CIRCANNUAL CLOCKS

ANNUAL BIOLOGICAL RHYTHMS

ACADEMIC PRESS RAPID MANUSCRIPT REPRODUCTION

Proceedings of a Satellite Symposium of the 140th Meeting of the American Association for the Advancement of Science Held in San Francisco, California, February 25, 1974

CIRCANNUAL CLOCKS

ANNUAL BIOLOGICAL RHYTHMS

EDITED BY

ERIC T. PENGELLEY

Department of Biology
University of California, Riverside

ACADEMIC PRESS, INC.

NEW YORK SAN FRANCISCO LONDON 1974

A Subsidiary of Harcourt Brace Jovanovich, Publishers

ACADEMIC PRESS, INC.
111 Fifth Avenue, New York, New York 10003

United Kingdom Edition published by
ACADEMIC PRESS, INC. (LONDON) LTD.
24/28 Oval Road, London NW1

LIBRARY OF CONGRESS CATALOG CARD NUMBER: 74-27489

ISBN 0-12-550150

PRINTED IN THE UNITED STATES OF AMERICA

Contents

Participants

Ivan Assenmacher, Laboratoire de Physiologie Animale, Faculté des Sciences, Place Eugene Bataillon, Université de Montpellier, 34 Montpellier, France.

Sigfried P. Berthold, Vogelwarte Radolfzell am Max-Planck-Institut für Verhaltensphysiologie, 7761 Schloss Moggingen, West Germany.

Mary Anne Brock, Gerontology Research Center, Baltimore City Hospitals, Baltimore, Maryland, 21224

Albert R. Dawe, Office of Naval Research, 536 S. Clark Street, Chicago, Illinois, 60605

James T. Enright, Scripps Institute of Oceanography, P. O. Box 1529, La Jolla, California, 92037

Richard J. Goss, Division of Biological and Medical Sciences, Brown University, Providence, Rhode Island, 02912

Bengt W. Johansson, Heart Laboratory, Department of Medicine, General Hospital, S-21401 Malmo, Sweden.

Helmut Klein, Max-Planck-Institut für Verhaltensphysiologie, Seewiesen und Erling-Andechs, 8131 Erling-Andechs, West Germany.

Michael Menaker, Department of Zoology, University of Texas, Austin, Texas, 78712

Nicholas Mrosovsky, Department of Zoology, University of Toronto, Toronto, Ontario, M5S IA5, Canada.

Eric T. Pengelley, Department of Biology, University of California, Riverside, California, 92502

Alain Reinberg, Laboratoire de Physiologie, Fondation A. de Rothschild, 29 Rue Manin, 75, Paris-19^{e}, France.

James T. Rutledge, Department of Animal Physiology, University of California, Davis, California, 95616

Robert Schwab, Department of Animal Physiology, University of California, Davis, California, 95616

Jerome B. Senturia, Department of Biology and Health Sciences, The Cleveland State University, Cleveland, Ohio, 44115

Wilma A. Spurrier, Department of Physiology, Stritch School of Medicine, 2160 South First Avenue, Maywood, Illinois, 60153

ABSENT

Sally J. Asmundson, Department of Biology, University of California, Riverside, California, 92502

Eberhard Gwinner, Max-Planck-Institut für Verhaltensphysiologie, Seewiesen und Erling-Andechs, 8131 Erling-Andechs, West Germany.

Preface

The purpose of this symposium is to summarize the present state of knowledge on endogenous annual rhythms (circannual clocks), and to point out their biological significance and importance. It is further hoped that the symposium will give the phenomenon the attention it deserves, particularly since many biologists are unaware of it. In addition we gathered our knowledge on the subject under one heading, so that future investigators, particularly the young, can use it as a starting point for future ideas and experiments. It is also hoped that there will be a meeting of minds on the subject, with careful analysis of each others work. This book is the outcome of these aims and objectives.

To the critical reader it will be perfectly obvious that our ignorance of the basic nature of the subject is profound. This is due primarily to two factors, namely that the discovery of the phenomenon is relatively new and that the demands of the time involved make it very difficult to study. It must also be stressed that we know of the existence of endogenous annual rhythms only in very few species of animals, and it seems remarkable that they have apparently never been explored or demonstrated in plants. Hopefully this will be undertaken, as it may prove of great importance. It will be of help also to point out that our terminology at this time has largely been borrowed from those people who study circadian rhythms. The adaptations and meanings of this terminology have been thoroughly discussed and explained in the papers by Dr. Alain Reinberg and by Pengelley and Asmundson. The inexperienced reader may wish to consult the introductions of these two papers first, in order to master the terminology.

In both the organization of the symposium and the preparation of this volume I have been greatly assisted by many people. First, I wish to acknowledge the initiative and leadership of Dr. Richard Goss. He was a constant source of encouragement and help when I needed it most. Second, I must thank Mrs. Elisabeth Zeutschel of the AAAS administrative staff for her excellent organization. She has somewhat restored my trust in administrators. I am indebted to Dr. William Fuller for supplying me with the background material on Professor Rowan, and to Mr. Robert Lister for writing his biography. I wish also to express my gratitude to Drs. Alain Reinberg, Ivan Assenmacher, Bengt Johansson, Helmut Klein, and Peter Berthold for traveling all the way from their respective European countries to be with us in San Francisco. Finally I thank Mrs. Mary Hickey who retyped the manuscripts for the final production process.

Professor William Rowan (1891-1957)
Karsh, Ottawa

This volume is dedicated, by the Participants of the Symposium, to the memory of Professor William Rowan, in honor of the fiftieth anniversary of his classic work on the Junco.

There follows a short biography of Professor Rowan, and his original paper of 1925.

CIRCANNUAL CLOCKS

ANNUAL BIOLOGICAL RHYTHMS

BIOGRAPHICAL SKETCH OF PROFESSOR WILLIAM ROWAN

by

Robert Lister
Formerly Research Assistant to
Professor Rowan
Edmonton, Alberta
January 1974

PROFESSOR WILLIAM ROWAN was born in Basle, Switzerland in 1891 of English-Swiss parents. He was educated in England and received his M.Sc. from University College, London. For a short time he taught at Bedford, but in 1919 accepted an appointment to the Zoology Department at the University of Manitoba. A year later he joined the Biology Department of the University of Alberta and when that department split in 1921, he became head of the new Department of Zoology, a post he held until his retirement in 1956.

Rowan's early field notes show his keen interest in birds while still a boy, but it was as a university student at Blakeney Point on the north Norfolk coast of England that he probably acquired his lifelong interest in bird migration.

The mud flats and saltings at Blakeney teemed with migrating shorebirds while sand and shingle annually hosted large colonies of breeding terns.

When he arrived in Alberta he found near Edmonton an area that was reminiscent of Blakeney. Here again, at Beaverhills Lake in spring, the lake flats swarmed with countless waders on passage to the Arctic and geese in their thousands rested to await the breakup of the northern ice.

It was the remarkable regularity with which the birds returned each spring that fired Rowan's imagination. The birds arrived within a day or two of the same date each year. Weather was variable and he looked for some constant that might regulate the spring migration. The lengthening days, it seemed to him, might be the governing factor. He decided to subject birds to increasing periods of light to simulate the lengthening daylight as a test of his hypothesis.

His classical experiments on juncos and crows are too well known to need elucidation. It is sufficient to note here that although we know today that light in itself does not have the profound effect on bird migration that Rowan thought it might, his work pioneered a new field of avian biology.

For his outstanding work he received his D.Sc. from the University of London and in 1945 was awarded the Flavelle medal of the Royal Society of Canada, Canada's highest honour in the field of Science.

William Rowan's ornithological interests were by no means confined to migration One of his earliest field studies was on the ecology and breeding behavior of Merlins *(Falco columbarius)* on the Yorkshire moors. During his early years in Alberta he contributed to "British Birds" a series of "Notes on Alberta Waders Included in the British List". These, as many other of his scientific and popular papers, were profusely illustrated with his delightful pen-and-ink sketches.

Rowan also had some interests in mammals. In the early 1920's he vigorously protested the federal government's plan to introduce plains bison into Wood Buffalo Park. He visited that park in 1925 and saw pure wood bison. Having seen European bison on the Duke of Bedford's estate in England, and being familiar with plains bison at Wainwright, Alberta, he was convinced that wood bison were intermediate between the other two.

The last years of his career were devoted largely to an attempt to understand the 10-year cycle in snowshoe hares and upland game birds. In an attempt to supplement meagre research grants for this work Professor Rowan designed, and attempted to sell, a set of Conservation Stamps. Unfortunately, he was too far ahead of public awareness in environmental concerns and sale did not produce the required funds.

He was an ardent collector and built up a large collection of Alberta birds which are now housed in the University of California. Perhaps the proximity of Beaverhills Lake fostered his abiding interest in the shorebirds. From this came his description of a new race of the short-billed Dowitcher, *Limnodromus griseus hendersoni*.

For many consecutive years he visited in spring the muskegs of central Alberta and became an authority on the breeding birds of these areas. He conceived the idea that if the Common Loon *(Gavia immer)* were a direct descendant of *Hesperornis* the embryo chicks might show evidence of tooth buds in their mandibles. Numerous incubated eggs were collected and the bills of the embryo sectioned, but nothing to substantiate this supposition was found. Unfortunately, he was disposed to disregard negative results and no notes on this work were published by him.

Through his many years of field work he became intensely interested in the periodic fluctuations in the numbers of certain birds and mammals in the northern hemisphere and published a paper entitled "The Ten Year Cycle". Unfortunately, his attempt to study a known population of Varying Hares *(Lepus americanus)* through one of these cycles was abandoned.

A most versatile man with many artistic as well as scientific qualities, Rowan will be remembered primarily for having discovered, as he himself puts it in his book, The Riddle of Migration, "a method of manipulating the annual rhythm of the reproductive organs....." of birds.

NOTE: Additional information may be obtained from Salt, W. Ray, Auk 75 (4):387-390, 1958.

RELATION OF LIGHT TO BIRD MIGRATION AND DEVELOPMENTAL CHANGES

That light is a factor of prime importance in the inauguration or stimulation of bird migration, has been suggested by many authors from the days of Seebohm onwards. While many of the suggestions will not bear close investigation, at least one very attractive view has been put forward by Sir E. Sharpey-Schafer. In an address delivered some years ago to the Scottish Natural History Society[1] he makes the following comments, " . . . the regularity with which migration occurs, indicates that the exciting cause must be regular. There is no yearly change, outside the equatorial zone, that occurs so regularly in point of time as the change in the duration of daylight. On this ground this may well be considered a determining factor in migration, and it has the advantage over other suggested factors that it applies to the northerly as well as to the southerly movement." He says further "That it [migration] is a result of developmental changes in the sexual organs is improbable."

Evidently inspired by the work of the botanists Garner and Allard on what they have termed "photoperiodism," an American author[2] has lately revived this theory and has apparently independently, come to the same conclusion as Sir Edward with regard to the absence of relation between developmental changes in the reproductive organs and migration.

On purely theoretical grounds it has always seemed to me that if the waxing and the waning of the days really in any way affect the migratory impulse, they must produce their effect through the gonads. This is not the place for theoretical discussion, and I merely wish to record an experiment that has just reached completion. Other corroborative work is still in progress and a critical histological examination of the experimental and normal material yet remains to be undertaken.

In September of last year I trapped a number of Juncos *(Junco hyemalis)* on their southward migration to the Middle States. These were turned into two large open-air aviaries removed from shelter of any kind. One, into which about a dozen birds were put for the experimental work, was fitted with two 50-watt electric lights. The other housed controls. Commencing on October 2, the lights were turned on at sunset (that is while the birds were still fully active) and kept on until five minutes after dark. Each day afterwards the time was lengthened by five minutes. Taking into consideration the differences in time of sunrise, the birds thus got about three minutes longer illumination daily. On account of the fact that they went to roost at their usual time on the first day in spite of the glaring lights and that attempts at educating them to keep awake were never wholly successful, and less so with some individuals than with others, it has proved impossible to estimate the effective light increases. For the same reason there is lack of uniformity in the results obtained.

Elimination of the warmth factor was unexpectedly successful--thanks to a severe winter--the lowest temperature to which the birds were exposed being 50^{0} below zero (Fahrenheit).

Birds were killed at intervals of approximately two weeks, with the following results:

Dates of Killing	Number Examined	Size of Testes[3]
Oct. 15	1	0.50 X 0.48
(A wild bird killed same date		.60 X .60)[4]
Oct. 29	1	.44 X .41
Nov. 13	1	.45 X .53
Nov. 26	1	.60 X .44
Dec. 11	1	.80 X .79
Dec. 27[5]	2	(A) .90 X ? (Part of ribbon destroyed)
		(B) 1.80 X 1.54

Catastrophe overtook my control birds and I had to find substitutes. Through the kindness of the Museum of Vertebrate Zoology University of California, I have been receiving fixed gonads of Juncos (a closely related species of approximately the same size) wintering in the Berkeley district. These are not strictly comparable with mine, therefore, but the samples include birds taken at intervals from November to early January. In spite of the California climate, the January testes are minute. My solitary female, killed also on December 27, as compared with the early January females from Berkeley, has an ovary two to two and a half times as large and with conspicuous follicles.

The two males and the female killed on December 27 were kept indoors for their last week at an average temperature of about 40°. (Their drinking water froze one night.) The marked difference in the size of the testes of the two males may probably be accounted for by their habits. A went to roost, in spite of the lights (the birds were together in the same cage), about an hour to an hour and a half each night before B. The female kept the latter company. A sang a good deal; B incessantly. All the birds were in excellent condition when killed.

It would, therefore, appear that whatever effect daily increases of illumination may or may not have on migration, they *are* conducive to developmental changes in the sexual organs. Comparison of the normal material from Riviera-like California with the experimental product from Alberta further suggests that favorable light conditions are more potent in this respect than favorable temperatures.

Footnotes

[1]"On the Incidence of Daylight as a determining Factor in Bird Migration,"

E. A. Schafer, NATURE, vol. 77, pp. 159-163 (December 19, 1907).

[2]"Is Photoperiodism a Factor in the Migration of Birds?" G. Eifrig. Auk, vol. 41, pp. 439-444.

[3]The whole series of testes was sectioned at 6u. The first column indicates the greatest diameter of the largest section in each series in millimetres. The second is arrived at by adding the total number of sections for each series and multiplying by 0.006.

[4]The testes of this bird, still on migration, had not yet reached the winter minimum and this accounts for their large size. Diminution in size during the initial stages of the experiment is very marked.

[5]A was an adult bird: B a bird of the year.

WILLIAM ROWAN.

11142 86th Ave., Edmonton, Alta., Canada, January 28.

Reproduced by kind permission of "Nature".
Reference - Volume 115, p. 494-495; April 4, 1925.

CIRCANNUAL RHYTHMICITY IN INVERTEBRATES

MARY ANNE BROCK

Laboratory of Cellular and
Comparative Physiology,
Gerontology Research Center,
National Institute of Child Health
and Human Development,
National Institutes of Health, PHS,
U.S. Department of Health, Education
and Welfare,
Bethesda and Baltimore City Hospitals,
Baltimore, Maryland, 21224

INTRODUCTION

Endogenous circannual rhythms are known for only two invertebrate species, an arthropod, the cave crayfish, *Orconectes pellucidus inermis* (Jegla & Poulson, 1970), and the colonial, marine cnidarian, *Campanularia flexuosa* (Brock, 1974a, 1974b, 1974c). The yearly cycles in reproduction of the cave crayfish and in the rate of aging, growth, and development of the cnidarian are similar to the circannual rhythms of more highly evolved birds and mammals (Benoit, *et al*, 1956; Davis, 1967; Gwinner, 1968; Pengelley & Fisher, 1963; Pengelley & Kelly, 1966). Although reports on other invertebrate species do include references to "circannual rhythms", seasonal variations, annual cycles, etc., in their titles, the observations were made on animals not kept under constant conditions for periods exceeding one year. Therefore the cycles should not be termed circannual rhythms. To establish that an annual cycle is an endogenous circannual rhythm, it must be shown that the system expressing a periodic variation is "capable of self-sustained oscillations" or is free-running (Aschoff, 1965). The absolute minimum required is the completion of one period and the initiation of a second oscillation under constant conditions, that is, in the absence of any known exogenous signals or zeitgeber from the environment. Under such constant conditions, the rhythms for different individuals of a given species will gradually go out of phase with each other, with other animals of the same species in their natural environment, and with the calendar year. It may then be assumed with the observation of two or more individuals whose rhythms progressively go further out of phase with each other that the zeitgeber is absent and the rhythms are free-running. For invertebrates, these criteria have been satisfied only in the studies on the arthropod and the cnidarian.

The existence of circannual rhythmicity in invertebrates, particularly less complex species, suggests that endogenous annual rhythms may have evolved as properties of individual cells, just as circadian or endogenous daily rhythms are expressed by unicellular organisms (Bunning, 1967; Hastings, 1970). From this point of view, annual rhythmicity in *C. flexuosa* is provocative. Cnidarians are considered to be at the tissue level of organization and have an unspecialized nerve net rather than a complex central nervous system. Therefore, the potential is provided for exploring whether control of the endogenous annual rhythms resides in cells of the unspecialized nerve net and/or other cell types.

The purpose of this review is to examine the circannual rhythms of an arthropod, the cave crayfish, *Orconectes pellucidus inermis*, and a cnidarian, *Campanularia flexuosa*, and to consider *C. flexuosa* as a model for studies of cellular aging.

CIRCANNUAL RHYTHMS IN ARTHROPODS

Endogenous, annual rhythms in reproduction and molting of the cave crayfish, *Orconectes pellucidus inermis*, were reported by Jegla and Poulson (1970). The authors chose this species for investigation since annual cycles of growth and reproduction did occur in nature, but in cave environments where neither light nor temperature changes were appreciable and which were considered "nearly seasonless". Male and female crayfish were collected from a cave in Breckinridge Co., Kentucky in November 1965 and placed in separate sections of aerated aquaria at 13.0 ± 0.5° C. This constant temperature was maintained throughout the observations. The animals were also kept in constant darkness except for 15-20 minute periods of light two to three times each month when they were fed and inspected for stages in the reproductive and molting cycles. The observations of 6 adult males and 6 adult females were continued for 2 to 2 1/2 years.

In earlier studies, Jegla analyzed the reproductive cycles of *O. p. inermis* and found that for males, the reproductive state changed alternately between reproductive competence (form I) and reproductive quiescence (form II) with each successive molt. For female cave crayfish, changes in the size of the ovary and oocytes were observed through the translucent exoskeleton and ranked from 0 to 4, with 0 representing a quiescent ovary and 4 one with large, yolky oocytes.

1. Seasonal Changes in Reproduction.

The male crayfish in the laboratory population were compared with those in a natural population described some time earlier, and, although the percentage of breeding males in the two groups corresponded well at the beginning of the observations, the laboratory population was out of phase with the natural population within 4 months (figure 1). After observations began in November 1965, molting did occur in the caves during February but was delayed in the laboratory until March 1966. Furthermore, the synchrony of molting among males in the laboratory population was lost by late summer of 1966 (figure 1).

If the individual male crayfish were considered, the free-running periods for each crayfish approximated one year in the constant laboratory environment (figure 2). Although the periods for the circannual

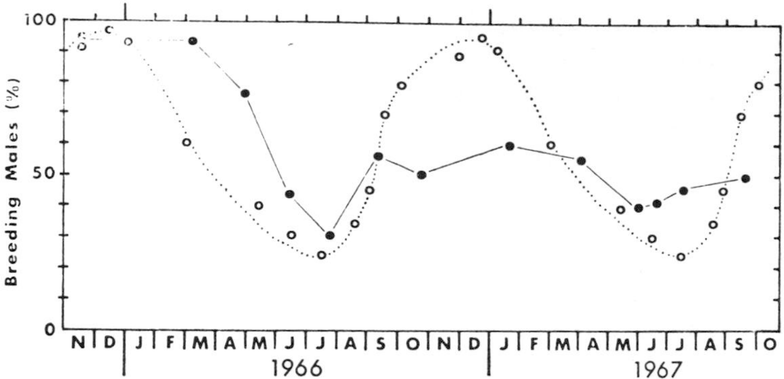

Fig. 1. Synchrony in the breeding condition of male crayfish with each other in the laboratory was lost by September 1966. The percentage of breeding males is indicated by the closed circles connected with solid lines. In contrast, the synchrony of crayfish with each other in the cave population persisted. The average monthly percentage of breeding males in Shiloh Cave, Indiana from 1960-1964 is shown by the open circles connected with dotted lines (from Jegla & Poulson, 1970).

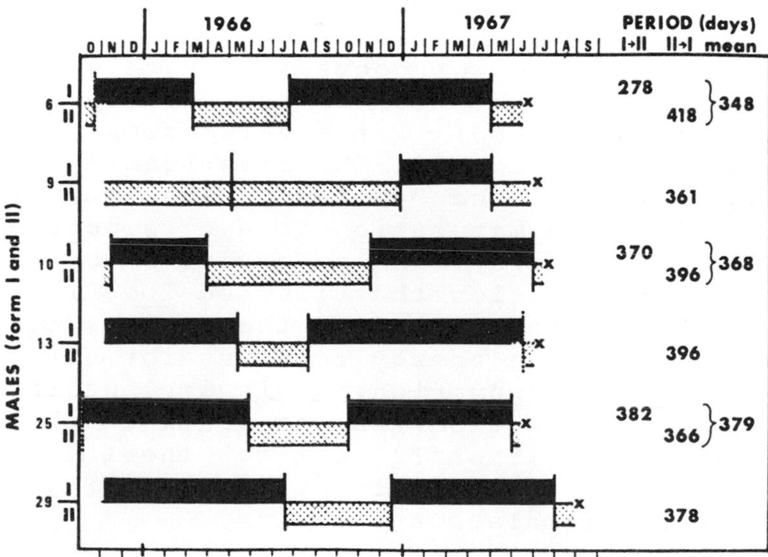

Fig. 2. The endogenous molting-reproduction rhythms for 6 male crayfish persisted for over 1½ years. The free-running period was defined by molting (vertical lines) either from breeding condition (form I, solid bars) to non-breeding condition (form II, cross-hatched bars) or vice versa. For one juvenile crayfish (no. 9), the period was defined by the 1966-1967 molt. An X indicates death (from Jegla & Poulson, 1970).

rhythms averaged 370 days for these 6 males, the authors noted that in the laboratory population, the proportion of the cycle spent as form I, or reproductive competence, varied considerably among the individuals. In view of the fact that there are differences in the length of the intermolt cycles in different size-classes of crayfish and other crustacea, the variability might well result from observations on a population of mixed size-classes (Travis, 1974).

The female crayfish also had reproductive cycles of about one year, but these were less well defined than those of the males. A precise reference for determining period length of the ovarian rhythms would be the time an individual lays eggs, however only two females laid eggs during the first spring and none in the laboratory population did so at the end of one annual cycle (figure 3). Rather, their ovaries regressed in size by resorption of oocytes, and this reabsorption of the oocytes was used to estimate the period. For the two females that can be considered, the free-running periods were 355 and 338 days. The lack of complete oocyte maturation in a cave population of female crayfish was explained by the authors' observation that limited food in caves results in failure of the females to accumulate enough energy reserves for both molting and reproduction. This implies that the two to three feedings per month in the laboratory were insufficient for the full expression of the annual reproductive rhythms of the female crayfish.

2. Adaptive Significance.

Jegla and Poulson (1970) further discussed 1) the advantages of circannual rhythms in anticipating future environmental

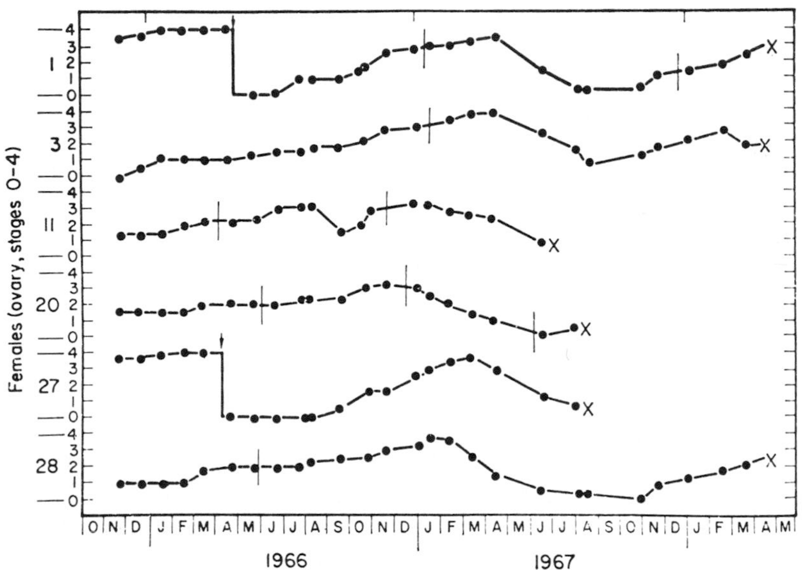

Fig. 3. The reproductive rhythms for 6 female crayfish were described by ovarian development, ranked 0 to 4. Vertical lines indicate molt; the two arrows indicate egg laying by females 1 and 27 in April 1966; and an X signifies death (from Jegla & Poulson, 1970).

events in the caves where seasonal changes are not marked and 2) the synchronization of individuals with each other and with the cycling of the environment. The latter must be of special importance in this species since the laboratory population became asynchronous within one period under constant conditions.

The synchrony of the reproductive cycles in nature with seasonal changes in the environment is shown in figure 4 and is from data collected over 4 years at Shiloh Cave, Indiana. The heavy precipitation, given in mm. of snow or rain, was followed by a time when surface water run-off above ground was presumed to be highest. At this same time, the cave water showed very subtle changes, the most abrupt being in volume flow and water temperature, with the latter in fact falling only 1-2° C over a week or less. Nevertheless, the female crayfish' eggs were laid at the time of maximum surface water run-off and influx of organic matter into the cave. The authors suggest that the rhythms remain in phase in nature due to subtle cues which trigger egg laying. The females carry their eggs and young for two to three months, and, after this lag, the young crayfish joined the population. The delay allowed the necessary time for micro-organisms in the cave to process the organic matter brought in at the time of maximum surface water run-off. Not only is sufficient organic matter available then for the young crayfish, but presumably no subtle changes in cave water occur that might be injurious to the young crayfish.

CIRCANNUAL RHYTHMS IN *CNIDARIA*

Free-running annual rhythms in growth

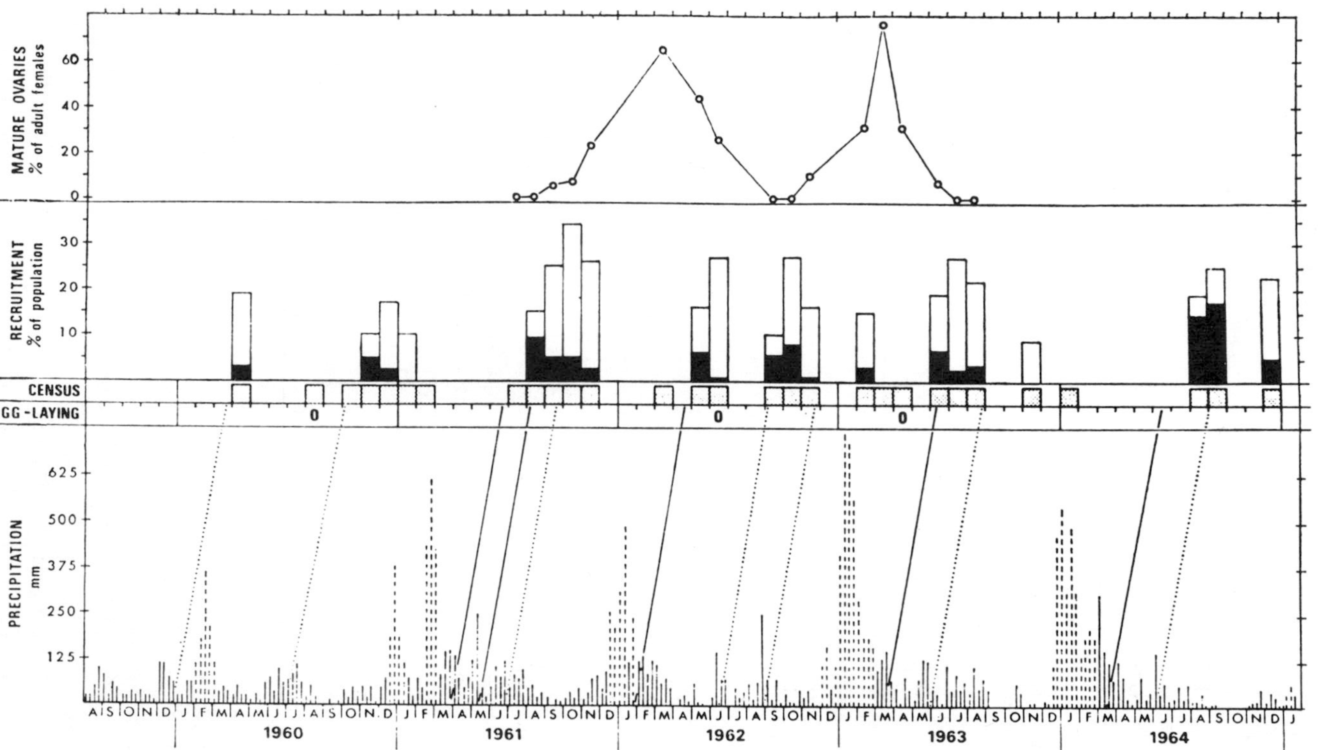

Fig. 4. Egg laying and addition of young crayfish to the population are related to the time of presumed peak water run-off above ground that followed maximum precipitation. The solid or dashed lines in the lower portion of this figure are drawn from the time of the presumed peak surface run-off to the time young crayfish joined the population. The lag was about 3 months. Egg laying (open circles signify female cave crayfish carrying eggs) occurred at the time of the peak run-off as recorded in 1960, 1962 and 1963 (from Jegla & Poulson, 1970).

of the colonies and in the development and longevity of hydranths exist in the colonial, marine cnidarian, *Campanularia flexuosa* (Brock, 1974a, 1974b, 1974c). Furthermore, 1) these endogenous rhythms of colonies obtained from nature at different times and cultured together in the same vessels were out of phase with each other and 2) the circannual rhythms were temperature-compensated.

The detailed descriptions of growth habit and the methods used to culture *C. flexuosa* were given in the three above cited papers. Briefly, the wild colonies of *C. flexuosa* were obtained from the Marine Biological Laboratory, Woods Hole, Massachusetts in May 1968. The colonies have common basal structures, stolons, that attach to a substrate, and many upright stems grow from the stolons. Alternately arranged branches from the upright stems bear either individual hydranths or several hydranths, also in an alternate branching pattern on secondary branches (figure 5). In the laboratory, *C. flexuosa* is cultured by fastening a portion of the wild colonies with thread to glass microscope slides. The growth of the stolons, which adhere to the surface of the slides, can be observed within a few days. The slides were kept in glass dishes containing aerated, circulating artificial sea water made with Neptune Salts (Westchester Aquarium Supply Co., Inc., White Plains, N.Y.) that was replaced with freshly prepared water weekly. Colonies of *C. flexuosa* were fed each day with an excess of brine shrimp (*Artemia*) that were hatched during the previous 24 hours.

A total of 6, 16 and 6 colonies were cultured at 10°, 17° and 24° C, respectively, and, of these, detailed daily observations were compiled for the number of colonies

Fig. 5. C. flexuosa colonies continuously cultured at 10° C. always had this lush appearance during the phases of luxuriant growth. The basal portions of the colony, the stolons (S) attach to glass microscope slides in laboratory cultures. Uprights (U) grow from the stolons at intervals and bear either hydranths or secondary branches in an alternate branching pattern that is clearly shown by the uprights marked with arrows (from Brock, 1974a).

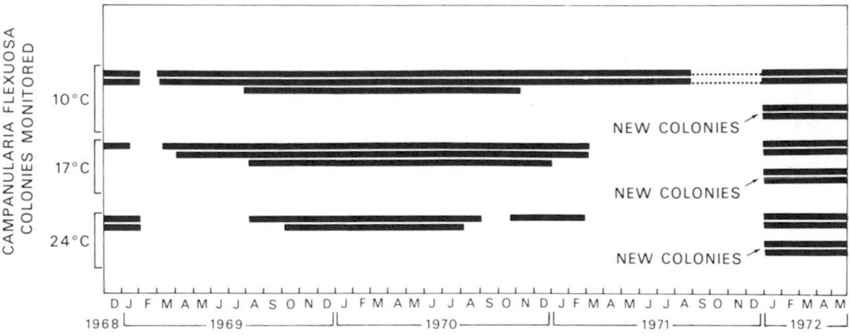

Fig. 6. The number of *C. flexuosa* colonies used for detailed daily observations of hydranth development and longevity is indicated by the number of lines at 10°, 17° and 24° C. The time span of uninterrupted daily observations of each colony is equivalent to the length of each line. Colonies were switched from 17° to 24° C. in October 1970 in order to complete the records through February 1971. The transitions from solid to broken to solid lines for 2 colonies at 10° C. indicate that from late August through December 1971 both colonies were monitored daily but detailed records were not kept. For daily observations in 1972, some of the colonies were switched from 10° to 17° C. in March 1971 and from 17° to 24° C. in November 1971. The new colonies that were obtained from Woods Hole, Massachusetts in November 1971 were cultured initially at 17° C. Some of these new colonies were switched to both 10° and 24° C. on December 21, 1971.

over the time intervals shown in figure 6. Stages in upright growth and in the development and longevity of, in most cases, over 100 hydranths in each colony were recorded on each successive day.

All of the colonies were kept in constant darkness except for brief (30-60 min.) unscheduled periods required for daily mapping and in constant temperature rooms at the ambient temperatures of $10 \pm 1^{\circ}$, $17 \pm 1^{\circ}$ and $24 \pm 1^{\circ}$ C for the duration of the experiments.

1. Seasonal Changes in Growth and Development.

The circannual rhythms of *C. flexuosa* were defined by seasonal changes in growth of the colonies and in development and longevity of the hydranths. During most of the year, the growth of the colonies was luxuriant and the hydranth life spans were long. However, this luxuriant growth was interrupted by phases of sparse growth each summer and by brief phases of sharply curtailed growth in mid-winter, both of which were accompanied by shorter hydranth life spans. Following the summer and mid-winter changes in growth habit, there was a reversal to luxuriant growth.

The rhythmic changes in the colony growth and hydranth development for each phase of the annual cycle have been given in detail previously (Brock, 1974a, 1974b, 1974c). Briefly, during the phases of luxuriant growth, hydranth differentiation was quite normal and followed the patterns described some time ago (Hammett, 1943; Berrill, 1950; Crowell, 1957). In contrast, there were marked differences in the number of hydranth buds initiated and in their differentiation during the summer and mid-winter phases of sparse growth. In these seasons, a 2- to 7-fold increase in the percentage of hydranth buds that were aborted in late stages of their differentiation was observed. The percentages of hydranth buds aborted at the three constant ambient temperatures are summarized for 1969-1970 in table 1, and the abnormal process can be compared to normal hydranth differentiation in figure 7. Colonies of *C. flexuosa* in luxuriant growth enlarged by adding new hydranths to the apices of the upright stems and the secondary branches; the large number of adult hydranths in the colony is also maintained by replacement of those hydranths which regress or die after their relatively brief life span. The colonies shown in figure 8 illustrate the great frequency both of the replacement and of initiation of new hydranths. However, during the phases of sparse growth and coincidental with the greater incidence of abnormal hydranth development, the initiation of hydranth buds both in apical growth zones and at sites of regression was markedly depressed (table 1).

As a result of all of these changes in the initiation and development of the hydranths as well as their decreased longevity during the phases of sparse growth (table 1), the number of feeding adult hydranths declined and the colonies were much reduced in size (figure 9). Often, because hydranths did not replace those that had regressed, only the skeletons of empty upright stems remained (figure 10). With the reversal to the luxuriant growth habit, either cellular components filled the empty upright stems (in colonies cultured at 10° C) or new upright stems sprouted from the stolons (more common in colonies cultured at 17° and 24°C). The rate of hydranth bud initiation rose as

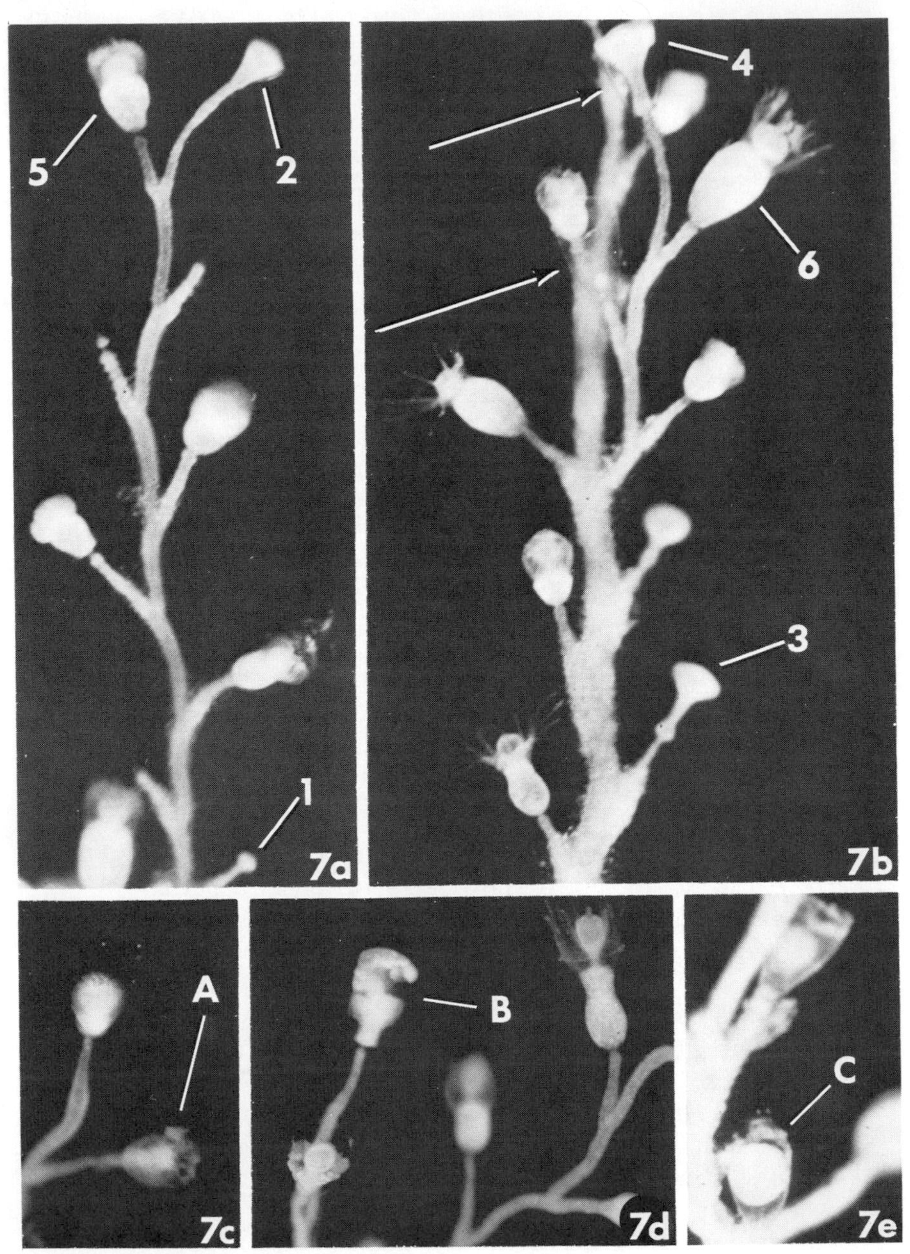
5
2
1
7a
4
6
3
7b
A
7c
B
7d
C
7e

Fig. 8. Hydranth buds of *C. flexuosa* in progressive stages of differentiation, cone, cylinder and short-tentacle stages, replace adult hydranths that have regressed at any position along the upright stems (arrows). New buds are also continuously initiated at the distal growth zones of the uprights, increasing the height of the uprights during phases of luxuriant growth (crossed arrows).

The normal regression of adult hydranths begins with a perceptible shortening of the tentacles (1, 2). The tentacles are further autolyzed and cellular fragments from them are propelled through the regressing hydranth to other hydranths in the colony where they are phagocytized. Eventually, only the structure of the hypostome is clear (3, 4), and, as it also is resorbed, an amorphous mass of cellular debris remains in the hydrotheca (5). The empty hydrotheca (6) eventually drops from the upright stem, and, after a few days, a new hydranth bud begins differentiation at the site of regression (from Brock, 1974a).

Fig. 7a and 7b. The differentiation of *C. flexuosa* hydranths begins with the enlargement of the tip of one of the alternate branches from an upright stem to form a cone-shaped structure (1, 2). The cones increase in diameter distally and at this stage resemble cyclinders (3). At a later stage, the anlage of a single ring of tentacles appear at the periphery of the distal portion of cylinders (4), and, with further differentiation, the tentacles elongate and surround the central extension of the hydranth, the hypostome (5). Finally, the tentacles extend completely, the mouth opens in the center of the hypostome, and the adult hydranth (6) may feed. The hydranths are radially symmetrical, and the entire colony is covered by a chitinous layer, the perisarc. The transparant perisarc can be seen surrounding one of the cylinders and remaining as an empty shell, the hydrotheca, at a site where an adult hydranth has regressed (arrows) (from Brock, 1974a).

Fig. 7c, 7d and 7e. During the phases of sparse growth, differentiation of the hydranth buds proceeded to the short-tentacle stage, but then the structure of the buds was gradually obliterated. This abortive process began with the loss of definition of the tentacles (A). The remaining structure became more amorphous, sometimes with a residue left distally (B), until finally only a compact, nearly spherical mass remained in the hydrotheca (C). The hydrotheca containing the cellular debris eventually dropped off the upright stem (from Brock, 1974a).

Table 1. **Seasonal Changes in the Initiation and Development of Hydranth Buds and in Hydranth Longevity of *C. flexuosa* at Three Constant Ambient Temperatures summarized for 1969-1970.**

		Luxuriant Growth	Summer Phase of Sparse Growth	Mid-Winter Phase of Curtailed Growth
Percent Hydranths Aborted	10° C.	0 - 4%	5 - 7%	19 - 30%
	17° C.	3 -10%	12 -23%	17 - 19%
	24° C.	2 - 8%	20 -33%	18 - 32%
Rate of Hydranth Bud Initiation; Apical Growth Zones	10° C.	4 every 18 days	1 in 18 days	
	17° C.	3 every 18 days	0 - 1 in 18 days	
	24° C.			
Average Time for Hydranth Replacement; Sites of Regression	10° C.	In 2 days	None Replaced	
	17° C.	In 1.5 days	None Replaced	
	24° C.	In 1.1 days	None Replaced	
Mean Hydranth Longevity	10° C.	15.6 - 16.7 days	7.4 - 9.7 days	9.7 - 9.9 days
	17° C.	7.1 - 10.1 days	4.5 - 6.7 days	4.4 - 6.4 days
	24° C.	5.9 - 7.3 days	3.5 - 3.7 days	3.9 days

Fig. 9. Colonies of *C. flexuosa* could reach this minimal size during seasons of sparse growth. This colony had been cultured at 24° C. and eventually died in February 1971. The cellular components have been lost from the distal segments of many upright stems, leaving only the empty perisarc.

Fig. 10. The empty perisarc at the apices of upright stems of *C. flexuosa* during the phases of sparse growth is shown in greater detail (from Brock, 1974a).

the percentage of buds developing abnormally fell, and hydranth life spans became much longer. Within a short time, the colonies were lush in appearance like those in figure 5.

The rhythmic changes in growth and development are shown for a representative colony that was cultured continuously at 10°C for over 3 years (figure 11). In each year, the phases of luxuriant growth were interrupted first in mid-winter and then again during the summer season. In mid-winter, a gradual change to the curtailed growth habit took place which required about 2 weeks; this was followed by a quiescent phase when few or no hydranth buds were initiated. Because the quiescent phase was so clearly expressed, the free-running periods could be determined precisely and are given in table 2. The quiescent phase lasted from 7 to 10 days, and then the reversal to luxuriant growth was clearly marked by the appearance of many hydranth buds. Figure 11 also shows that a somewhat different growth pattern was observed in 1972, for the curtailed growth phase began on March 20 and continued until June when detailed records were discontinued.

The second interruption of the luxuriant growth during each summer season also began and ended gradually, and the midpoints for this phase were used to calculate the free-running periods (table 2). In 1969, the first year that the colonies were cultured at 10° C, only one of the features which characterized the sparse growth habit during the summer season was expressed (figure 11). For one month, June 13 through July 12, there appeared to be abnormalities in the differentiation of some hydranth buds, although only five buds were actually

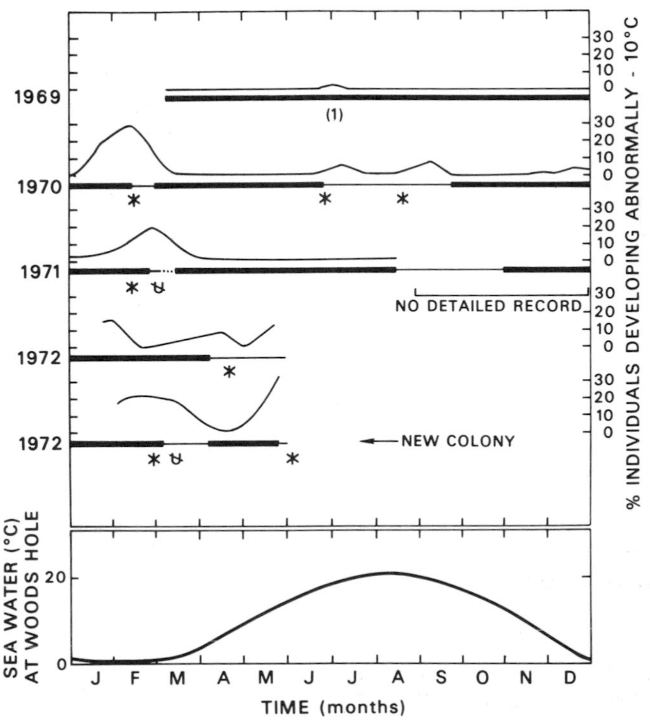

Fig. 11. Free-running circannual rhythms in a *C. flexuosa* colony at 10° C. are defined by seasonal changes from phases of luxuriant growth (bars) to phases of mid-winter quiescence and of summer sparse growth (lines). Severely curtailed growth was observed in March 1971 as shown by the dotted line. The percentage of individuals, including adults, that appeared abnormal is given at (1) (see text). An asterisk (*) indicates few or no distal hydranth buds initiated on uprights and branches, little replacement of regressed hydranths, and empty distal positions on uprights; a crossed U (Ʉ) signifies death of uprights. In April 1972, this colony which had been monitored at 10° C. for over 3 years began the curtailed growth phase and did not reverse to luxuriant growth. The opposite pattern was observed in a representative of the new colonies obtained from nature in November 1971; the curtailed growth phase was expressed from late February until April 1972. The yearly fluctuations in coastal water temperature at Woods Hole, Massachusetts were calculated from daily observations made over 3 years (1967-1969) by the U. S. Coast and Geodetic Survey (from Brock, 1974a).

Table 2. Periodicity of the endogenous changes in colony growth and hydranth longevity of *C. flexuosa*.

	WINTER		SUMMER	
	Termination of Mid-Winter Curtailed Growth	Free-running Period (days)	Summer Lull Mid-point	Free-running Period (days)
10° C.	February 18, 1970		June 28, 1969	
		375		403
	March 1, 1971		August 5, 1970	
		403 (1)		411
	April 6 →, 1972		September 20, 1971	
	March 19, 1972 *			
17° C.	February 26, 1970		June 20, 1969	
		369		407
	March 2, 1971		August 1, 1970	
		382		
	March 18, 1972			
	May 3 →, 1972 *			
24° C.	February 1, 1969			
		389		
	February 25, 1970			
		370		
	March 2, 1971			
		385 (2)		
	March 21 →, 1972			
	February 27, 1972 *			

	WINTER		SUMMER	
	Mid-point in Phase of Decreased Hydranth Longevity	Free-running Period (days)	Mid-point in Phase of Decreased Hydranth Longevity	Free-running Period (days)
10° C.	February 2, 1970		June 4, 1969	
		372		432
	February 8, 1971		August 25, 1970	
17° C.	February 14, 1970		July 31, 1969	
		353		315
	February 3, 1971		June 12, 1970	
24° C.	February 4, 1970			
		366		
	February 5, 1971			

* - New colonies obtained from Woods Hole, Massachusetts late in 1971.

The arrows following dates in 1972 indicate that the phase of curtailed growth was observed on these dates and no switch to the luxuriant growth habit took place prior to May 1972 when records were discontinued. The periods, 403 days at 10° C. (1) and 385 days at 24° C. (2), were calculated using the dates, April 6, 1972 and March 21, 1972, respectively.

aborted. In addition, many 1- to 4-day old hydrants appeared to be in early stages of regression. However, after 2 to 3 days, either normal hydranth differentiation was completed or the adults resumed feeding. A total of 4% of the differentiating or adult hydranths were involved. Interestingly, a sharp decrease in hydranth mean life spans coincided with this first summer season (see figure 14).

Temperature-compensation of the circannual rhythms (uniformity of the oscillations despite changes in ambient temperature) in *C. flexuosa* was evident when colonies cultured at 17° and 24° C were compared with those at 10° C. The seasonal changes in growth and development were nearly parallel at all three ambient temperatures, and, like the colonies at 10° C, the free-running periods of those at 17° and 24° C were slightly longer than one year (table 2). However, some of the rhythms were amplified at the elevated temperatures and the increases in amplitude, summarized from the detailed data, are shown in table 1. Figure 12 illustrates the periodicity of the rhythms for representative colonies at the three constant ambient temperatures. The colonies at both 17° and 24° C grew continuously for a limited time, in contrast to those cultured at 10° C. This shortened survival, 2 years for colonies cultured at 17° C and 11 months for those at 24° C, may reflect more accelerated colonial aging at the higher temperatures. Elevated ambient temperatures have a well-known effect in reducing the longevity of solitary-living poikilotherms (Comfort, 1964), and also generally shortened the life spans of *C. flexuosa* hydranths, irrespective of the phase of the annual rhythm (Brock, 1974c).

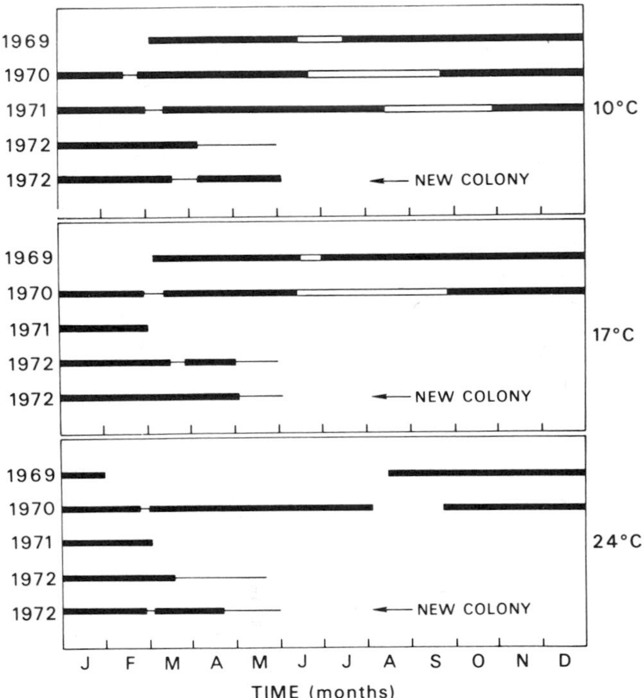

Fig. 12. The mid-winter phase of sharply curtailed growth in *C. flexuosa* was usually followed by a brief quiescent phase, the change shown by the transition from solid bars to narrow lines for three representative colonies cultured at 10°, 17° and 24° C. The breaks in the records during 1969 and 1971 for the colonies cultured at 17° and 24° C. signify colony death. The end of the quiescent phase and the reversal to the luxuriant growth habit are shown by the transitions from narrow lines to solid bars. Open bars extend through the summer phases of sparse growth. Colonies were switched from 10° to 17° C. in March 1971 and from 17° to 24° C. in November 1971 in order to complete the observations in 1972. The new colonies obtained from Woods Hole, Massachusetts in November 1971 were cultured in the same vessels with the older colonies. Despite their common environment, the new and the older colonies were out of phase with each other at each of the three ambient temperatures. The detailed records were discontinued in June 1972 (from Brock, 1974b).

2. Free-running Behavior: Phase Changes With Time.

The endogenous nature of the circannual rhythms in growth and development was further established by the observation that colonies obtained from Woods Hole, Massachusetts in different years, 1968 and 1972, and cultured at 10°, 17° and 24° C together in the same vessels were out of phase with each other. Figures 11 and 12 illustrate that at each of the three ambient temperatures, the colonies cultured under constant conditions in the laboratory for over 3 years were in a phase of sparse growth at the same time that the luxuriant growth habit was expressed by the new colonies. From these observations, one may assume not only the absence of any known zeitgeber but also the absence of unrecognized variables in the culture methods themselves that could bring the rhythms into synchrony.

3. Seasonal Changes In Hydranth Longevity.

The fact that the mean life spans of *C. flexuosa* hydranths doubled during seasons of luxuriant growth despite constant culture conditions (table 1) is, if the hydranths are considered individual organisms, a contradiction to the accepted concept of "specific age". Specific age is defined by Comfort (1964) as the "age at death which is characteristic of the species" in "organisms which undergo senescence". *C. flexuosa* hydranths do senesce but also exhibit circannual rhythmicity in the rate of aging. These rhythmic changes from long to short mean life spans of the hydranths were nearly synchronous with the reversals from the luxuriant to sparse growth habit. The detailed mortality curves and tables showing the contrasts between phases have been reported

(Brock, 1974c), and figure 13 illustrates that hydranths in a representative colony cultured continuously at 10^{0} C have a 50% survivorship of about 16 days during phases of luxuriant growth in contrast to only 7 to 10 days in the summer phases of sparse growth. At the two higher temperatures, similar endogenous changes in longevity were observed, and the maximum and minimum values for hydranth life spans in the colonies cultured at three constant ambient temperatures are given in table 3. The rhythmicity of the changes in hydranth longevity for over 2 years is shown in figure 14a. It is clear that the longevity curves at three different temperatures are approximately parallel but with longer hydranth life spans at lower temperatures. As already indicated, longevity of colonies was similarly extended at lower temperatures. The parallelism in longevity curves of *C. flexuosa* hydranths at three different temperatures definitively indicates that hydranth aging is controlled by both ambient temperature and circannual rhythms. Furthermore, the annual cycles in hydranth longevity are temperature-compensated just as were the rhythms in colony growth and in hydranth development.

The specific values for the mean life spans on each of the curves shown in figure 14a are placed at the mid-points of shorter, repeating rhythms (circa-lunar rhythms) in hydranth longevity (figure 14b). The mid-points of the circa-lunar rhythms during the summer and mid-winter phases when hydranth life spans were the shortest were used in calculating the free-running periods (table 2). These free-running periods are in close agreement with the free-running periods determined from the growth and developmental cycles.

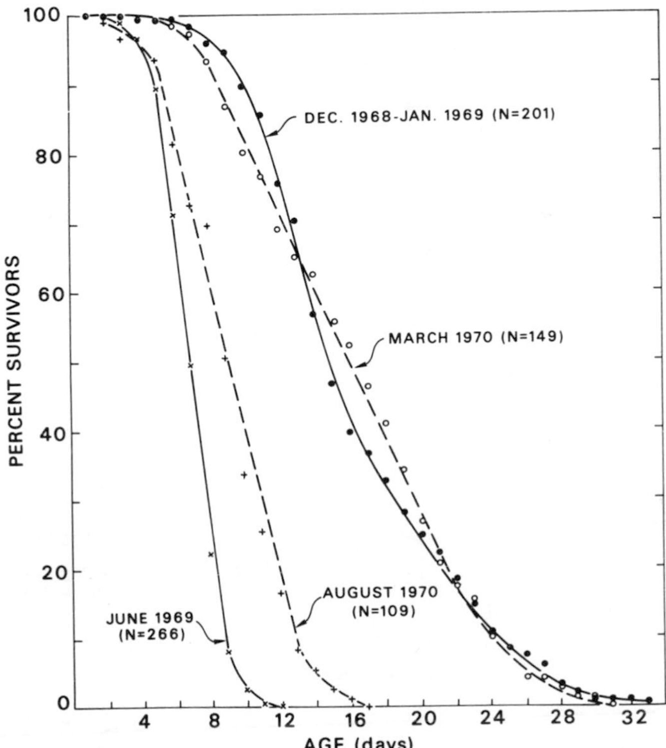

Fig. 13. The longer life spans of the hydranths of *C. flexuosa* during December 1968-January 1969 and March 1970, phases of luxuriant colony growth, contrast with the diminished hydranth life during June 1969 and August 1970, phases of sparse colony growth. The animals were cultured continuously in a constant 10° C. environment (from Brock, 1974c).

Table 3. **Maximum and Minimum Average Life Spans for Hydranths of *C. flexuosa* Cultured at Three Constant Ambient Temperatures (from Brock, 1974c).**

Date	Number of Hydranths	Longest Hydranth Life Span (Days)	Mean Life Span (Days ± S.E.)
Colonies at 10°			
Dec. 1968-Jan. 1969	201	34	16.7 ± 0.4
June 1969	266	12	7.4 ± 0.1
Oct.-Nov. 1969	361	42	15.7 ± 0.4
Jan.-Feb. 1970	133	17	9.7 ± 0.4
March 1970	149	31	16.7 ± 0.5
August 1970	109	17	9.7 ± 0.3
Oct.-Nov. 1970	214	32	15.6 ± 0.3
Feb. 1971	144	20	9.9 ± 0.3
May-June 1971	374	29	13.3 ± 0.2
Colonies at 17°			
Dec. 1968-Jan. 1969	67	10	7.4 ± 0.2
July-Aug. 1969	231	12	4.5 ± 0.1
Aug.-Sept. 1969	147	9	4.5 ± 0.2
Nov.-Dec. 1969	221	22	7.1 ± 0.2
Oct.-Nov. 1969	128	21	7.6 ± 0.3
Feb. 1970	94	12	4.4 ± 0.2
Feb. 1970	312	19	6.4 ± 0.2
Apr.-May 1970	111	27	9.6 ± 0.4
Apr. 1970	142	20	10.1 ± 0.3
May-June 1970	99	15	6.7 ± 0.3
May 1970	15	14	6.3 ± 0.9
Nov.-Dec. 1970	33	16	9.8 ± 0.5*
Sept.-Nov. 1970	134	23	8.0 ± 0.3
Jan.-Feb. 1971	94	22	8.1 ± 0.4*
Jan.-Feb. 1971	115	20	6.5 ± 0.3
Colonies at 24°			
Dec. 1969-Jan. 1969	244	18	7.2 ± 0.2
Sept. 1969	112	8	3.7 ± 0.1
Nov. 1969	365	21	7.2 ± 0.2
Feb. 1970	224	9	3.9 ± 0.1
Mar.-Apr. 1970	59	10	7.3 ± 0.2
June 1970	39	5	3.5 ± 0.2
Oct.-Nov. 1970	86	15	5.9 ± 0.2*
Jan.-Feb. 1971	109	13	5.0 ± 0.2*

Student's t-test and tables of P values, two-tailed, were used for evaluation of statistical significance. The differences between successive values for the mean life span were highly significant, $P < 0.001$, except for the two pairs marked by asterisks where $P < 0.01$. The two sets of data for colonies at 17° C. represent colonies kept in two separate culture systems but in the same constant temperature room.

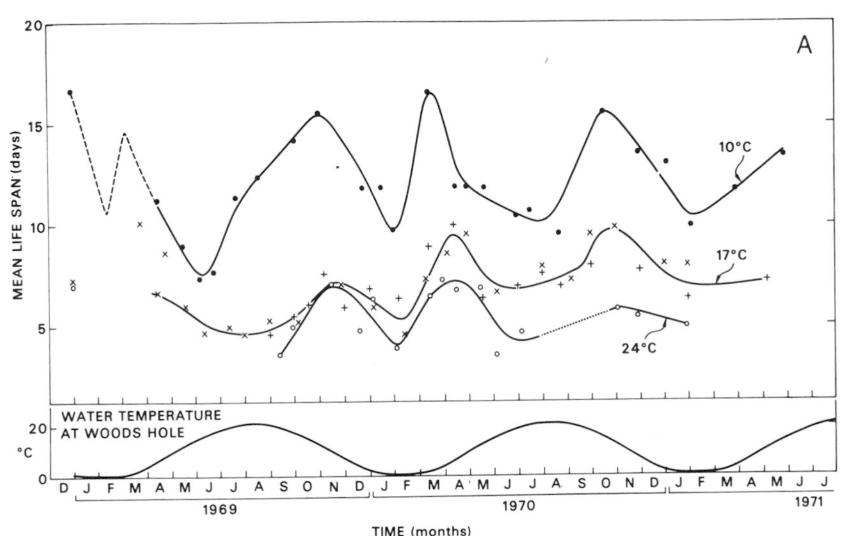
A
MEAN LIFE SPAN (days)
20
15
10
5
10°C
17°C
24°C
WATER TEMPERATURE
AT WOODS HOLE
°C
20
0
D J F M A M J J A S O N D J F M A M J J A S O N D J F M A M J J
1969
1970
1971
TIME (months)

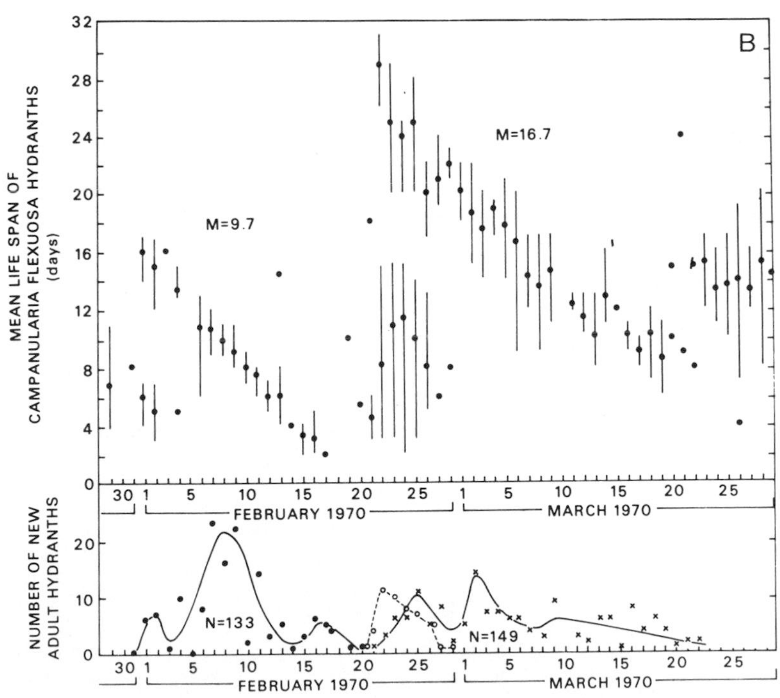
B
MEAN LIFE SPAN OF
CAMPANULARIA FLEXUOSA HYDRANTHS
(days)
32
28
24
20
16
12
8
4
0
M=9.7
M=16.7
30 1 5 10 15 20 25 1 5 10 15 20 25
FEBRUARY 1970
MARCH 1970
NUMBER OF NEW
ADULT HYDRANTHS
20
10
0
N=133
N=149

Fig. 14a. The rhythmicity of the seasonal changes in the longevity of hydranths of *C. flexuosa* persisted for over 2 years in colonies grown continuously at 10°, 17° and 24° C. The dashed line from January through April 1969 for the curve describing the colonies at 10° C. indicates that data from the new colonies obtained from Woods Hole in late 1971 were used to fill a gap in the initial studies. In the constant 24° C. environment, the colonies died in July 1970, and others were switched from 17° to 24° C. in October 1970. The dotted segment of the curve for the colonies at 24° C. connects these two points. In the lower portion of the figure, the yearly changes in the coastal water temperature at Woods Hole, Massachusetts are shown. These were calculated from daily records (1967-1969) provided by the U. S. Coast and Geodetic Survey, Woods Hole, Massachusetts (from Brock, 1974c).

Fig. 14b. Daily mortality data for February-March 1970 that corresponds to the mid-winter phase of curtailed growth followed by a reversal to the luxuriant phase. The mean life spans (closed circles) and the range of life spans for adults that matured on each successive day are shown. Initially, hydranth life spans were long, and they progressively became shorter from February 1-17. The mean life for hydranths in this entire cycle was 9.7 days. A quiescent phase followed with few adults maturing. Then a transitional phase was observed when 2 groups of adults matured, those with intermediate life spans and those with very long life spans. The latter initiated another cycle that began with adults having extremely long life spans that progressively became shorter. The mean life for hydranths in this entire cycle, February 22-March 22, was 16.7 days.

The number of hydrants that became adults on each successive day is also plotted in the lower portion of the figure. For each cycle, there were 3 peaks. If a large porportion of hydranths matured mid-way in the cycle, a shorter average life was recorded for all the hydranths included in the cycle. If more hydranths matured early in the cycle, then the average life for all the hydranths included in the cycle was long. The open circles connected by dotted lines indicate the number of adults that matured each day in the transitional group having life spans intermediate between those of the first and second cycles.

Circannual rhythmicity in the mean life span of hydranths of *C. flexuosa* whether they are considered individual organisms or parts of a colony has no parallels. On the contrary, gerontologists have emphasized that a fixed longevity is characteristic of each animal species (Comfort, 1964). For poikilothermic invertebrates, it must be added that the mean life spans are thought to be invariable so long as the environment remains constant. In view of the general acceptance of the concept of specific age, it is not surprising that contradictions either have not been raised previously or have been dismissed. The usual methods for collecting mortality data would, in themselves, obscure longevity rhythms because the data for life spans is a composite determined over timespans of varied lengths. Consequently, there is no discrimination of possible circa-lunar or circannual cycles in longevity. Furthermore, even in studies of short-lived invertebrates where successive measurements of longevity could be made, seasonal differences have not been reported. Illustrative of this is a previous report (Brock & Strehler, 1963) which gave a mean life span of *C. flexuosa* hydranths in colonies cultured at 17° C from April through June as 7.0 days. If the later data that was collected on a daily schedule (Brock, 1974c) is lumped for April through June of 1969 and 1970, an average life span of 7.1 days results. Although the second value determined after a seven year interval is reassuringly close to the first mean, complete blurring of the rhythmicity in hydranth longevity is accomplished by combining daily records. Just how much data is dismissed because it is in disagreement with the specific age concept is difficult to estimate. One published report, in part on the effects of different ambient temperatures on the longevity of rotifers, *Euchlanis dilatata*, suggests another contradiction (King, 1970). The rotifers were collected from fresh water ponds in May and June and subsequently cultured at 27° and 19° C. Although King reported that the mean life spans were similar at both temperatures, his data shows that there was actually a marked reduction in the longevity of rotifers collected on May 1, 1968 and then grown at 27°C (King's clones A and B). In contrast, the higher ambient temperature did not decrease the longevity of rotifers collected either June 5 or June 12, 1968 (King's clones C, D, E and F). This suggests that the two groups of rotifers were in different phases of an annual cycle which was expressed in their responses to different ambient temperatures. Additional contradictions, relative to changes in invertebrate life spans when experiments are repeated under constant laboratory conditions, have been mentioned in conversations but never documented in published papers.

4. Adaptive Significance.

The endogenous rhythms in *C. flexuosa* hydranth longevity can be related to the seasonal changes in coastal water temperature in the animals' native environment, Woods Hole, Massachusetts (figure 14a). Although the longevity rhythms appear biphasic, for most of the annual cycle with the exception of the mid-winter phases of sparse growth, there is roughly an inverse relationship of hydranth mean life spans to the cyclic changes in the water temperature. Furthermore, the endogenous rhythms anticipate the yearly variations in the coastal water temperature (figure 15). It may be assumed that by anticipating the annual changes in their environment, the hydranths are prepared physiologically and biochemically for the approaching season.

CAMPANULARIA FLEXUOSA, A MODEL FOR STUDIES OF CELLULAR AGING

Many animal models, predominantly mammalian, have been proposed and are used to investigate age-related changes in dividing and non-dividing cell types. *C. flexuosa*, although a more simply organized poikilotherm, is far more responsive than mammals to experimental manipulation of the rate of aging and has other advantages including 1) the similarity of many cell types to those of complex mammalian species, including man; 2) the parallels in the age-related changes of cellular organelles in long-lived cells of *C. flexuosa* and mammalian species; 3) the accelerated rate of aging resulting in marked cellular alterations within days; and 4) the endogenous rhythmic changes in the rate of aging of *C. flexuosa* hydranths. The last, from the author's point of view, is the most intriguing for it raises the question of whether cellular changes associated with aging are simply accelerated in those organisms that have shorter life spans during certain phases of the annual cycle or whether different patterns of change exist in hydranths with increased longevity. And even more basic is the question of the nature of the control of the endogenous rhythms in the rate of cellular aging.

The known age-related cellular changes in *C. flexuosa* hydranths have been reported for only one phase of the annual cycle. In

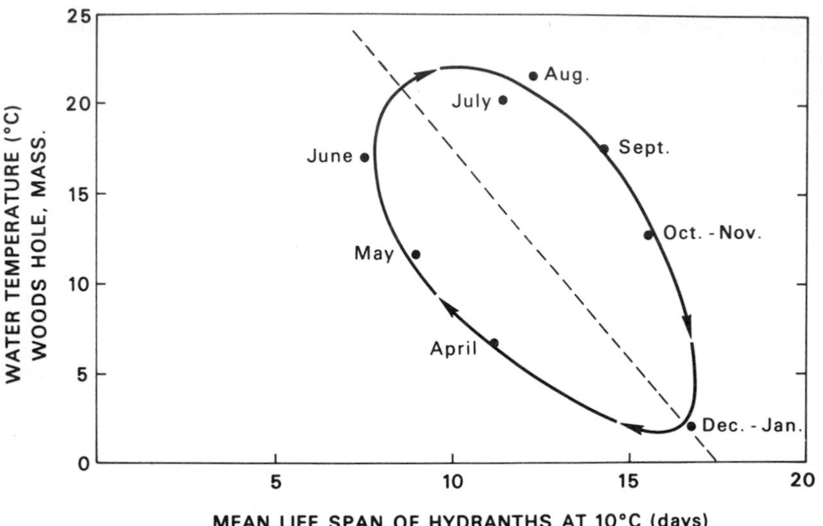

Fig. 15. The programmed changes in the longevity of *C. flexuosa* hydranths in colonies that were cultured continuously at 10° C. from December 1968 through November 1969 are related to the seasonal changes in the coastal water temperature at Woods Hole, Massachusetts. The lowest and highest water temperatures occur in January-February and August, respectively. If the rhythms in longevity evolved in synchrony with the yearly variations in water temperature, a direct correlation between values for the mean life spans of *C. flexuosa* hydranths and water temperature would be expected, and the data for mean life spans would lie on a straight line. The dashed straight line in this figure connects the point for the longest average life span with the point on the ellipse approximating the shortest average life span. The changes in longevity, however, anticipate the variations in water temperature. The hydranth mean life spans increased (July) prior to the seasonal fall in water temperature. Only one year is considered because rhythms with free-running periods different from 365 days will, in time, run further out of phase with the calendar year (from Brock, 1974c).

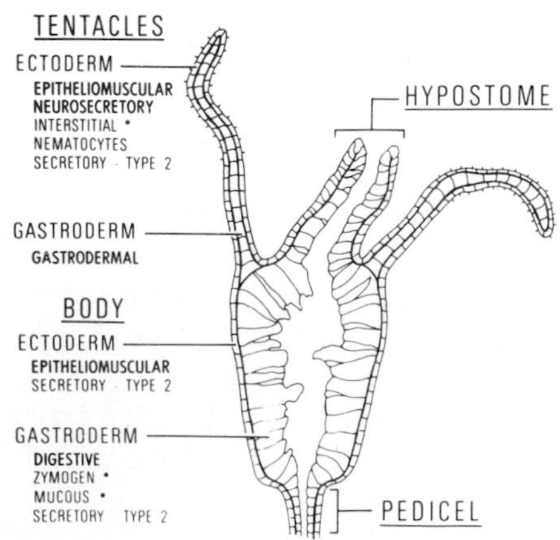

Fig. 16. C. flexuosa hydranths have a single ring of tentacles that surround a central projection, the hypostome, and the hydranth body is attached to the colony by the pedicel. Two cellular layers, the ectoderm and gastroderm, are separated by an acellular mesoglea, shown by the heavier line in this drawing. All of the ectodermal and gastrodermal cell types are listed. The boldfaced letters indicate the kinds of cells which display prominent age-associated ultrastructural changes. The asterisks mark the cell types in older hydranths with less marked alterations (from Brock, 1970).

ultrastructural studies, all of the cell types of the hydranths and those which were markedly different in older hydranths were described (figure 16) (Brock *et al*, 1968; Brock, 1970). In epitheliomuscular, neurosecretory, gastrodermal and digestive cells, an increased incidence of lysosomes with age was apparent, just as has been observed in fixed post-mitotic mammalian cells, such as muscle and nerve which are considered to be long-lived. These organelles in *C. flexuosa* and in the mammalian cell types are predominantly representative of the final stages of lysosomal activity, residual bodies, which contain indigestible residues of degraded cell organelles. The intracellular processes responsible for residual body formation begin with the biosynthesis of primary lysosomes that contain a variety of acid hydrolases (de Duve & Wattiaux, 1966). The enzymes at this point are potentially capable of causing vast degradation but are inactive. The primary lysosomes may coalesce either with cellular organelles that have been segregated from the remainder of the cytoplasmic matrix in autophagic vacuoles or with exogenous material phagocytized by the cell. Activation of the acid hydrolases occurs after this fusion and digestion ensues in the secondary lysosomes until only non-degradable material remains. This often very dense residue has no resemblance to cellular organelles and is the structural component commonly observed in aging cells. The changes with age reflected in epitheliomuscular cells of *C. flexuosa* are shown in figures 17 and 18, with stages in the degradation of segregated cellular organelles illustrated further in figures 19, 20 and 21.

The marked accumulation of residual bodies in older long-lived cells implies

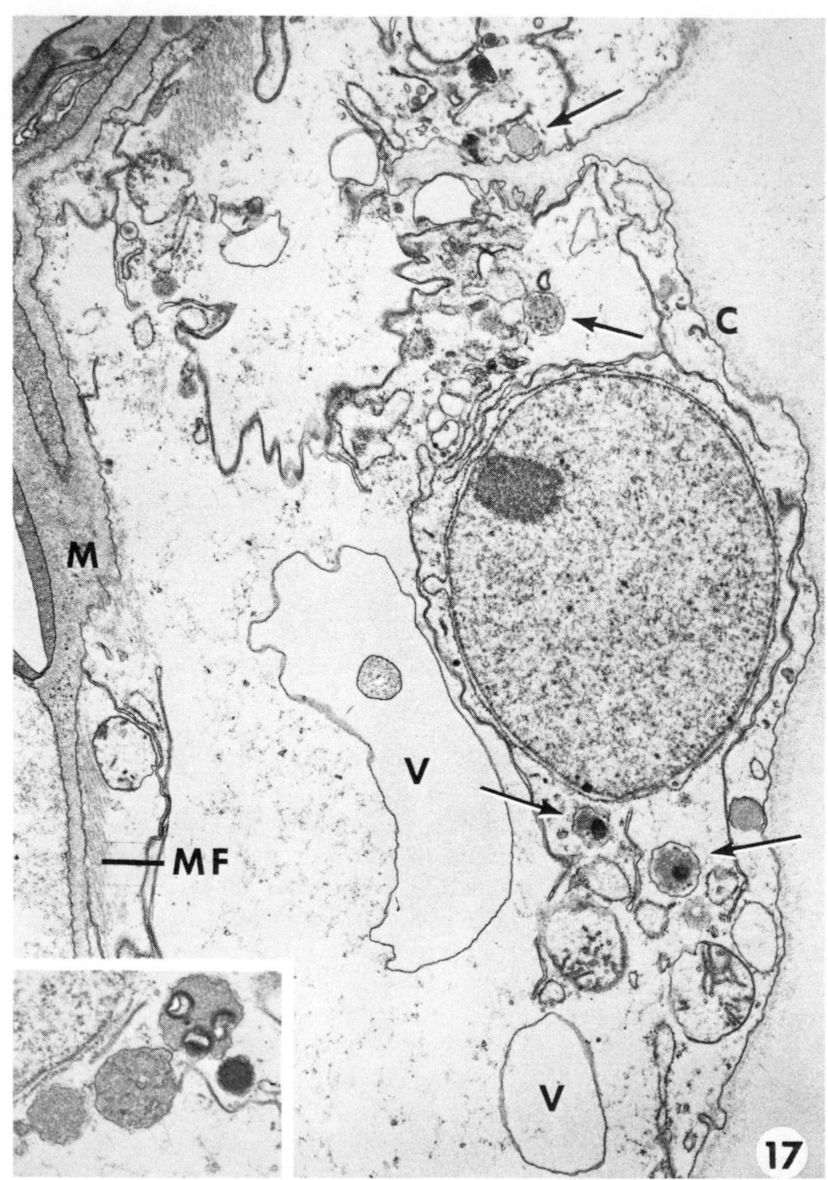

Fig. 17. An epitheliomuscular cell from a 9-day old hydranth has prominent primary lysosomes (arrows and inset) but otherwise is not altered appreciably. In this longitudinal section, the exterior surface is covered by the cuticle (C) and the basal area containing myofibrils (MF) lies next to the mesoglea (M). Conspicous vacuoles (V) that contain electron lucent material usually occupied the central portions of these cells, with the nucleus excentric. Mitochondria, sparse cisternae of the RER (rough endoplasmic reticulum), free ribosomes and smooth ER are scattered in the amorphous appearing cytoplasmic matrix. In other sections, scattered Golgi complexes, lipid droplets and vesicles were seen. X 14,000 (from Brock, 1970).

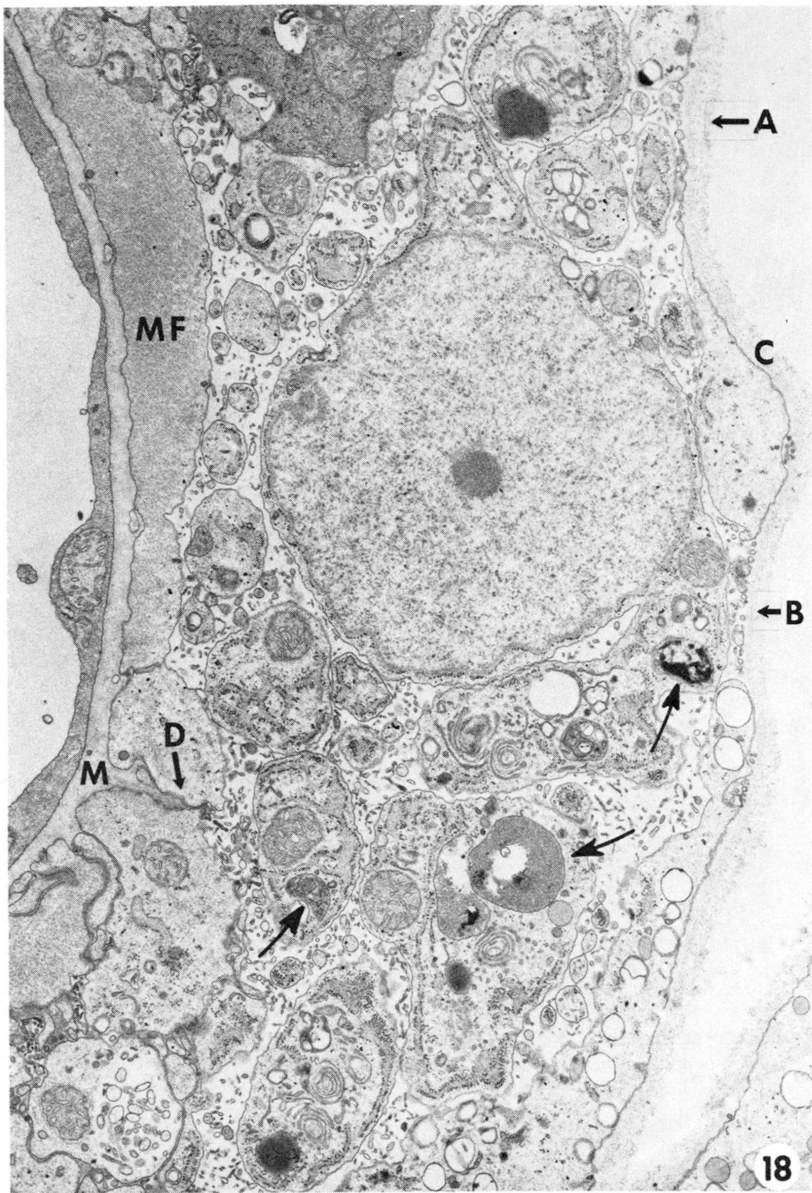

Fig. 18. The conspicous age-related changes in this epitheliomuscular cell reflect increased lysosomal activity. The large central vacuole is packed with autophagic vacuoles which contain both cellular components and material in variable stages of degradation (arrows). Interruptions in the plasma membrane appear at A and B where freed vacuolar material lies next to the cuticle (C), although the cuticle is uninterrupted. A channel (D) appears to connect the mesoglea (M) with the intravacuolar space. The basal cellular processes containing myofibrils (MF) and the mesoglea appear similar to corresponding areas in younger cells. X 15,600 (from Brock, 1970).

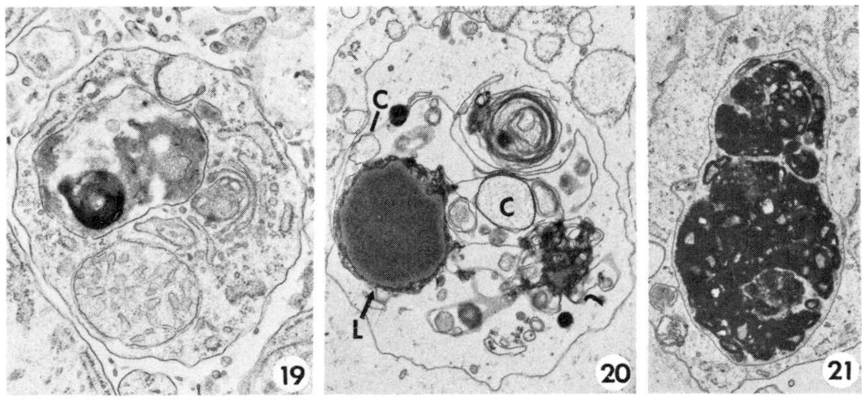

Fig. 19. This autophagic vacuole contains partially degraded, amorphous material and electron dense, laminar whorls in the homogeneous, electronlucent matrix and separated from the other contents by a single membrane. The mitochondrion, cisternae of the RER, smooth ER and ribosomes appear unaltered. The stages in degradation shown in this and the next two figures were observed in 8-day old hydranths. X 31,500 (from Brock, 1970).

Fig. 20. Membranous residues presumably the result of hydrolytic activity are conspicous in this autophagic vacuole. The membranous configurations around the lipid droplet (L) may indicate the beginning of lipid degradation. Other areas (C) resemble the cytoplasmic matrix of epitheliomuscular cells. X 25,100 (from Brock, 1970).

Fig. 21. This residual body with its electron-opague contents represents the final stage in the degradative events. X 24,100 (from Brock, 1970).

that some or all facets of lysosomal function may change with age. One may postulate 1) accelerated, programmed synthesis of primary lysosomes; 2) programmed alteration or deterioration with time of cell organelles that marks them for segregation and eventual degradation; and 3) changes in the nature of the intracellular membrane system that may provoke increased segregation activity. Even though specific organelles of some long-lived cells are segregated, as exemplified by the mitochondria of old rat myocardium (Travis & Travis, 1972), there is no convincing evidence which implicates either the mitochondria or some facet of the lysosomal system in inducing the selective degradation, although it seems plausible that subtle structural and/or functional alterations occurred. Little evidence exists even in non-aging systems relating to the mechanisms of selective degradation. One example is the discriminative autophagy of the rough endoplasmic reticulum (RER) in rat prostatic epithelial cells undergoing remodeling (Helminen & Ericsson, 1971). In these cells alterations of the RER cisternae preceeded their segregation in autophagic vacuoles, and, because the cellular remodeling followed castration, an indirect hormonal control was hypothesized. It is clear now that the interpretation of the numerous observations of enhanced lysosomal activity in aging cells demands definitive answers to the critical question of the nature of the cellular controls. In this regard, a system that expresses endogenous rhythmicity in the rate of aging provides a regularly changing program of cellular aging with each phase of the cycle.

SUMMARY

1. Endogenous circannual rhythms are known for two invertebrate species, an arthropod, the cave crayfish, *Orconectes pellucidus inermis*, and the colonial, marine cnidarian, *Campanularia flexuosa*. Only in the observations reported for these two species were the absolute minimum requirements for circannual rhythms met, the completion of one period and the initiation of a second oscillation under constant conditions.

2. The free-running periods for the cycles in reproduction of the cave crayfish and in the rate of aging, growth and development of *C. flexuosa* slightly exceeded one year.

3. Individual cave crayfish drifted out of phase with each other as did different colonies of *C. flexuosa* under constant laboratory conditions, eliminating the possibility that unknown zeitgebers or the culture conditions phased the rhythms.

4. The circannual rhythms of *C. flexuosa* were temperature-compensated, persisting in near synchrony at 10°, 17° and 24° C.

5. The endogenous rhythms reported for both of these invertebrate species allow anticipation of seasonal variations in the natural environment.

6. Yearly rhythms in *C. flexuosa* were defined by seasonal changes in the growth of the colonies and in development and longevity of the hydranths: luxuriant growth was interrupted each summer and mid-winter by phases of sparse growth, and, coincidentally, the hydranth mean life spans decreased to

approximately half the values observed during luxuriant growth phases.

7. The possibility of the existence of endogenous rhythms in longevity in other invertebrate species and the use of *C. flexuosa* as a model for studies of cellular aging are discussed.

8. Annual rhythmicity in a highly organized species like the cave crayfish and one with an unspecialized nerve net rather than a complex central nervous system, *C. flexuosa*, argues for the universality of circannual rhythms.

REFERENCES

ASCHOFF, J. (1965). Circadian vocabulary. Circadian Clocks. North Holland Publ. Co., Amsterdam. pp. x-xix.

BENOIT, J., ASSENMACHER, I. & BRARD, E. (1956). Apparition et maintien de cycles sexuels non saisonniers chez le Canard domestique place' pendant plus de trois ans a l'obscurite' totale. J. Physiol., Paris 48, 388-391.

BERRILL, N.J. (1950). Growth and form in calyptoblastic hydroids II. Polymorphism within the Campanularidae. J. Morphol. 87, 1-26.

BROCK, M.A. (1970). Ultrastructural studies on the life cycle of a short-lived metazoan, *Campanularia flexuosa* II. Structure of the old adult. J. Ultrastruct. Res. 32, 118-141.

BROCK, M.A. (1974a). Circannual rhythms I. Free-running rhythms in growth and development of the marine cnidarian, *Campanularia flexuosa*. Comp. Biochem. & Physiol., in press.

BROCK, M.A. (1974b). Circannual rhythms II. Temperature-compensated free-running rhythms in growth and development of the marine cnidarian, *Campanularia flexuosa*. Comp. Biochem. & Physiol., in press.

BROCK, M.A. (1974c). Circannual rhythms III. Rhythmicity in the longevity of hydranths of the marine cnidarian, *Campanularia flexuosa*. Comp. Biochem. & Physiol., in press.

BROCK, M.A. & STREHLER, B.L. (1963). Studies on the comparative physiology of aging. IV. Age and mortality of some marine cnidaria in the laboratory. J. Gerontol. 18, 23-28.

BROCK, M.A., STREHLER, B.L. & BRANDES, D. (1968). Ultrastructural studies on the life cycle of a short-lived metazoan, *Campanularia flexuosa* I. Structure of the young adult. J. Ultrastruct. Res. 21, 281-312.

BUNNING, E. (1967). The Physiological Clock. Springer-Verlag, New York.

COMFORT, A. (1964). Aging: The Biology of Senescence. Holt, Rinehart and Winston, Inc., New York.

CROWELL, S. (1957). Differential responses of growth zones to nutritive level, age, and temperature in the colonial hydroid Campanularia. J. Exp. Zool. 134, 63-90.

DAVIS, D.E. (1967). The annual rhythm of fat deposition in woodchucks (*Marmota monax*). Physiol. Zool. 40, 391-402.

DE DUVE, C. & WATTIAUX, R. (1966). Functions of lysosomes. Ann. Rev. Physiol. 28, 435-492.

GWINNER, E. (1968). Circannuale periodik als Grundlage des jahrzeitlichen Funktionswandels bei Zugvogeln: Untersuchungen am Fitis (*Phylloscopus trochilus*) und am Waldlaubsanger (*P. sibilatrix*). J. Ornithologie 10, 70-95.

HAMMETT, F.S. (1943). The role of the amino acids and nucleic acid components in developmental growth. Part One. The growth of an *Obelia* hydranth. Chapter One. Description of *Obelia* and its growth. Growth 7, 331-399.

HASTINGS, J.W. (1970). Cellular-biochemical clock hypothesis. The Biological Clock, Academic Press, New York, pp. 61-91.

HELMINEN, H.J. & ERICSSON, J.L.E. (1971). Ultrastructural studies on prostatic involution in the rat. Mechanism of autophagy in epithelial cells, with special reference to the rough-surfaced endoplasmic reticulum. J. Ultrastruct. Res. 36, 708-724.

JEGLA, T.C. & POULSON, T.L. (1970). Circannian rhythms - I. Reproduction in the cave crayfish, *Orconectes pellucidus inermis*. Comp. Biochem. Physiol. 33, 347-355.

KING, C.E. (1970). Comparative survivorship and fecundity of mictic and amictic female rotifers. Physiol. Zool. 43, 206-212.

PENGELLEY, E.T. & FISHER, K.C. (1963). The effect of temperature and photoperiod on the yearly hibernating behavior of captive golden-mantled ground squirrels (*Citellus lateralis tescorum*). Canad. J. Zool. 41, 1103-1120.

PENGELLEY, E.T. & KELLY, K.H. (1966). A "circannian" rhythm in hibernating species of the genus *Citellus* with observations on their physiological evolution. Comp. Biochem. Physiol. 19, 603-617.

TRAVIS, D. (1974). Personal communication.

TRAVIS, D.F. & TRAVIS, A. (1972). Ultrastructural changes in the left ventricular rat myocardial cells with age. J. Ultrastruct. Res. 39, 124-148.

CREDITS

Figures 1,2,3 and 4. Reproduced by permission of Pergamon Publishing Co. and Dr. Thomas Jegla from Reproduction in the cave crayfish, *Orconectes pellucidus inermis*. Comp. Biochem. Physiol. 33: 347-355; 1970.

Figures 5,7a,7b,7c,7d,7e,8,10 and 11. Reproduced by permission of Pergamon Publishing Co. from Circannual rhythms I. Free-running rhythms in growth and development of the marine cnidarian, *Campanularia flexuosa*. Comp. Biochem. Physiol., in press, 1974.

Figure 12. Reproduced by permission of Pergamon Publishing Co. from Circannual rhythms II. Temperature-compensated free-running rhythms in growth and development of the marine cnidarian, *Campanularia flexuosa*. Comp. Biochem. Physiol., in press, 1974.

Figures 13,14a,15 and table 3. Reproduced by permission of Pergamon Publishing Co. from Circannual rhythms III. Rhythmicity in the longevity of hydranths of the marine cnidarian, *Campanularia flexuosa*.

Figures 16,17,18,19,20 and 21. Reproduced by permission from Ultrastructural studies on the life cycle of a short lived metazoan, *Campanularia flexuosa* II Structure of the old adult. J. Ultrastruct. Res. 32: 118-141, 1970.

CIRCANNUAL RHYTHMS IN BIRDS WITH DIFFERENT MIGRATORY HABITS

PETER BERTHOLD

(Instructor at the University of Constance)

Max-Planck-Institut für
Verhaltensphysiologie,
Vogelwarte Radolfzell,
D-7761 Schloss Moeggingen,
WEST GERMANY

INTRODUCTION

As far back as 1702 von Pernau already suspected that in the fall a migratory bird of higher latitudes is not so much driven away by "hunger and coldness" but rather by a "hidden urge for migration" which is located in the bird itself. Though relatively vague, this is probably the earliest statement on endogenous factors involved in the control of annual biological rhythms. Only in this century however, did the question of endogenous factors in annual cycles achieve special importance, and particularly in plants and birds endogenous processes and circannual rhythms have been postulated, chiefly from observations of free living organisms (e.g., Sperlich, 1919; Okada, 1930; for review on birds, see Gwinner, 1971; Berthold *et al*, 1972a). After the existence of endogenous annual biological clocks had been shown in the snail *Limax flavus* (Segal, 1960) and in the mammalian hibernator *Citellus lateralis* (Pengelley & Fisher, 1963), the first clear-cut demonstrations of circannual rhythms in birds were shown by Gwinner (1967) for *Phylloscopus trochilus*, Berthold *et al* (1971) for *Sylvia atricapilla* and *S. borin*, and Schwab (1971) for *Sturnus vulgaris*. At present, circannual rhythms are well established for nine species of birds and can be assumed for a further seven species (Berthold, 1974a).

A comparative study of circannual rhythms in two warbler species by Gwinner (1971) revealed differences in the "rigidity" of the endogenous control and in the persistence of circannual rhythms in identical constant photoperiodic conditions, recalling the obviously different manifestation of annual clocks in closely related species of mammals of the genus *Eutamias* (Heller & Poulson, 1970). From Gwinner's study the intriguing question arose, as to whether circannual rhythms in birds are widespread or restricted to specialists, especially to migratory birds wintering in equatorial regions and to species living in other seasonally quasi-stable environments. In addition, one wants to know, whether poor or negative findings are conclusive with respect to the existence of annual clocks, or may merely be due to inadequate experimental conditions. According to Pengelley & Asmundson (1971), "it seems highly likely that circannual clocks are widespread in the animal world....circannual clocks may, in fact, be almost as universal as circadian ones." On the other hand, Farner (in Gwinner, 1971) has pointed out that "we should bear in mind that two thirds of the evolution of life was over before a circannual periodicity could have any adaptive significance...". Hence, not only from Gwinner's experiments, but also theoretically, the question of how widespread circannual rhythms in birds may be, is entirely open. To approach the question, within the warbler program of our institute (Vogelwarte 24: 320, 1968), a larger group of closely related bird species, with very different migratory habits, was tested for the existence, "rigidity", quality and importance of circannual rhythms. As experimental birds the genus *Sylvia* was chosen. The following overview presents our present knowledge on the occurrence, manifestation and importance of circannual rhythms in birds with different migratory habits in a comparative way.

THE SPECIES INVESTIGATED

Comparative studies of circannual rhythms in birds, with different migratory habits, have been performed in two genera of old-world warblers: in the genus *Phylloscopus* (two species) by Gwinner (1967, 1968a, b, 1971, 1972a, b) and in the genus *Sylvia* (six species) by Berthold (1973a, b, 1974b), Berthold & Berthold (1973a) and by Berthold *et al* (1971, 1972a, b, c). In the following short account some main features of the species of the two genera are given:

Sylvia borin, the garden warbler, *S. cantillans*, the subalpine warbler and *Phylloscopus trochilus*, the willow warbler; typical long-distance migrants (trans-

Sahara migrants); European populations winter exclusively in Africa; pre and postnuptial molt; large body weight increase during migration in consequence of heavy fat deposition.

Sylvia atricapilla, the blackcap and *Phylloscopus collybita*, the chiffchaff: less typical middle-distance migrants; European populations winter in Europe and Africa; pre and postnuptial molt; minor body weight increase during migration by means of moderate fat deposition.

Sylvia melanocephaia, *Sardinian warbler* and *S. undata*, the Dartford warbler; less typical partial migrants wintering within the Mediteranean breeding area; post and (at least in *melanocephala*) obviously prenuptial molt; body weight fairly constant throughout the year, at least in *melanocephala* with slight increase during the cold season.

Sylvia sarda balearica, Marmora's warbler; resident and endemic on the Balearic Isles and on the Pithyuses in the Mediteranean Sea; post and possibly prenuptial molt; possibly slight body weight increase in the cold season.

For more details on the annual cycle of the species treated see Gwinner (1968a, b, 1971); Berthold (1973a, 1974b); Berthold & Berthold (1973b) and Berthold *et al* (1972a, b, c).

As a rule, the experimental birds were taken from the nest at an age of a few days, raised by hand, and were always transferred to constant conditions within the course of their first year of life. Only a few blackcaps and garden warblers have been trapped during the period of their juvenal molt. The birds were housed individually in registration cages in which their locomotor activity was continuously recorded. Body weights and data on molting were regularly taken twice a week, and in groups of blackcaps and garden warblers the length of the left testis was measured *in situ* following approximately monthly unilateral laparotomy.

CIRCANNUAL RHYTHMS

1. Occurrence, persistence, properties. Garden warbler: Garden warblers have been kept in four different but constant photoperiodic conditions; LD 10:14, 12:12 and 16:8 (400:0, 01 lux, 20 ± 1.5^{0} C), in LD 12:12 with and without twilight, for up to 34 months.

29 out of 33 southwest German individuals showed clear circannual rhythms throughout the course of the experiment (figs. 1-3). Circannual rhythms could be established for four different annual processes: (1) molt, (2) body weight changes, (3) nocturnal (migratory) restlessness, and (4) changes in testis size (figs. 1-3). As fig. 1 shows, pre and postnuptial (juvenile) molts, body weight increases and periods of nocturnal restlessness during autumn and spring migratory periods occurred at intervals of approximately half a year, peaks of testis size at about a yearly interval. With some exceptions, the vast majority of the experimental birds showed circannual rhythmicity in all processes tested, with period lengths clearly deviating from that of the calendar year (figs. 1-3; Berthold *et al*, 1971, 1972a, b). In addition, an annual clock may be also involved in seasonal changes of food preference in the garden warbler (Berthold & Berthold, 1973a): In a one-year experiment, changes of food preference could be detected which are corre-

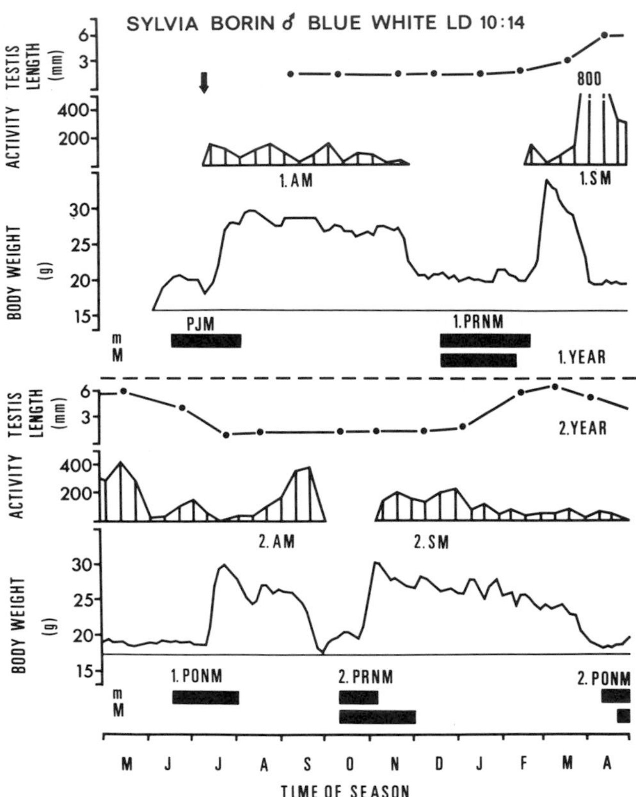

Fig. 1. Circannual rhythms of testis length, nocturnal restlessness, body weight changes and molt in *Sylvia borin* in LD 10:14 during a two-year experiment (above: data obtained in the first year, below: data obtained in the second year). Arrow indicates date of transfer of the handraised bird to constant photoperiodic conditions, m = molt of body feathers, M = molt of wing and tail feathers, PJM = postjuvenile molt, PRNM, PONM = pre and postnuptial molt, AM, SM = autumn, spring migration. From Berthold *et al.* (1971).

Fig. 2. Circannual rhythms of body weight, molt, and nocturnal restlessness in *Sylvia borin* in groups of experimental birds kept for a three-or two-year period in three different constant photoperiodic conditions. Numbers in parenthesis: experimental groups; years in parenthesis: experimental years of group 4. White bars: nocturnal restlessness, black bars: molt (upper row: molt of body feathers, lower row: molt of wing and tail feathers), KG (g): body weight in grams, thin lines: standard deviations, * indicates dates of hatching of the hand raised birds, arrow indicates dates of transfer to constant photoperiodic conditions. From Berthold *et al.* (1972 a).

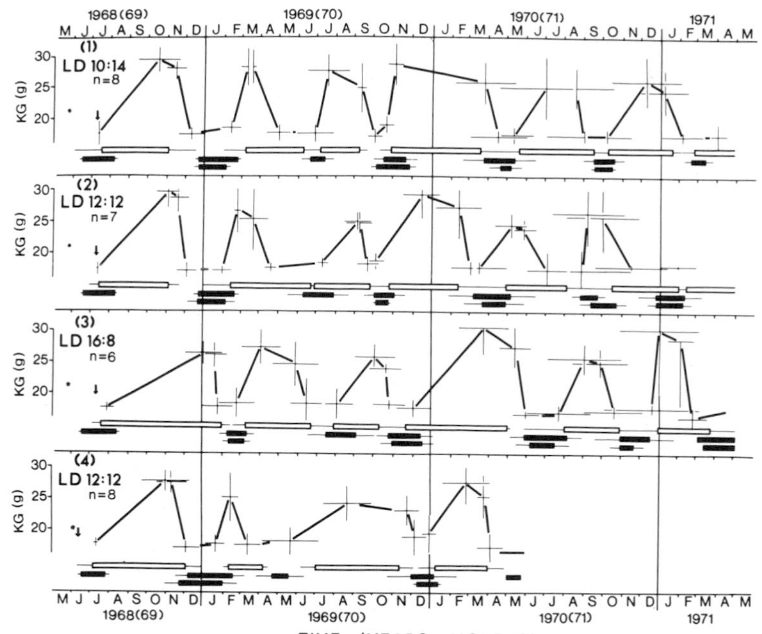

Fig. 2. See legend on preceeding page.

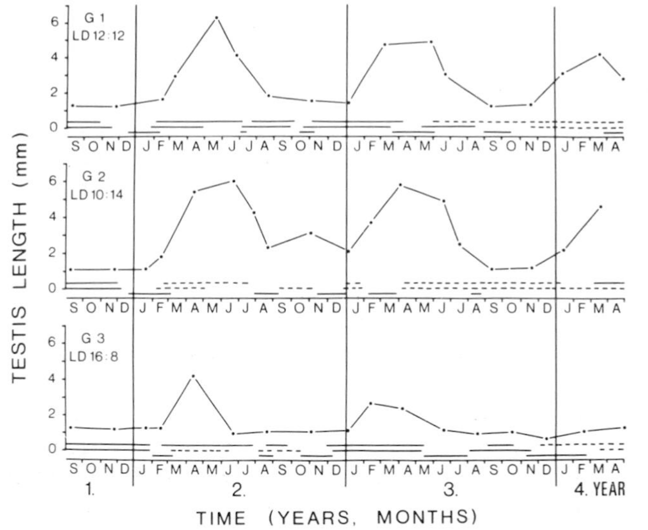

Fig. 3. Circannual rhythms of testis length in *Sylvia borin* kept for a three-year period under three constant photoperiodic conditions. From below to above, the solid bars represent molt, fat deposition and nocturnal restlessness. Broken bars indicate the onset and / or completion of an event which were not established exactly. From Berthold *et al.* (1972 b).

lated with seasonal changes of body functions, for which circannual rhythmicity is well established. During winter molt, plant food (berries) is preferred but during the main period of fat deposition (i.e. in the fall migration period) animal food (meal worms) is selected. Such seasonal changes in food preference have been interpreted in terms of nutritional adaptations to bird migration (for review, see Berthold, 1974c).

The results obtained from the study of the garden warbler are in aggrement with those on the willow warbler by Gwinner (*loc. cit.*): The willow warbler as a typical migrant shows clear circannual rhythms in molt, seasonal body weight changes and nocturnal (migratory) restlessness in different but constant conditions; changes in gonad size have not been tested.

The garden warbler is at present the bird species in which the largest number of circannual rhythms have been shown, and also over the longest experimental time. Only in the mammalian species *Citellus lateralis* has an annual clock been demonstrated for more parameters: i.e. hibernation, seasonal changes in body weight, food and water consumption, and changes in reproductive conditions (for review, see Pengelley & Asmundson, 1971).

Subalpine warbler: Subalpine warblers, (a total of eight birds from southern France, have been investigated in a two-year experiment under LD 10:14 conditions (400:0, 01 lux, 20 ± 1.5^{o}C). Because of the small size of the species, no laparotomy was performed, and therefore gonad size was not measured.

In the subalpine warbler, there was more interindividual variation in the occurrence of circannual rhythms than in the garden warbler. All experimental birds showed clear rhythms of molt (fig. 4, 17). "Summer" and "winter" molts occurred at about annual intervals. Only six out of eight birds, however, had persisting circannual rhythms of nocturnal restlessness and only three showed rhythms in body weight changes (fig. 4). Mean values could therefore only be calculated for the data on molting (fig. 17). Moreover, the maximum body weights gained during periods of migratory fattening continuously decreased in successive migratory periods (fig. 5) (see Berthold, 1973a, b, 1974b). These striking differences in comparison with the findings in the garden warbler will be discussed in section 6.

Blackcap: Blackcaps have also been kept under the same four conditions as the garden warblers: in LD 10:14, 12:12 and 16:8 (400:0, 01 lux, 20 ± 1.5°C), in LD 12:12 with and without twilight, for up to 34 months.

23 out of 32 southwest German individuals showed clear circannual rhythms through out the course of the experiment with period lengths clearly deviating from that of the calendar year (figs. 6-8). The behavior of the blackcaps kept in the same conditions as the garden warblers differed, however, in several respects from the latter species: First, in contrast to the garden warbler, the blackcap's circannual rhythms in LD 16:8 only persisted until the end of the second experimental year, and rhythmicity disappeared after that (fig. 6). Second, circannual rhythms could be established only for three different annual processes: (1) molt, (2) nocturnal (migratory) restlessness, and (3) changes in testis size (figs. 6-8), but not for body weight changes. Except for some increases of body weight during the first periods of nocturnal activity, body

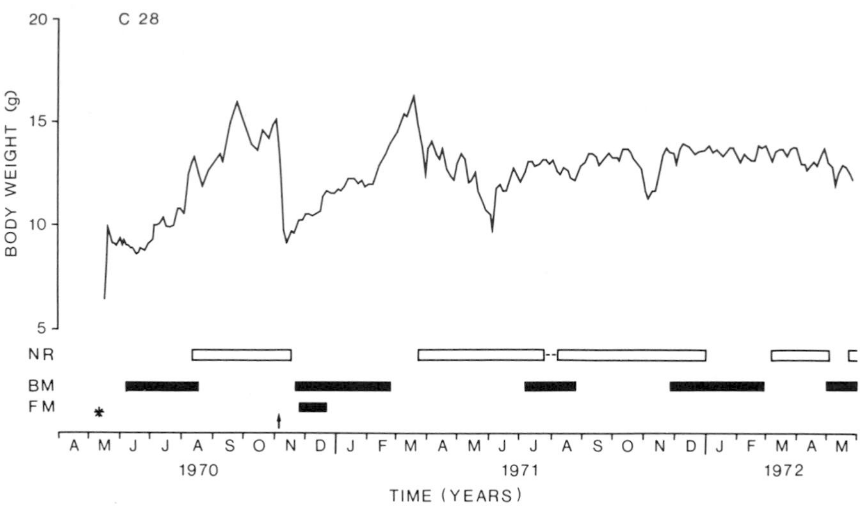

Fig. 4. Circannual rhythms of molt, nocturnal restlessness and body weight changes of *Sylvia cantillans.* FM black bars = molt of flight feathers; BM black bars = molt of body feathers; NR white bars = nocturnal restlessness; * date of hatching; arrow indicates date of transfer to constant photoperiodic conditions (LD 10:14). From Berthold (1974 b).

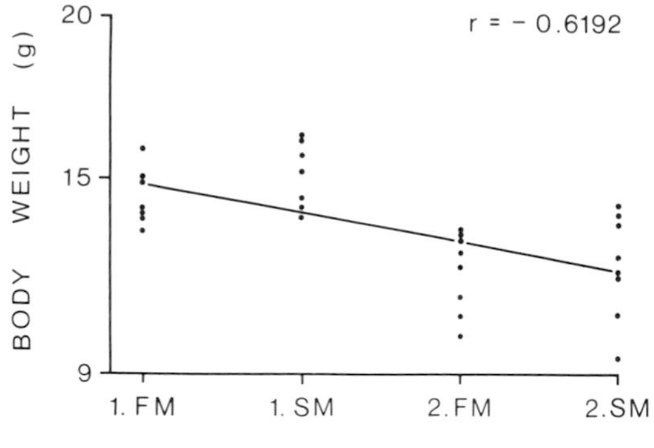

Fig. 5. Decreases of mean maximum body weight in successive migratory periods in *Sylvia cantillans.* FM = fall migration, SM = spring migration. From Berthold (1974 b).

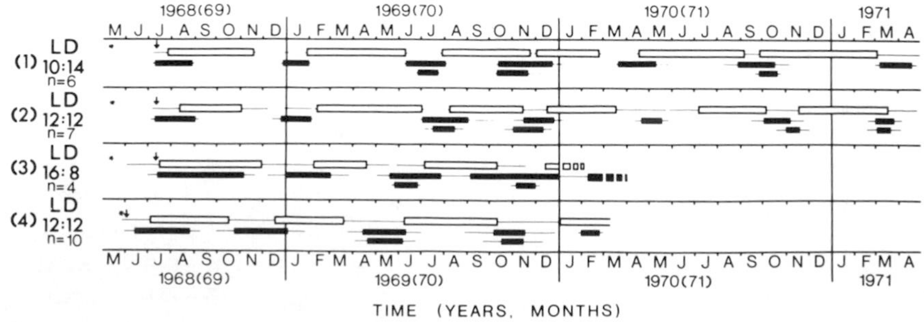

Fig. 6. Circannual rhythms of molt and nocturnal restlessness in *Sylvia atricapilla* kept for a three or two year period in three different constant photoperiodic conditions. For further explanations see fig. 2. From Berthold *et al.* (1972 a).

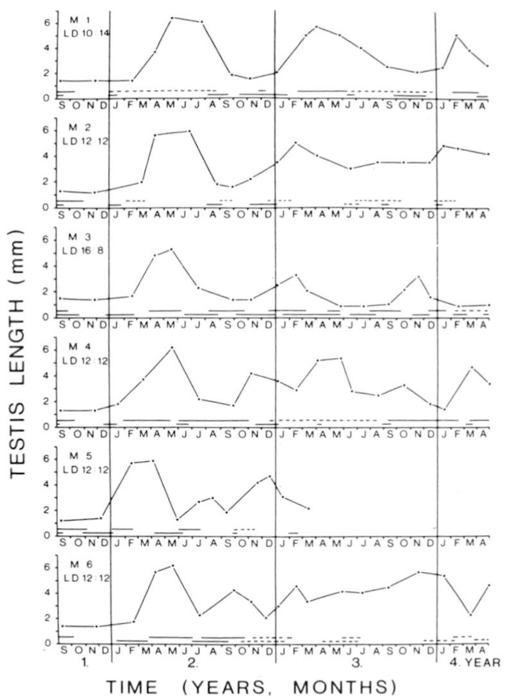

Fig. 7. Circannual rhythms of testis length of *Sylvia atricapilla*. From below to above, the solid bars represent molt and nocturnal restlessness. For further explanations see fig. 3. From Berthold *et al.* (1972 b).

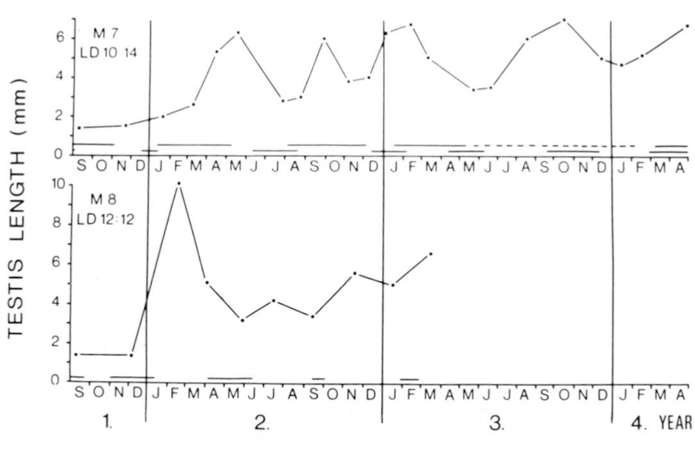

Fig. 8. As for fig. 7.

weight was rather constant. In freeliving conspecifics of the same population, however, there is a regular body weight increase of several grams at least during the fall migratory period. Third, in contrast to the garden warbler, there was greater interindividual variation with respect to the gonadal cycles. In LD 10:14 and 12:12, some blackcaps showed two peaks of testis size per year (figs. 7, 8). These circasemiannual rhythms will be discussed in detail in section 5. With the exceptions of these three differences, circannual rhythms in the blackcap were equal to those of the garden warbler (Berthold *et al*, 1971, 1972a, b).

The results obtained from the study of the blackcap differ considerably from those reported by Gwinner (1971, 1972a, b) for another middle-distance migrant, the chiffchaff. In contrast to the blackcaps, chiffchaffs kept in LD 12:12, with initially existing rhythms of molt, body weight changes and nocturnal restlessness became increasingly irregular and rhythmicity disappeared in part within an experimental period of only 15 months.

Sardinian warbler: A total of six birds from southern France have been investigated in a two-year experiment at LD 10:14 (400:0, 01 lux, 20 ± 1.5^{o}C). Because of the small size of the species, no laparotomy was performed, and therefore gonad size was not measured (Berthold, 1973a, b, 1974b).

Two birds showed clear circannual rhythms in "summer" and "winter" molt with period lengths considerably deviating from twelve months (fig. 9). In the other birds, at first sight no rhythmicity in the molt could be seen. However, if molt intensity was taken into account, circannual rhythms of molt intensity could easily be detected (fig. 10). This interesting discovery will be discussed in section 3.

In caged Sardinian warblers, no rhythms of body weight could be observed (figs. 9, 10), whereas in freeliving conspecifics body weight slightly increases during the cold season (Berthold, 1974b). There was considerable variation in the occurrence of nocturnal restlessness: Not a single bird exhibited nocturnal restlessness throughout the course of the experiment. One bird was restless in three or four different periods (fig. 9), another bird in two different periods, and four birds only once. The absence of a regular occurrence of nocturnal (migratory) restlessness on the one hand, and of regular cyclic body weight changes on the other is in good agreement with the bird's habits as a partial migrant and, when migrating, as a short-distance migrant (see section 2). The problem of intraspecific variation in the occurrence of nocturnal restlessness is discussed in section 4.

Dartford warbler: Six birds from southern France were investigated in a two-year experiment under the same conditions and in the same way as the Sardinian warblers; i.e. in LD 10:14, for molt, body weight, and restlessness cycles (Berthold, 1973a, b, 1974b).

The results obtained for that species are highly congruent with those for the Sardinian warblers: One bird (fig. 11) showed clear circannual rhythmicity in molt: in approximately annual intervals "summer" and "winter" molts occurred, but period length deviated considerably from twelve months. In all other specimens a circannual periodicity in molt could only be detected by considering changes in molt intensity (figs. 12, 13). One bird displayed periodical

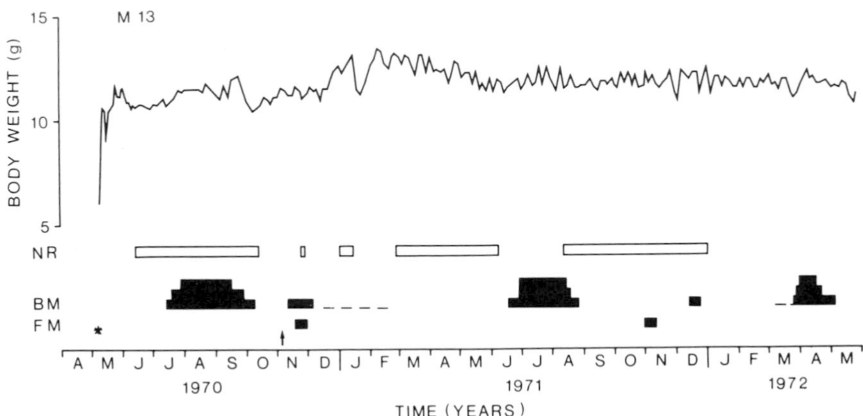

Fig. 9. Circannual rhythm of molt as well as nocturnal restlessness and body weight in *Sylvia melanocephala.* BM, single black bar: up to one third of the bird's plumage parts are molting; double black bar: up to two thirds and threefold bar more than two thirds of the plumage parts are molting. Broken line: some few body feathers are molting. For further explanations see fig. 4. From Berthold (1974 b).

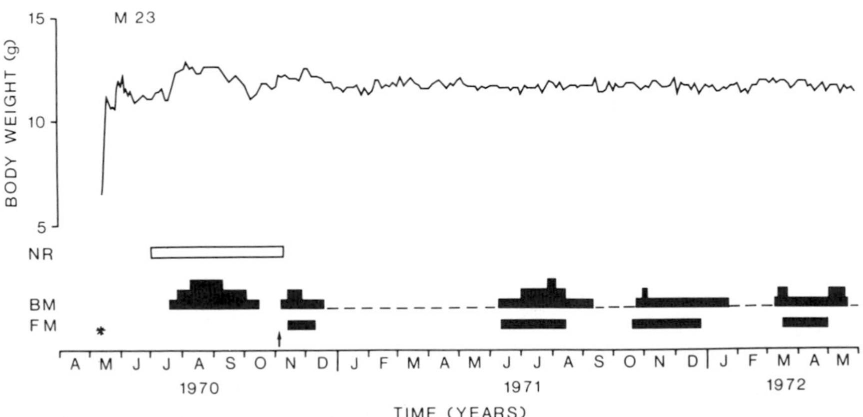

Fig. 10. Circannual rhythm of molt as well as nocturnal restlessness and body weight in *Sylvia melanocephala.* For further explanations see figs. 4 and 9. From Berthold (1974 b).

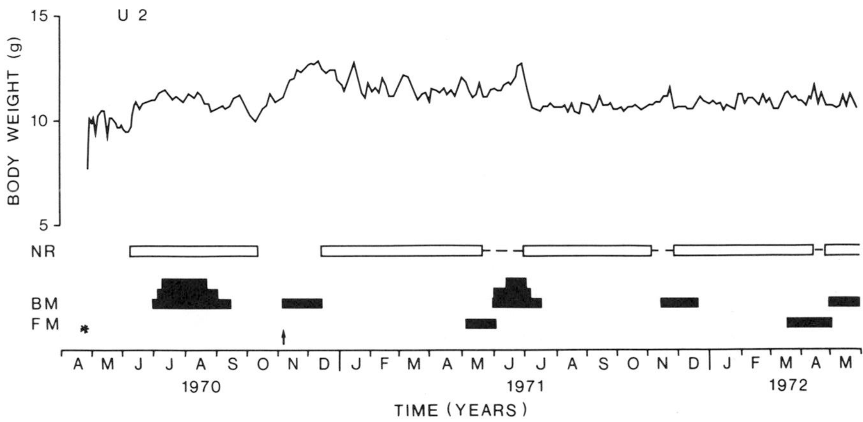

Fig. 11. Circannual rhythms of molt and nocturnal restlessness as well as body weight of *Sylvia undata.* For further explanations see figs. 4 and 9. From Berthodl (1974 b).

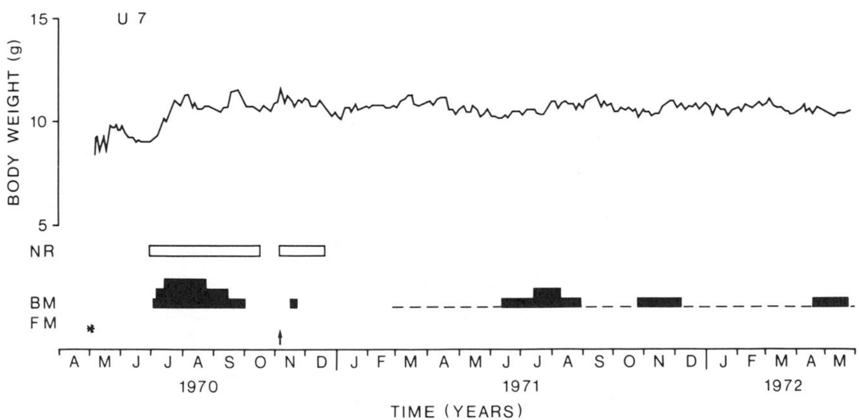

Fig. 12. Circannual rhythm of molt as well as nocturnal restlessness and body weight of *Sylvia undata.* For further explanations see figs. 4 and 9. From Berthold (1974 b).

nocturnal restlessness in regular intervals throughout the experiment (fig. 11), and the increasing deviation of the onsets of this restlessness from the calendar year clearly indicates that this restlessness was based on a circannual rhythm. In the other birds, the occurrence of nocturnal restlessness varied from none to irregular appearance until the end of the experiment. Body weight did not show any rhythmicity, but was fairly constant throughout the experiment (figs. 11, 12) as is obviously the case in freeliving conspecifics throughout the year (Berthold, 1974b). In some birds, however, occasional peaks of body weight occurred (fig. 13), which mainly occurred during the winter months but which were also related to periods of nocturnal activity and are, therefore, difficult to interpret. In the Dartford warbler, as in the Sardinian warbler, the general absence of a regular exhibition of nocturnal (migratory) restlessness and of regular cyclic body weight changes is in good agreement with the bird's habits as a partial migrant and a short-distance migrant (see section 2).

Marmora's warbler: A total of 9 birds from Formentera, Pithyuses, were investigated in a two-year experiment under the same conditions and in the same way as the other Mediteranean species, i.e. in LD 10:14, for molt, body weight, and restlessness cycles (Berthold, 1973a, b, 1974b).

Three birds showed clear circannual rhythms of molt with corresponding molts at approximately yearly intervals (fig. 14). In the other specimens (fig. 15), cyclic changes in molt intensity at clearly circannual intervals for corresponding molts showed up. Maximum occurrence of nocturnal restlessness was seen in six birds in two

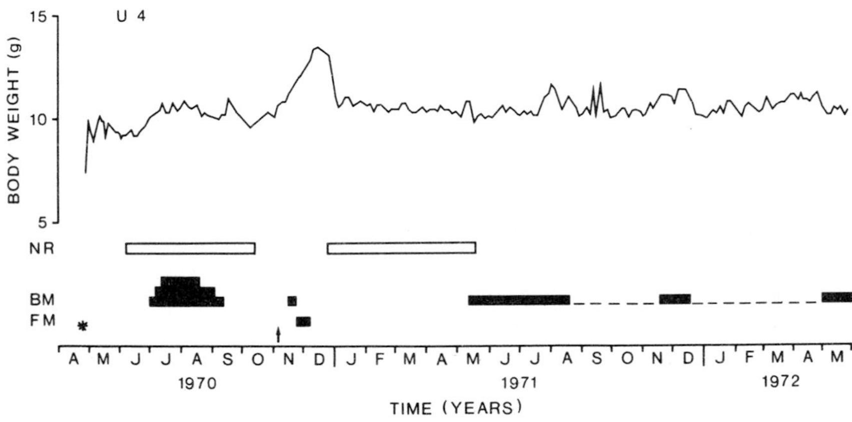

Fig. 13. Circannual rhythm of molt as well as nocturnal restlessness and body weight of *Sylvia undata.* For further explanations see figs. 4 and 9. From Berthold (1974 b).

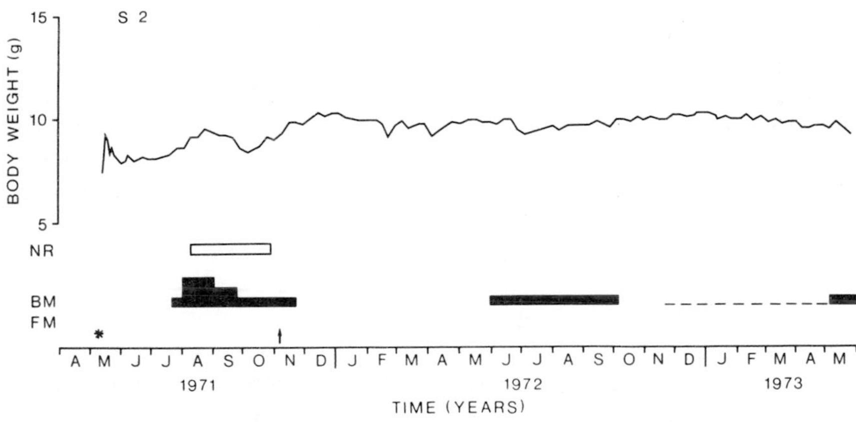

Fig. 14. Circannual rhythm of molt as well as nocturnal restlessness and body weight of *Sylvia sarda.* For further explanations see figs. 4 and 9. From Berthold (1974 b).

successive periods, three birds were restless during only one period. Body weight was fairly constant or showed some occasional peaks (figs. 14, 15).

Besides Marmora's warbler as a quasi-resident bird, and the race *balearica* as an obviously completely resident race (Berthold & Berthold, 1973b), a fully resident species has been tested for circannual rhythms: i.e. the crested tit *Parus cristatus* (Berthold, 1973b), which is resident in the whole area of their European distribution, including the northernmost parts of Europe. In a two-year experiment in LD 10:14, eight birds, tested for molt, body weight changes, and restlessness cycles, showed constant body weight as did conspecifics in the wild, displayed no restlessness at all, but showed clear circannual rhythms in molt, at least in the intensity of the molt of the body feathers (fig. 16).

In short, all seven species of the genera *Sylvia* and *Parus* investigated possess circannual rhythmicity; and at least molt or molt intensity in partial migrants and resident birds is controlled by circannual rhythms. In the genus *Phylloscopus*, circannual rhythms have been demonstrated at least in the willow warbler.

2. Period length, τ.

In *S. atricapilla* and *S. borin*, mean period lengths of circannual rhythms were calculated for rhythms of molt and nocturnal restlessness, and in *S. Borin* for body weight changes also (Berthold, *et al*, 1972a). For the Mediteranean warblers, mean values for the molt could be calculated (fig. 17). The mean values of τ for all *Sylvia* species are presented in Table 1.

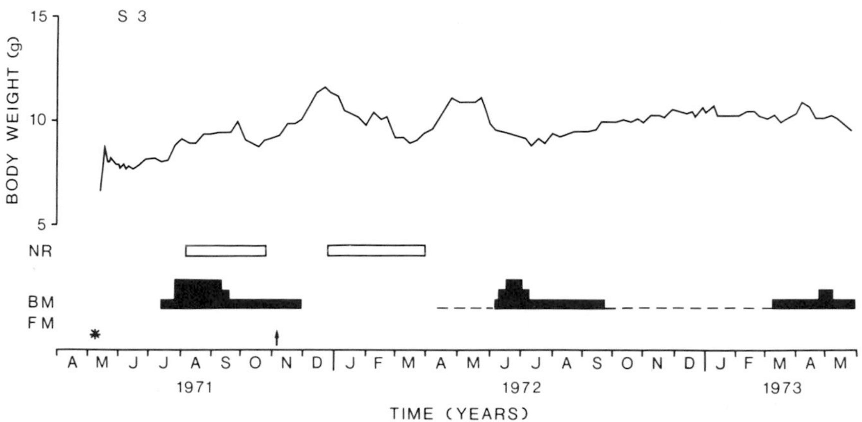

Fig. 15. Circannual rhythm of molt as well as nocturnal restlessness and body weight of *Sylvia sarda.* For further explanations see figs. 4 and 9. From Berthold (1974 b).

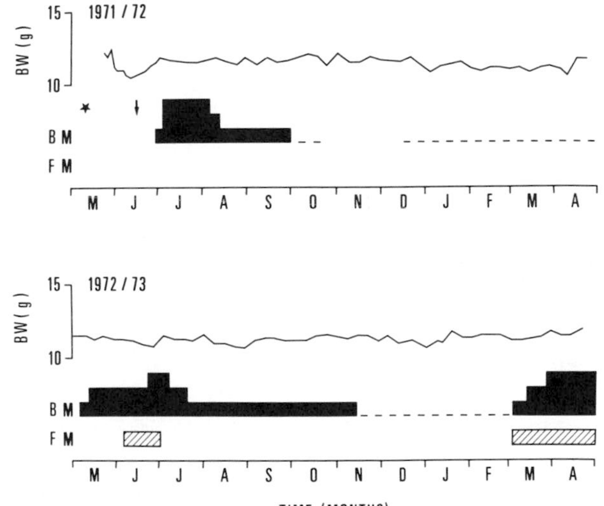

Fig. 16. Circannual rhythm of molt as well as body weight in a *Parus cristatus.* For further explanations see figs. 4 and 9. From Berthold (1973 b).

Table 1

Mean values of τ (in days) and standard deviations (s) averaged for several annual processes in *S. atricapilla* and *S. borin*, and for molt in the other species (Berthold, 1974b, Berthold *et al*, 1972a).

Species	τ (days)	s
Sylvia atricapilla	319	± 48.2
S. borin	322	± 55.2
S. cantillans	346	± 64.4
S. melanocephala	320	± 54.1
S. sarda	316	± 40.0
S. undata	338	± 18.8

There were no statistically significant interspecific differences in the period length τ. That holds also true when values of τ are only compared for corresponding annual processes - molt - and for equal experimental durations (Berthold, 1974b). Hence period length τ in the genus *Sylvia* is uniform, and it is generally shorter than the calendar year. Recent experiments by Gwinner (1971) and Schwab (1971) also established period lengths of less than one year for *Phylloscopus* and *Sturnus*, and in *Parus cristatus* cycles are also less than twelve months (Berthold, in preparation). A relative shortness of the period length of the endogenous rhythm in comparison with that of the coordinated environmental cycle has been considered to be advantageous for organisms, due to the fact that a clock which is fast makes its owner continuously disposed in advance to the environmental and necessary physiological changes (Berthold, 1974a). In this connection the following is interesting: within the Mediteranean warbler species which have been treated in entirely the same way, and can therefore be directly and mutually compared with one another, *S. cantillans* is a typical migrant with minor variations from year to year in the occurrence of annual events, and shows a tendency to a somewhat larger τ than the other species. In these there is greater variability from year to year in the course of annual processes. The resident *S. sarda balearica* shows a tendency to the shortest τ. Again in the temperate-zone species, the blackcap, as a less typical migrant, tends to have a shorter τ than the garden warbler as a more typical migrant. Keeping the previous interpretation of short period lengths in mind, it might be assumed that in *S. atricapilla, S. melanocephala, S. sarda* and *S. undata* it is the shorter period length which provides greater readiness for reactions to changing environmental conditions by a higher degree of preparedness in advance to these reactions. On the other hand, an annual clock which runs with a period length closer to that of the calendar year may be advantageous for typical migrants when wintering in quasi-stable environments.

3. Molt of long duration, changes of molt intensity.

In the less typical migrants and in the resident species of the genus *Sylvia*, (apart from normal circannual rhythms of molt with onsets of corresponding pre- and postnuptial

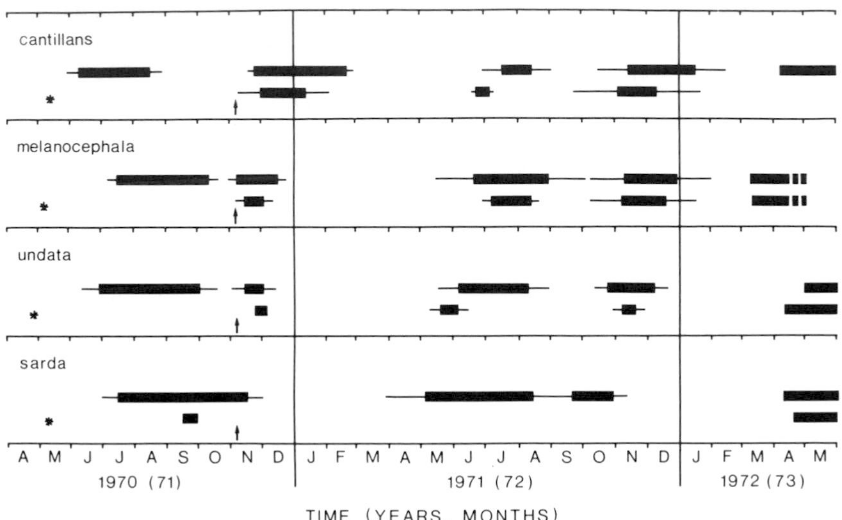

Fig. 17. Circannual rhythms of molt for experimental groups. Black bars: mean values, thin lines: standard deviations. Molt of very low intensity (of very few body feathers) is not depicted. For further explanations see fig. 4. From Berthold (1974 b).

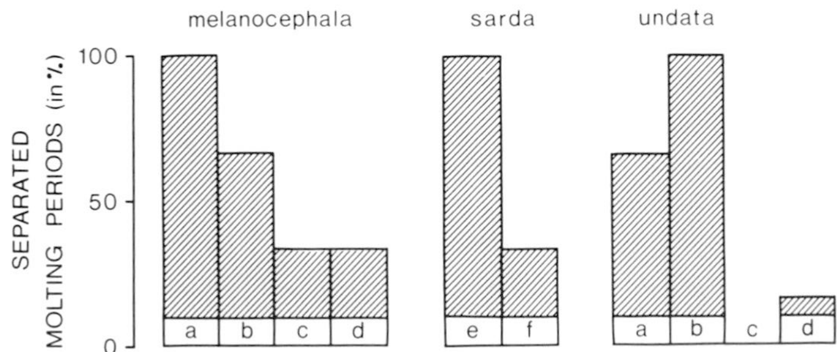

Fig. 18. Percentages of separated molts of body feathers in successive molt periods. a = juvenile molt - 1. winter molt, b = 1. winter molt - 2. summer molt, c = 2. summer molt - 2. winter molt, d = 2. winter molt - 3. summer molt, e = 1. summer molt - 2. summer molt, f= 2. summer molt - 2. winter or 3. summer molt. From Berthold (1974 b).

molts at approximately annual intervals) there occurred a molt of body feathers of long duration with periodic changes of molt intensity (section 1). This extended molt increased with longer experimental time (fig. 18). From these observations the intriguing question arises as to whether the molt of long duration is normal, i.e., a good reflection of the event as it occurrs in freeliving conspecifics, or whether it is an artefact caused by the experimental treatment. Fortunately, data of extensive studies on molt in one species, *S. melanocephala* from southern France, are available (fig. 19). In this species molting has been found all the year round except in January. In that month however, very slight molting like that in *S. sarda* (Berthold, 1974b) may simply have been overlooked in routine molt checks. From these data, it may be cautiously concluded that in individual birds also, very extended molting periods occur; even periods longer than one year cannot be excluded, so that the molt of long duration in the experimental birds should not necessarily be considered as an artefact.

Let us now consider another interesting observation: molt of long duration has only been observed when nocturnal restlessness was not present or was no longer displayed (figs. 10, 12-16), and it never occurred in species and individual birds regularly exhibiting nocturnal restlessness (figs. 1, 2, 4, 6, 9, 11). These data indicate that molt of long duration and nocturnal (migratory) restlessness are obviously mutually exclusive. If there is a cause-and-effect relationship between both events, the manifestation of circannual rhythmicity of molt in partial migrants, either in separated molt periods or in intensity

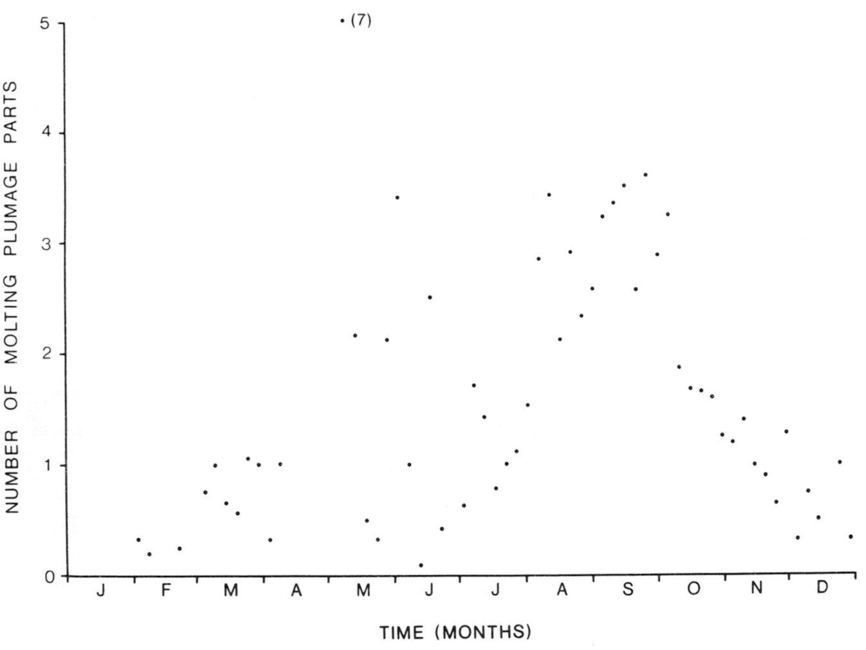

Fig. 19. Intensity of molt of body feathers of *Sylvia melanocephala* in the Camargue, S-France, in the course of the year according to data from Station Biologique for 1952-1970. From Berthold (1974 b).

fluctuations, would only or mainly depend on the display of restlessness. It is hypothesized that restlessness suppresses molt like those in the well known cases in which molt is interrupted by actual migration (e.g., Stresemann & Stresemann, 1966).

It is common knowledge that in many partial migrants migratory activity decreases with increasing age (e.g., Schuz *et al*, 1971). If this also holds true for the warbler species investigated, and if migratory activity in these birds is permanently preprogrammed (see section 4), then in addition the increased occurrence of long duration molt with increasing duration of the experiment, (i.e., with an increasing age of the birds), could be normal.

4. Nocturnal restlessness.
For all six species of the genus *Sylvia* (fig. 20), and for *Phylloscopus trochilus* and *P. collybita* (e.g., Gwinner, 1968b, 1972a) it has been shown that the first display of nocturnal restlessness is essentially an expression of true migratory activity. In *Sylvia* species, the relative ratio of the amount of nocturnal activity displayed under identical conditions corresponds closely to that of the average migratory distance. Furthermore, for *Phylloscopus* species it has been shown by comparing distances travelled by freeliving ringed and recovered birds on the one hand, and of amounts of restlessness displayed by caged birds in corresponding periods on the other, that nocturnal restlessness displayed as migratory activity in the wild, would just enable the birds to reach the species-specific winter quarters. Since we know that at least in the two *Phylloscopus* species and in *S. atricapilla* and *S. borin* the amount of restlessness in the

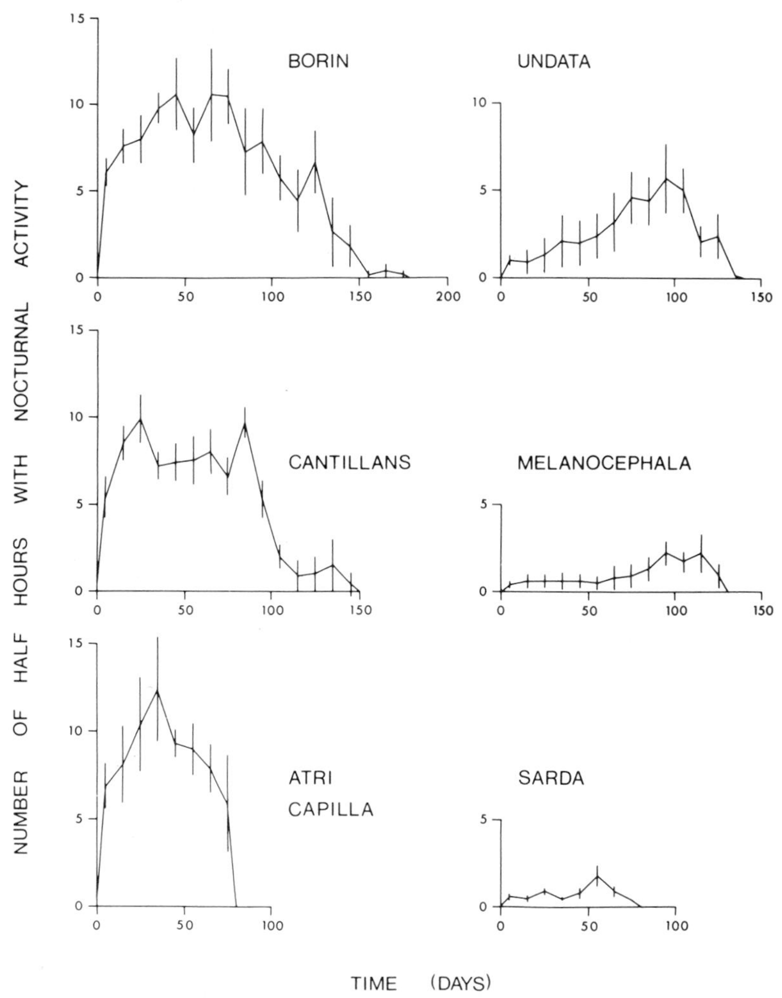

Fig. 20. Patterns of migratory restlessness of *Sylvia* species. Mean values and standard errors of means for 10-day periods. From Berthold (1973 a).

first migratory period is chiefly inborn, we can conclude that endogenous factors not only control the onset of migratory activity, but duration and pattern as well. The same holds true for the fat deposition during hibernation in the mammal (Heller & Poulson, 1970), and for the temporal organization of juvenile development in warblers (Berthold *et al*, 1970; Gwinner *et al*, 1971, 1972). These observations on nocturnal restlessness led to the hypothesis that in warblers an endogenous time-program is essentially involved in finding their winter quarters at least during their first fall migration (Gwinner, 1968b; Berthold *et al*, 1972c--vector navigation; Schmidt-Koenig, 1973). The predictions of this hypothesis have been confirmed by numerous observations on migratory behavior of several species in the wild and by uncompleted experiments with different races of *Zonotrichia leucophrys* (Gwinner, 1972a).

The development of small amounts of nocturnal activity in caged resident *S. sarda balearica* has been interpreted in terms of an atavistic remnant of an ancestral migratory behavior (Berthold, 1974b), as has been formerly done for *Zonotrichia leucophrys nuttalli* (Smith *et al*, 1969).

In the Mediteranean partial migrating *Sylvia* species, there was a surprisingly high variation with respect to the number of periods in which nocturnal activity was displayed throughout the experiment (section 1). This variation can be explained in different ways: (1) The disappearance of nocturnal activity in the course of the experiment could be due to artefacts caused by the experimental treatment - for instance, by internal desynchronisation (e.g., Berthold *et al*, 1972a) - and consequently could be caused by a fading of the circannual rhythm. However, the persistence of circannual rhythms of molt in all birds investigated argues strongly against the assumption of a general fading of the rhythms. (2) The large interindividual variation in nocturnal restlessness could be due to different endogenous programs, i.e., to genetic differences. If this is true and also applicable to other partial migrants, polymorphism would generally be the reason for different migratory behavior in individuals of populations of partial migrants, as has been assumed repeatedly (for review see Schuz *et al*, 1971; Berthold, 1974c).

In the Mediteranean *Sylvia* species except for *S. cantillans*, there was a conspiciously high degree of overlap between nocturnal restlessness and juvenile molt. In the extreme cases, restlessness began before the onset of molt, and in any case very early in comparison with the onset of migration of freeliving conspecifics (figs. 9-15). In Mediteranean *Sylvia* species, early summer movements within the breeding areas are known which may lead to settlement in another area (e.g., Berthold, 1973a). Our results suggest that in the Mediteranean partial migrating and resident *Sylvia* species, at least part of the nocturnal restlessness displayed, is possibly not true migratory activity but some type of "dismigration restlessness" following the terminology of Berndt & Sternberg (1966). This interpretation would mean that not only migration, but also dismigration (in the form of dispersal--Berndt & Sternberg, 1966) is preprogrammed at least to some extent. Many more investigations are necessary in order to ascertain whether these assumptions are valid.

5. Gonadal cycles.

In the two *Sylvia* species, for which gonadal cycles under constant conditions were investigated, and for which proof was obtained that they are controlled by circannual rhythms, considerable interspecific differences and differences between caged and freeliving birds were found (Berthold *et al*, 1972b). Most striking was the occurrence of two testis cycles (circasemiannual rhythms) within one calendar year in the blackcap. Since we know that (1) in contrast to the garden warbler the blackcap undergoes regular autumnal courtship display, and (2) the blackcap in its resident population on Cap Verde Islands obviously breeds in spring and in autumn (Bannerman & Bannerman, 1968), it is not unlikely that the two-peaked testis cycle of the experimental southwest German blackcaps is an expression of an original disposition. In the wild, the autumnal peak of gonadal development in the tested migratory southwest German birds could easily be suppressed by environmental or internal factors, such as migratory behavior, which exerts strong effects on the gonadal cycle (Berthold, 1969).

The interpretation proposed for the circasemiannual gonadal rhythm in the blackcap might also be valid for the endogenously controlled gonadal cycles with high frequencies in the Pekin duck (Benoit *et al*, 1955).

6. Existence, manifestation and rigidity of circannual rhythms.

Proof that circannual rhythms are involved in the control of annual cycles was obtained for all *Sylvia* species in relation to their different migratory behavior, and to the various annual and environmental cycles they display or experience respectively. But proof was of different quality. In *S. atricapilla*, periodicity disappeared entirely in LD 16:8 and never occurred under various constant photoperiodic conditions for body weight. In *S. borin*, testis cycles ceased in LD 16:8, and in *S. cantillans*, body weight cycles faded away in LD 10:14. In *Phylloscopus*, only in *P. trochilus* (but hardly for *P. collybita*) could circannual rhythms be established (Gwinner, *loc. cit.*). Furthermore, as a rule, interindividual variation in circannual rhythmicity was greater in less typical migrants than in typical migrants (e.g., Gwinner, 1971, 1972b; Berthold, 1974a, b; Berthold *et al*, 1972a, b, c). What support can the different findings in different species give to our general concept of circannual rhythms in birds?

At present, the importance of constant experimental conditions as permissive factors for the development of circannual rhythms is abstruse (e.g., Gwinner, 1971; Berthold, 1974a), and Gwinner has concluded from a review of several experiments: "The range of permissive conditions for the expression of circannual rhythms is evidently held within narrow limits...". Consequently as a first conclusion, all negative findings in experiments on circannual rhythms, (above all in experiments with only one or a few different constant conditions), are extremely inconclusive and must be very cautiously interpreted. One and the same constant photoperiodic condition may play quite a different role as permissive factors for different species. A constant daylength of 16 hours, for instance, may for *S. atricapilla* (living the year round in shorter daylengths than *S. borin*) be much less permissive for the development of circannual rhythms than for *S. borin*. Hence, as a second conclusion from the different manifestations of circannual rhythms in different species and in only a few different constant conditions, one cannot simply argue in favor of different amounts and rigidity of endogenous control in the species investigated. The same applies to interspecific differences in the variability in timing of events based on circannual rhythms. Less uniform timing may be due to less rigid endogenous control and may cause a stronger control by environmental factors. Less uniform timing can, however, also reflect another type of endogenous program in which a greater interindividual variability is genetically preprogrammed. It might be advantageous for a population to offer a wider range of more different individual endogenous programs to a more variable environment. Thus, real interspecific differences in circannual rhythms of birds with different migratory habits can only be assumed, but are not fully proven. That *S. atricapilla* shows no rhythmic changes in body weight in any of the three different constant photoperiodic conditions, suggests of course, a mild, if any, endogenous control in the body weight cycle of this species. However, it cannot at present be excluded that the body weight cycle of this species is highly sensitive even to the slightest internal desynchronisation (Berthold *et al*, 1972a), and therefore hardly becomes apparent; or it may well be expressed in another fourth constant photoperiodic condition. Moreover, we must even face the possibility that in some species with circannual rhythms we will not be able to demonstrate them in whatever constant conditions we use. Nor can we breed many bird species because of hormonal deficiencies caused by caging (von Tienhoven,

1961).

It is obvious from this discussion, that demands for proof of endogenous circannual rhythms by maintaining the animals under constant light or constant darkness (Evans, 1970; Hamner, 1971) are unrealistic, since such environmental circumstances may be too unnatural for birds.

To sum up: Species with very different migratory habits, even two resident temperate-zone forms[1], show clear circannual rhythms. Interspecific differences occur, and they may either be caused by different degrees of the rigidity of endogenous programs (due to different types of endogenous programs) or may simply reflect experimental artefacts. Finally, from the results obtained so far, I agree with Pengelley & Asmundson (1971), that circannual rhythms may be widespread, and in my opinion there will be a rapidly increasing body of experimental evidence for circannual rhythms in birds.

[1]Whether the American European Starlings, once introduced from England, for which circannual rhythms are demonstrated (Schwab, 1971), derive from the resident British population or from migrating continental population, is not known (Berthold, 1968).

SUMMARY

1. Comparative studies on circannual rhythms in birds with different migratory habits have been performed in two genera: in *Phylloscopus* (two species: *collybita, trochilus*) and in *Sylvia* (six species: *atricapilla, borin, cantillans, melanocephala,*

sarda, and *undata*). Individual birds were investigated for up to 34 months and the same species were tested under four different conditions. Data from the resident *Parus cristatus* are also discussed.

2. All species except *Phylloscopus collybita* showed clear circannual rhythms: *Sylvia borin* in molt, body weight changes, nocturnal restlessness, testis size, and probably in seasonal changes of food preference. *Phylloscopus trochilus* and *S. cantillans* in molt, nocturnal restlessness, and body weight changes. *S. atricapilla* in molt, nocturnal restlessness and testis size, and the other *Sylvia* species and *Parus cristatus* in molt (at least in fluctuations of the intensity of the molt of body feathers), and to some extent in nocturnal restlessness in the *Sylvia* species.

3. As a rule, events which are known to be periodic in freeliving birds, were also periodic in caged conspecifics with some exceptions: In *S. atricapilla*, no changes of body weight occurred, and in *S. cantillans* the rhythm of body weight changes disappeared.

4. In *Sylvia atricapilla*, some birds showed circasemiannual rhythms of testis size, possibly due to an atavistic expression of an original disposition.

5. The mean period length τ was in all species considerably shorter than the calendar year, as it is generally the case in all bird species investigated so far. It follows that annual processes deviate the more from the season in which they normally occur the longer an experiment lasts, so

that the emerging rhythms are circannual.

6. With respect to τ, there were no significant differences between the species investigated. However, within the Mediterranean and temperate-zone *Sylvia* species, more typical migrants showed a tendency to a longer τ, i.e., closer to the period length of the calendar year. These differences may be adaptive and establish in less typical migrants a greater preparedness for reactions in advance to more variable environmental conditions; they may also provide an extremely exact annual, instead of a circannual, clock for typical migrants wintering in quasi-stable environments.

7. In partial migrants and resident forms, a molt of body feathers of very long duration, possibly also occurring in free-living conspecifics, is mutually exclusive with nocturnal restlessness. It is hypothesized that restlessness suppresses molt and influences molt rhythms.

8. Endogenous factors do not only initiate annual events, they also control duration and pattern of annual processes. In *Phylloscopus* and *Sylvia*, the temporal organization of juvenile development and the amount of nocturnal restlessness is preprogrammed. An endogenous time-program of migratory activity enables young inexperienced birds to find the species-specific winter quarters by vector navigation.

9. In partial migrants, interindividual differences in migratory behavior may be due to polymorphism in the genetic fixation of migratory activity. Some type of dismigration-restlessness may also be

preprogrammed.

10. It is unknown whether interspecific differences in the manifestation of circannual rhythms (differences in the persistence of rhythms and in the interindividual variation of timing events) are due to different impacts of experimental conditions (as permissive factors), or are caused by real species-specific differences in circannual rhythms.

ACKNOWLEDGEMENTS

These studies have been supported by grants from the Deutsche Forschungsgemeinschaft and from the Max-Planck-Gesellschaft. Dr. Werner Loher, University of California, Berkeley, kindly corrected my English.

REFERENCES

BANNERMAN, D. A. & BANNERMAN, W. M. (1968). History of the birds of the Cape Verde Islands. Oliver & Boyd, Edinburgh.

BENOIT, J., ASSENMACHER, I. & BRARD, E. (1955). Evolution testiculaire du Canard domestique maintenu a l'obscurite totale pendant une longue duree. C. R. Acad. Sci. 241, 251-253.

BERNDT, R. & STERNBERG, H. (1966). Dispersion bei Vogeln. Abstr. XIV Congr. Internat. Ornithol., Oxford, 33-36.

BERTHOLD, P. (1968). Die Massenvermehrung des Stars, *Sturnus vulgaris*, in fortpflanzungsphysiologischer Sicht. J. Orn. 109, 11-16.

BERTHOLD, P. (1969). Uber Populationsunterschiede im Gonadenzyklus europaischer *Sturnus vulgaris*, *Fringilla coelebs*, *Erithacus rubecula* und *Phylloscopus collybita* und deren Ursachen. Zool. Jb. Syst. 96, 491-557.

BERTHOLD, P. (1973a). Relationships between migratory restlessness and migration distance in six *Sylvia* species. Ibis 115, 594-599.

BERTHOLD, P. (1973b). Circannuale Periodik bei Teilziehern und Standvogeln. Naturwiss. 60, 522-523.

BERTHOLD, P. (1974a). Endogene Jahresperiodik. Konstanzer Universitatsreden. Universitatsverlag, Konstanz (in press)

BERTHOLD, P. (1974b). Circannuale Periodik bei Grasmucken (*Sylvia*). III. Periodik der Mauser, der Nachtunruhe und des Korpergewichtes bei mediterranen Arten mit unterschiedlichem Zugverhalten. J. Orn. 115 (in press)

BERTHILD, P. & BERTHOLD, H. (1973a). Jahreszeitliche Anderungen der Nahrungspraferenz und deren Bedeutung bei einem Zugvogel. Naturwiss. 60, 391-392.

BERTHOLD, P. & BERTHOLD, H. (1973b). Zur Biologie von *Sylvia sarda balearica* und *S. melanocephala*. J. Orn. 114, 79-95.

BERTHOLD, P., GWINNER, E., & KLEIN, H. (1970) Vergleichende Untersuchung der Jugendentwicklung eines ausgepragten Zugvogels, *S. borin*, und eines weniger ausgepragten Zugvogels, *S. atricapilla*. Vogelwarte 25, 297-331.

BERTHOLD, P., GWINNER, E. & KLEIN, H. (1971). Circannuale Periodik bei Grasmucken (*Sylvia*). Experientia 27, 399.

BERTHOLD, P., GWINNER, E. & KLEIN, H. (1972a) Circannuale Periodik bei Grasmucken. I. Periodik des Korpergewichts, der Mauser und der Nachtunruhe bei *Sylvia atricapilla* und *S. borin* unter verschiedenen konstanten Bedingungen. J. Orn. 113, 170-190.

BERTHOLD, P., GWINNER, E. & KLEIN, H. (1972b) Circannuale Periodik bei Grasmucken II. Periodik der Gonadengrose bei *Sylvia atricapilla* und *S. borin* unter verschiedenen konstanten Bedingungen. J. Orn. 113, 407-417.

BERTHOLD, P., GWINNER, E., KLEIN, H. & WESTRICH, P. (1972c). Beziehungen zwischen Zugunruhe und Zugablauf bei Garten- und Monchsgrasmucke (*Sylvia borin*) und (*S. atricapilla*). Z. Tierpsychol. 30, 26-35.

GWINNER, E. (1967). Circannuale Periodik der Mauser und der Zugunruhe bei einem Vogel. Naturwiss. 54, 447.

GWINNER, E. (1968a). Circannuale Periodik als Grundlage des jahreszeitlichen Funktionswandels bei Zugvogeln. Untersuchungen am Fitis (*Phylloscopus trochilus*) und am Waldlaubsanger (*P. sibilatrix*). J. Orn. 109, 70-95.

GWINNER, E. (1968b). Artspezifische Muster der Zugunruhe bei Laubsangern und ihre mogliche Bedeutung fur die Beendigung des Zuges im Winterquartier. Z. Tierpsychol. 25, 843-853.

GWINNER, E. (1971). A comparative study of circannual rhythms in warblers. In: Biochronometry (M. Menaker ed.). Nat. Acad. Sci., Washington, D.C., 405-427.

GWINNER, E. (1972a). Endogenous timing factors in bird migration. In: Animal orientation and navigation (Galler, S. R., Schmidt-Koenig, K., Jacobs, G. J. & Belleville, R. E. eds.). NASA, Washington, D.C., 321-338.

GWINNER, E. (1972b). Adaptive functions of circannual rhythms in warblers. Proc. XV internat. Ornithol. Congr., Den Haag, 218-236.

GWINNER, E., BERTHOLD, P. & KLEIN, H. (1971). Untersuchungen zur Jahresperiodik von Laubsangern. II. Einflus der Tageslichtdauer auf die Entwicklung des Gefieders, des Gewichts und der Zugunruhe bei *Phylloscopus trochilus* und *Ph. collybita*. J. Orn. 112, 253-265.

GWINNER, E., BERTHOLD, P. & KLEIN H. (1972). Untersuchungen zur Jahresperiodik von Laubsangern. III. Die Entwicklung des Gefieders, des Gewichts und der Zugunruhe sudwestdeutscher und skandinavischer Fitisse (*Phylloscopus trochilus trochilus* und *Ph. t. acredula*). J. Orn. 113, 1-8.

HELLER, H. C. & POULSON, T. L. (1970). Circannian rhythms - II. Endogenous and exogenous factors controlling reproduction and hibernation in chipmunks (*Eutamias*) and ground squirrels

(*Spermophilus*). Comp. Biochem. Physiol. 33, 357-383.

OKADA, Y. (1930). Study of *Euryale ferox Salisb*. V. On some features in the physiology of the seed with special respect to the problem of the delayed germination. Sci. Rep. Tohok Univ. 5, 42-116.

PENGELLEY, E. T. & ASMUNDSON, S. J. (1971). Annual biological clocks. Sci. Amer. 224, 72-79.

PENGELLEY, E. T. & FISHER, K. C. (1963). The effect of temperature and photoperiod on the yearly hibernating behavior of captive golden-mantled ground squirrels (*Citellus lateralis tescorum*), Can. J. Zool. 41, 1103-1120.

VON PERNAU, F. A. (1702). Unterricht, Was mit dem lieblichen Geschopff, denen Vogeln, auch ausser dem Fang, nur durch Ergrundung deren Eigenschafften und Zahmmachung oder anderer Abrichtung man sich vor Lust und Zeitvertreib machen konne. Nurnberg.

SCHMIDT-KOENIG, K. (1973). Uber die Navigation der Vogel. Naturwiss. 60, 88-94.

SCHUZ, E., BERTHOLD, P., GWINNER, E. & OELKE, H. (1971). Grundris der Vogelzugskunde. Parey, Berlin & Hamburg.

SCHWAB, R. G. (1971). Circannual testicular periodicity in the European starling in the absence of photoperiodic change. In: Biochronometry (M. Menaker ed.) Nat. Acad. Sci., Washington, D.C., 428-447.

SEGAL, E. (1960). Discussion to A. J. Marshall. Cold Spring Harbor Symp. Quant. Biol. 25, 504-505.

SMITH, R. W., BROWN, I. L. & MEWALDT, L. R. (1969). Annual activity patterns of caged non-migratory white-crowned sparrows. Wilson Bull. 81, 419-440.

SPERLICH, A. (1919). Uber den Einflus des Quellungszeitpunktes von Treibmitteln und des Lichtes auf die Samenkeimung von *Alectorolophus hirsutus All.*, Charakterisierung der Samenruhe. S.B. Akad. Wiss. Wien, Math.-naturwiss. Kl. I, 128, 477-500.

STRESEMANN, E. & STRESEMANN, V. (1966). Die mauser der Vogel. J. Orn. 107, Sonderheft.

VAN TIENHOVEN, A. (1961). Endocrinology of reproduction in birds. In: Sex and internal secretion (W.C. Young ed.). Baltimore, 1088-1169.

CREDITS

Figure 1. Reproduced by permission of Birkhauser Verlag from Experimentia Vol. 27.

Figures 2, 3, 4, 5, 6, 7, 8, 9, 10, 11, 12, 13, 14, 15, 17, 18 and 19. Reproduced by permission of Deutsche Ornithologen-Gesellschaft from J. Ornithologie.

Figure 16. Reproduced by permission of Springer-Verlag from Naturwiss Vol. 60.

Figure 20. Reproduced by permission of Great Glemham House from Ibis Vol. 115.

CIRCANNUAL RHYTHMICITY IN HIBERNATING MAMMALS

ERIC T. PENGELLEY
(Professor of Biology)

and

SALLY J. ASMUNDSON
(Staff Research Associate)

Department of Biology,

University of California,

Riverside, California, 92502

INTRODUCTION

Throughout recorded history, and no doubt before it, men have observed the changing patterns of behavior of almost all animals and plants which are easily related to two fundamental geophysical phenomena, namely the daily 24 hour rotation of the earth on its axis and the yearly 365 1/4 day orbit of the earth around the sun. Associated with the former is what we term day and night, and with the latter the coming and going of the seasons. These are of course important events in the overall environment of virtually all living organisms, and unless adaptions took place, the organism would undoubtedly have become extinct. The survival value of such adaptations has recently been well documented by Aschoff (1964, 1965), Pengelley (1967), Heller and Poulson (1970) and many others, and is also thoroughly treated elsewhere in this current symposium.

As a result of the evolutionary process, animals and plants exhibit many varieties of adaptations to the environment, but from a scientific point of view there would seem to be only three basic ways of synchronizing behavioral physiology to a changing environment (Pengelley, 1967). These are (a) a direct response to various changing geophysical stimuli; (b) an endogenous rhythm which programs the organism's behavior to the exogenous temporal period, i.e. usually 24 hours and 365 1/4 days; or (c) a combination of both. There seems little doubt that animals and plants from single celled protists (Sweeney, 1969) to complex multicellular animals and plants (Aschoff, 1965; Bunning, 1964, 1967) have in fact evolved the mechanism of a combination of both.

It should be clearly understood that in order to demonstrate an endogenous rhythm certain precise criteria must be applied. Firstly, the rhythm must exhibit a frequency which is not exactly synchronous with any known environmental periodic signal. The latter usually consists of light, ambient temperature or both; but there may be other variables as well (Pittendrigh, 1958; Pengelley and Asmundson, 1974). Weihaupt

(1964) presents an impressive number of geophysical variables not normally considered by most biologists. Secondly, the period of the endogenous rhythm must deviate at least slightly from the time being measures, i.e. 24 hours or 365 1/4 days; and it was because of this that Halberg *et al* (1959) proposed the term 'circadian' (*circa* = about; *dies* = day) for the 24 hour rhythm rather than diurnal. Similarly Pengelley (1967) proposed the word 'circannian' (*circa* = about; *annum* = year) for the 365 1/4 day rhythm; this word was subsequently changed to 'circannual' (Pengelley and Asmundson, 1970a) in order to conform with the word used by Gwinner (1968) and his German associates. Thirdly, the rhythm must be relatively temperature independent. If all these three criteria apply then we may be reasonably certain that the organism exhibits an endogenous rhythm, and the use of the word circadian or circannual implies that the rhythm is endogenous.

Endogenous rhythms in the strict sense are of fairly recent discovery, i.e. perhaps about 35 years ago. However they have been speculated upon for well over 200 years. The French scientist De Mairan (1729) clearly suspected their existence and Charles Darwin (1880) was puzzled by phenomena in plants which we now explain as a result of endogenous rhythms. The German botanist Pfeffer (1875) was also a pioneer in the field and actually performed experiments in which plants were kept under constant environmental conditions. In this century the number of scientists engaged in investigations of endogenous rhythms has rapidly increased and there is now a vast and ever expanding literature on the subject. A most important event in this respect was the

Cold Spring Harbor Symposia (1960) on Biological Clocks, which brought together the then existent knowledge and emphasized the extreme importance of the phenomena.

Since, as a part of the definition of an endogenous rhythm, the period deviates from the geophysical time being measured, it is obvious that the organism must have a means of synchronizing its endogenous rhythm with the exogenous one, and a great deal of work has been done on this in circadian rhythms, though little in circannual rhythms. Aschoff (1960) has coined the word zeitgeber for the environmental agent which entrains the organisms behavior to the changing environment. As Aschoff (1963) has pointed out, entrainment requires not simply synchrony, i.e. equal speed of the driving (environment) and driven (organism) oscillators, but phase control: "a clearly defined and stable phase angle difference between the biological oscillation and the zeitgeber". Now to achieve such a situation, it is necessary that the organism undergo a periodically changing sensitivity to the zeitgeber, "which then corrects the phase of the oscillation at least once during each period". In organisms with circadian oscillators, and which respond to light, this zeitgeber is usually the photoperiod and there is a complex terminology and literature expanding on this (Aschoff, 1965). It is only necessary to note here that there are three important time periods, i.e. the activity time, rest time, and the entire period (activity + rest) which is referred to as the free-running period. It has been assumed that the same criteria apply to circannual rhythms, though Pengelley, Bartholomew and Licht (1972) have stressed that the problems in circannual rhythms are likely to be much more complex.

From an annual point of view, the timing of the reproductive activities of animals, as well as the various migrations and behavioral phenomena such as dormancy and hibernation, are probably the major physiological events of the year, and have of course been observed by man since before recorded history and indeed made use of by him in such areas as animal husbandry and agriculture. However, once again it is only fairly recently that any scientific understanding of these phenomena has developed. A most important discovery in this respect was that of Rowan (1925, 1926, 1938) who clearly demonstrated for the first time that photoperiodism, reproductive periodism and the annual migrations of birds were interrelated, though he also noted that photoperiodism could not be the only controlling factor involved. Rowan was clearly far in advance of his time and visualized photoperiodism as a means of manipulating what he termed "the annual rhythm of the reproductive organs", though it is doubtful that he thought of this rhythm as entirely endogenous. Similarly, Bissonette (1935) showed that modification of some mammalian sexual cycles was possible by manipulation of the photoperiod. However, apart from increasing emphasis on the photoperiod as the apparently major controlling environmental aspect in seasonal reproductive periodism, little more work had been attempted until the 1960's on a synthesis of the various other possible factors involved in the timing of mammals' physiological activities and behavior with the seasons of the year. Nevertheless it should be noted that a productive attempt at this was first made by the French workers, Benoit, Assenmacher and Brard

(1955, 1956a, 1956b).

HIBERNATING MAMMALS

The phenomenon of hibernation in mammals has come under intensive investigation in recent years (Lyman and Chatfield, 1955; Eisentraut, 1956; Kalabukov, 1960; Kayser, 1957, 1960, 1961, 1964; Mammalian Hibernation I, 1960; Mammalian Hibernation II, 1964; Mammalian Hibernation III, 1967; Hibernation-Hypothermia, 1972). Hibernating mammals undergo a dramatic change on a yearly basis alternating between the active and hibernating physiological states, thus making them well suited for the study of long-term biological rhythms. It was as a result of this fact that Pengelley and Fisher (1957) first observed what they interpreted to be an endogenous annual rhythm. It should be noted that Dubois (1896), Kayser (1940) and later Lyman (1948, 1954) all recognized the temporal nature of hibernation, and that this was not a simple response of an animal to an environmental stimulus. However, it is in fact only within the last 10-15 years that the temporal nature of hibernation has received much attention (Bartholomew, 1972; Hoffman, 1964; Kristoffersson and Soivio, 1964; Kristoffersson and Suomalainen, 1964; Licht, 1972; Mrosovsky, 1970; Pengelley, 1964, 1967, 1968, 1969; Pengelley and Asmundson, 1969, 1970a, 1970b, 1971, 1972, 1974; Pengelley and Fisher, 1957, 1961, 1963, 1966; Pengelley and Kelly, 1966a; Richter, 1965; Scott and Fisher, 1970; Strumwasser, 1959a; Twente and Twente, 1964, 1967).

There is a great deal of confusion as to what is meant by hibernation, and it will be well to thoroughly clarify this before proceeding to its control in some species by a circannual rhythm. Lyman and Chatfield (1955) and Hock (1958, 1960) have proposed definitions of hibernation and Bartholomew and Hudson (1960) and Hudson and Bartholomew (1964) have compared hibernation and estivation. There is little doubt from the work of the latter authors that in mammals hibernation and estivation are qualitatively alike, if not the same phenomena at the physiological level. However, they differ in that they are induced in response to different environmental conditions. For the purposes of precise scientific terminology hibernation in mammals has been accurately defined as those mammals which "at rest are capable of maintaining a high and constant body temperature, usually close to 37°C, against the normal range of environmental temperatures; but which, at certain times, under natural conditions, abandon this homothermic state and permit the body temperature to fall close to, if not to, the environmental level, with a low limit of about 0°C, but which are able to regain the homothermic condition at any time against the environmental gradient" (Pengelley and Fisher, 1961). This definition is not all inclusive but in mammals it confines it to a relatively few species which characteristically exhibit this reversible physiological change of state at certain times of the year, or perhaps as a direct result of being exposed to a particular environment for a period of time. When hibernating mammals are in the active state, they are said to be homothermic as is the case with most other mammals, but when they are in the hibernation state they are said to be heterothermic, i.e. their temperature varies but at the same time they have internal control over those variations, as distinct from poikilothermic animals whose body temperature simply varies with the environmental temperature without any apparent internal control.

It should also be clearly understood that there is a great deal of both physiological and behavioral variation between species of hibernating mammals. Thus Morrison (1960) points out that some hibernating mammals have an "option", so to speak, as to whether they hibernate or not. These are referred to as "permissive" hibernators in contrast to others called "obligate" hibernators which apparently at certain times must hibernate regardless of the external conditions. Furthermore, it must be clearly understood that not all hibernating mammals exhibit a circannual rhythm. Cade (1963, 1964) has ably reviewed the great variety of behavior in mammals from temporary torpor of a few hours, to some form of temporary hypothermia, to periods of true hibernation and estivation. He has also related these to the evolution of the various species and it is highly probable that hibernation in some mammals is of polyphylogenetic origin (Pengelley, 1967). However, it is important to note that while a mammal in hibernation differs from an active one in having a body temperature lower than the normal homothermic level of 37°C, with subsequent torpor and incapacity to move, there are nevertheless many other profound physiological differences between the homothermic and heterothermic states (Aloia, Pengelley, Bolen and Rouser, 1974; Fawcett and Lyman, 1954; Johansson and Senturia, 1972; Kayser, 1953, 1960, 1961; Lyman, 1958, 1961; Lyman and

O'Brien, 1960; Pengelley, Asmundson and Uhlman, 1971; Pengelley and Chaffee, 1966; Pengelley and Kelley, 1966; Popovic, 1960, 1964; Riedesel, 1960, *et al.* 1964; Strumwasser, 1959a, 1959b, 1959c). It may be noted also that just as all mammals which have been studied, exhibit a circadian rhythm, so also do hibernating mammals (Folk, 1960; Menaker, 1961, 1964; Strumwasser, Gilliam and Smith, 1964; Strumwasser, Schlechte and Streeter, 1967). However any circadian rhythms will be only incidentally treated here.

1. The golden-mantled ground squirrel.

By far the greatest proportion of our knowledge on circannual rhythms in hibernating mammals, comes from studies on the golden-mantled ground squirrel, *Citellus lateralis*, and it is to these that considerable attention will now be given. This small rodent is found in western North America at high altitudes (5000-12,000 ft.) from northern British Columbia to southern California. It was first studied systematically in the field by Mullally (1953), but Pengelley *et al.* (1957 et seq.) have elucidated its behavioral physiology to some extent, though much remains unknown.

Both young and adult squirrels captured in July and kept in the laboratory under constant environmental conditions, behave as indicated in figs. 1 and 2 which are representative of many animals in the experiments. In these experiments all external factors were kept as constant as possible consistent with observation, care and maintenance of the animals. All disturbances were kept to a minimum, the cages were not even cleaned while the animals were hibernating and environmental temperature, photoperiod, food and water supply and cage conditions were all constant, in this case for over 700 days. Thus the animals had no contact with any known periodic environmental stimulus. The only difference in the environment of the two animals represented here is that the ambient temperature of the animal in figure 1 was 22^{0}C and in figure 2 it was 0^{0}C. In these experiments, three parameters were measured, the physiological state, i.e. active (homothermic) or hibernating (heterothermic), the body weight and mean daily food consumption. Whether or not the animal was in the homothermic or heterothermic state was determined each day using the method of Pengelley and Fisher (1961). As indicated in the figures, the hibernating periods are shown by solid horizontal black bars, and it is obvious that these alternate with active periods in a distinct rhythmic manner. It is important however to note that within these periods of heterothermy, there are yet other rhythms in which the animal periodically arouses from hibernation, remains homothermic for a few hours and then enters hibernation again. This phenomenon was first noted and studied by Pengelley and Fisher (1961), and it is only necessary to point out that these periodic arousals (which are apparently common to all hibernators) are not shown here. What the black bars represent are entire heterothermic periods, while the distances in between the black bars represent the entire homothermic periods. The body weight, and mean daily food consumption also clearly show a distinct rhythm, and calculating the free-running period from the onset of each heterothermic period to the onset of the next, it is about 300 days.

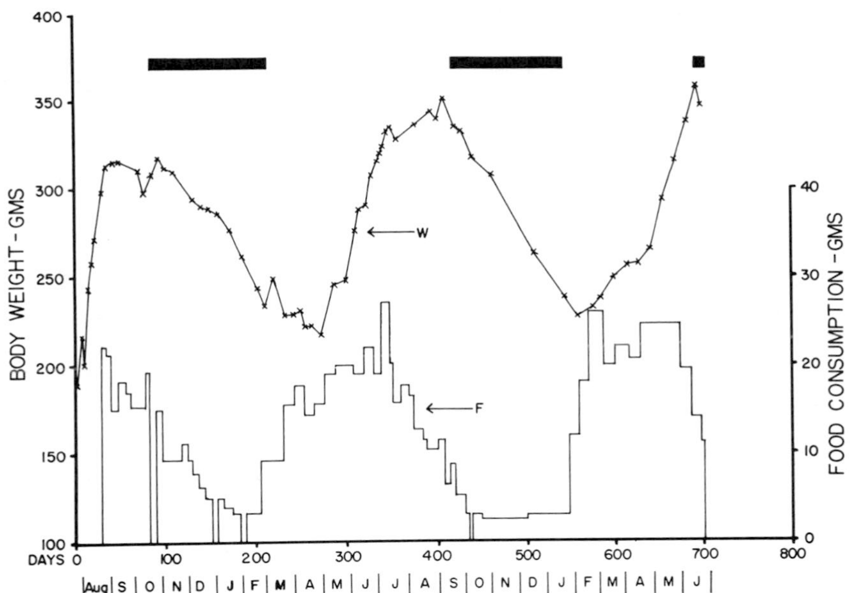

Fig. 1. Body weight (W), mean daily food consumption (F), hibernating periods (solid horizontal bars), active periods (clear spaces) of an individual *Citellus lateralis* observed daily for two years. Environmental temperature 22°C, photoperiod 12 hrs, food (Purina Chow) and water *ad libitum,* ample bedding. Rhythmic arousals not indicated. (From Pengelley and Fisher, 1963.)

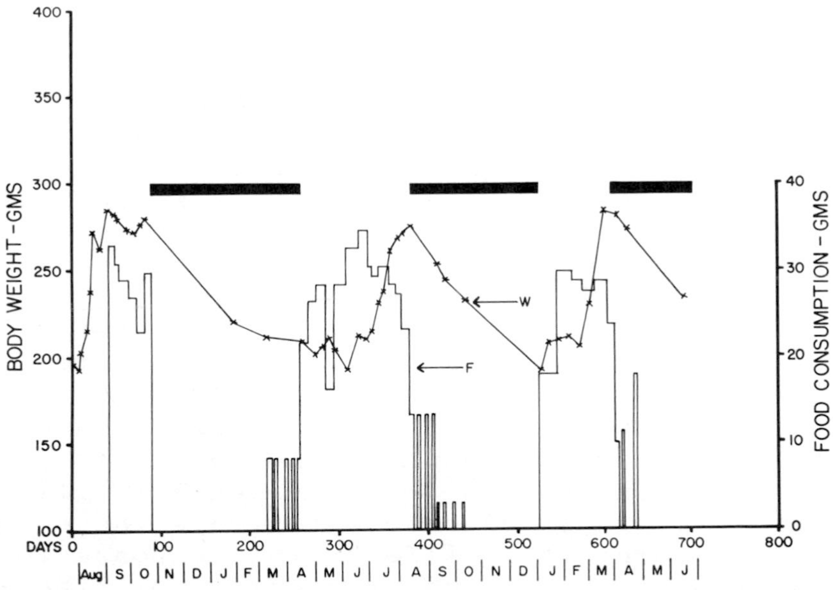

Fig. 2. Same as fig. 1, except environmental temperature 0°C. (From Pengelley and Fisher, 1963.)

It is quite clear from the experiments just described that the criteria necessary to demonstrate an endogenous circannual rhythm have been met. That is the rhythm is not synchronous with any known environmental periodic signal, since these were excluded by the constant environment; the period deviates from 365 1/4 days, and it is relatively temperature independent, i.e. it is more or less unaffected by an environmental temperature of 22°C or 0°C. Since the rhythm is obviously closely related to the geophysical period of a year, it seems entirely justified to refer to it as a circannual rhythm.

Since the free-running period of the circannual rhythm in these cases is considerably less than 365 1/4 days, it is entirely pertinent to ask what is the nature of the zeitgeber which in the natural state entrains the rhythm. No simple answer is possible at this point, but as we proceed with the experimental evidence we will discuss various possibilities. In figs. 1 and 2 it can be seen that at the lower environmental temperature the maximum weight was attained earlier, but was always less than in a matched animal at 22°C. It is obvious also that the onset of the heterothermic period is closely correlated with the attainment of the maximum body weight, and indeed the animals at the environmental temperature of 0°C did in fact hibernate earlier than those at 22°C. However, there was no significant statistical difference between the two groups in the length of the free-running period. The evidence is clear that over a period of 300-400 days these animals exhibit a differential response to temperature, i.e. they are active or hibernate, and the exact time at which they hibernate or become active is also responsive to temperature. All this suggests that ambient temperature is a potential circannual zeitgeber, though its exact mechanism of operation is far from clear.

From these experiments it is also clear that both the mean daily food consumption and the body weight are under the overall influence of an endogenous rhythm, since the animals hibernate and become active again, and gain or lose weight even with food and water *ad lib*. Nevertheless, food supply can be used to manipulate the length of the hibernation period. Fig. 3 shows the results of an experiment (Pengelley, 1967) in which animals had food and water removed from the cages 2-3 weeks after the hibernation period had begun. No food was available therefore, either during the rhythmic arousals or when hibernation would normally terminate the homothermic state be reestablished. All these animals (group2) hibernated approximately 100 days longer (then dying without arousing from hibernation) than did controls (group 1) with food and water *ad lib*. On the other hand if food and water are removed only after the animals have regained the homothermic state (group 3), then it is found that denial of food within the first two weeks of homothermism causes the animal to resume the heterothermic condition, but if food is denied after this time then there is no return to the heterothermic state. What solid evidence there is, therefore, seems to indicate that food supply is also a potential circannual zeitgeber at least for this species of hibernation mammal.

Fig. 4 represents two animals out of twenty-three in a laboratory experiment which was kept going for nearly four years.

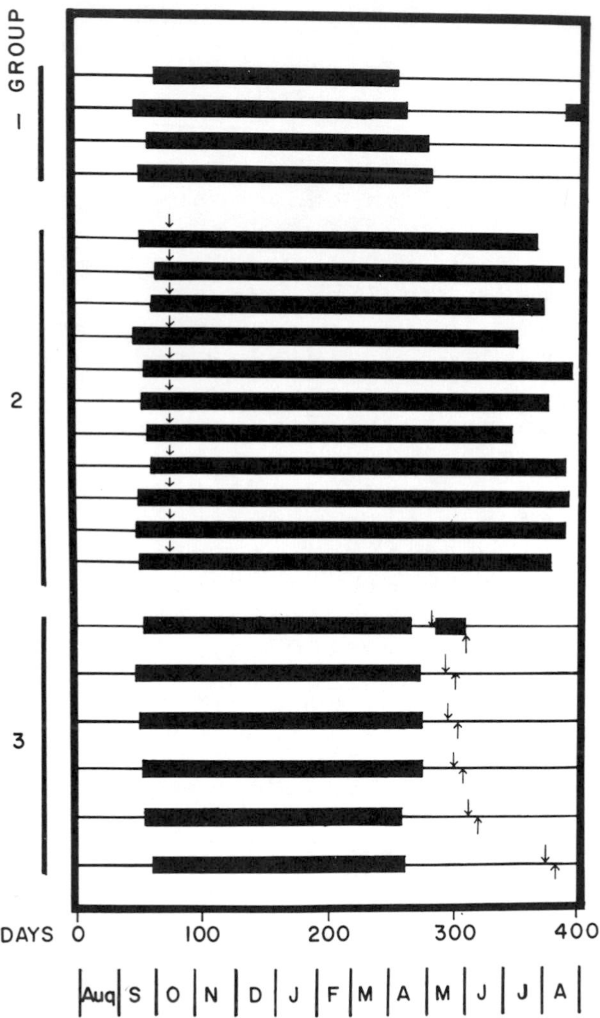

Fig. 3. Effect of food denial on hibernation period (solid horizontal bars) of *C. lateralis.* Environmental temperature 0°C, photoperiod 12 hours. Down arrow indicates food stopped, up arrow indicates food reintroduced. (From Pengelley, 1967.)

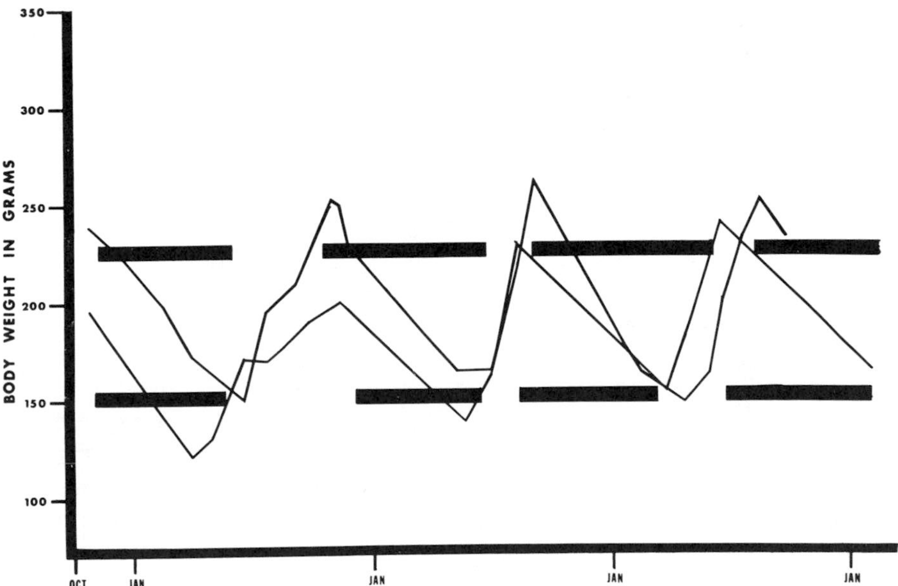

Fig. 4. Interrelationship of body weight (g), and whole hibernation periods (black bars), of two representative animals (*C. lateralis*) for nearly 4 years. Upper animal at 12°C ambient, lower at 3°C ambient, both with artificial photoperiod of 12 hrs. (From Pengelley and Asmundson, 1969.)

These animals were caught in the wild in August and consisted of both adults and juveniles. They were divided into two matched groups of eleven and twelve each according to sex and age (juveniles/adults), and in addition one half of all males and females were castrated. From this figure it is obvious that the body weight rhythm is closely synchronous with the homothermic-heterothermic rhythm and in fact is probably phase locked to it, as the weight peak is just prior to the onset of hibernation, and the nadir is just after its termination. It is also obvious that the animals at the higher environmental temperature have an annual weight cycle at a higher level with regard to both peak and nadir than those at the lower environmental temperature. Thus the data in figs. 1 and 2 are confirmed over the much longer period of four years in which all the requirements to demonstrate an endogenous rhythm have been met. Further more, the experiment tends to support our view that ambient temperature is a potential zeitgeber. Table 1 summarizes the results found for the free-running circannual homothermic-heterothermic periods for all animals in the experiment. However, we were unable to detect any significant differences in the free-running periods between sexes, adults and juveniles, or castrated and controls, and there is also no significant difference between those at ambient temperatures of 12°C and 3°C. It is also worthy of note, though conclusions should be avoided at this point, that of the 61 complete free-running periods (onset to onset of hibernation) which it was possible to measure, only 13 were longer than the annual period of 365 1/4 days. The longest period was 445 days and the shortest 229.

Table 1--Free running circannual periods (onset to onset of hibernation) in days of *C. lateralis* under constant environmental conditions. (From Pengelley and Asmundson, 1969)

	Circannual periods (days)			
Animal	1st	2nd	3rd	4th
1	346	357		
2	356	324		
3	331			
4	303	374	346	338
5	299	279	277	229
6	290	337	390	
7	345	318	348	358
8	387	231	234	288
9	333	318	345	
10	380	345	393	
11				
12	385	366		
13	438			
14	388	312	325	
15	386	262	325	
16	353	379	413	
17	365	267	310	325
18	348	360		
19	322	445	354	
20	409	248	309	319
21	332	343	319	
22	325			
23	340	312		

Animals 1-11, 12°C 12 hr photoperiod; animals 12-23, 3°C 12 hr photoperiod.

Similar long term experiments have been carried out in an attempt to determine whether light is in fact also a potential circannual zeitgeber. Fig. 5 represents graphically the results from such an experiment. All the animals involved were laboratory born from wild caught pregnant females (Pengelley, 1966). Shortly after weaning they were divided into four matched groups, and set up under the following experimental conditions. Group 1 comprised the controls with an artificial photoperiod of 12 hours. Group 2 were the same except that they had an artificial photoperiod of 20 hours. Group 3 were castrated and then placed with an artificial photoperiod of 12 hours. Finally group 4 were bilaterally enucleated. All the animals were kept at an ambient temperature of 3°C for nearly 4 years, and food and water were always *ad lib*. In this experiment the free-running period was measured from termination to termination of each heterothermic period. The results of this experiment indicate quite clearly that the animals in group 4 (bilaterally enucleated) have more accurately timed successive free-running periods than any of the other groups or those in other experiments. Thus while the time of termination of the heterothermic period varies considerably between each animal in group 4, the successive variations in the free-running periods of individual animals are remarkably small. For the fourth animal in this group it is possible to measure three complete free-running periods. They are 313, 314, and 310 days. The largest variation of the free-running period of an individual animal in this group is 29 days, whereas variations up to 75 days in the other three groups are common. From this experiment it is possible

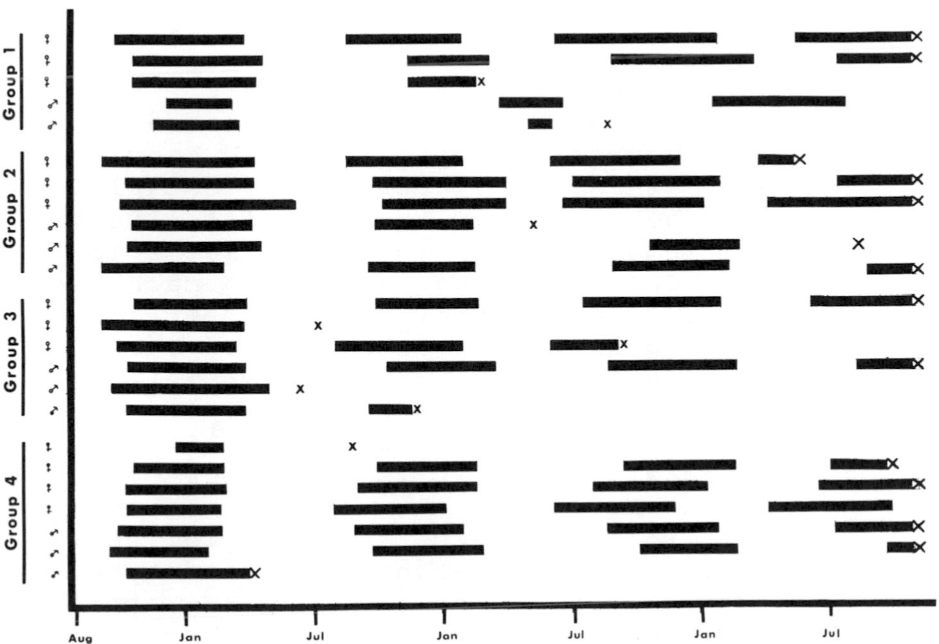

Fig. 5. Graphical representation of free running circannual periods of 24 animals (*C. lateralis*) for 4 yr. Black bars indicate heterothermic period, clear space homothermic period, X = death. All groups at 3°C ambient temperature. Group 1, controls with artificial photoperiod 12 hr. Group 2, normal with artificial photoperiod 20 hr. Group 3, castrated with artificial photoperiod of 12 hr. Group 4, bilaterally enucleated. M = Male, F = Female. (From Pengelley and Asmundson, 1970.)

to conclude several things with regard to the endogenous rhythm. Firstly, light is a factor involved in the determination of the free-running circannual rhythm of this species, and is therefore a potential zeitgeber. However, it is certainly not possible at present to amplify on how it actually acts as a zeitgeber. Secondly, laboratory born juveniles establish circannual rhythms without detectable differences from wild caught animals. This includes the fact that the homothermic-heterothermic and body weight rhythms are synchronously established from the beginning of the animals' lives, and continue until death. This seems to be clear evidence that no imprinting mechanism is involved, but rather that the circannual rhythm is genetically determined. Furthermore, it would appear that the mediating influence of light does not exert its effect via any gonadic cycle, since in this and other experiments no differences between castrated and normal animals can be detected. Thirdly, it seems that the influence of light on the free-running period takes some time (over a year) to show any appreciable effect, for it is easily discernible from fig. 5 that in groups 1, 2, and 3, it is not until the third and fourth years that large discrepancies occur in the temporal sequence of events. In this connection it is interesting to recall that nearly twenty years ago the late Raymond Hock (1955) after many years studying the artic ground squirrel, *Citellus undulatus*, speculated that the photoperiod was probably a stimulus for the onset of hibernation in the wild, a fact since confirmed by Drescher (1967). However, despite attempts in similar experiments, Pengelley and Fisher (1963) were unable to demonstrate this in the laboratory in *C. lateralis*. However, their experiments were for only one year, and it is clear now that it takes longer for the effect of light to have its impact. Even though it is still quite unknown how light really acts as a zeitgeber we can say that it has that potentiality. Table 2 summarized the data for group 4 in fig. 5. In part A the free-running circannual period is measured from onset to onset of heterothermy, while in part B it is measured from termination to termination of heterothermy. It can be seen that in this group the method of measurement makes little difference in the length of the free-running period, and it is not easy to decide which is the best method of measurement. Since it is known that all these particular animals were born within a period of only a few weeks it is clear that their endogenous temporal physiological differences would be slight in relation to their initial entrance into hibernation.

New data on the circannual rhythm in *C. lateralis* have recently come from Heller and Poulson (1970), and are summarized in figs. 6 and 7. These not only confirm the basic phenomenon of an endogenous circannual rhythm, but also add new parameters and another means of measuring the length of the free-running period. In fig. 6 three parameters are plotted, namely body weight, daily water consumption and reproductive competence. These data are for an animal in its second year under the constant ambient temperature of 16^{0}C and a photoperiod of 12 hrs. Their method of judging reproductive competence consisted of continuous observations on the testes in males and the vulvae in females. During most of the year males had abdominal testes and were

Table 2--Free-running circannual periods in days of *C. lateralis* bilaterally enucleated and kept at 3°C. Data from figure 5, group 4. (From Pengelley and Asmundson, 1972.)

A. Circannual periods in days (onset to onset of heterothermy)

Animal	1st	2nd	3rd
2	338	341	293
3	329	324	324
4	293	293	305
5	334	354	324
6	358	369	353

B. Circannual periods in days (termination to termination of heterothermy)

Animal	1st	2nd	3rd
2	348	377	
3	349	329	
4	313	314	310
5	337	359	
6	376	361	

(Animals 1 and 7 are not shown as neither completed an annual cycle before death).

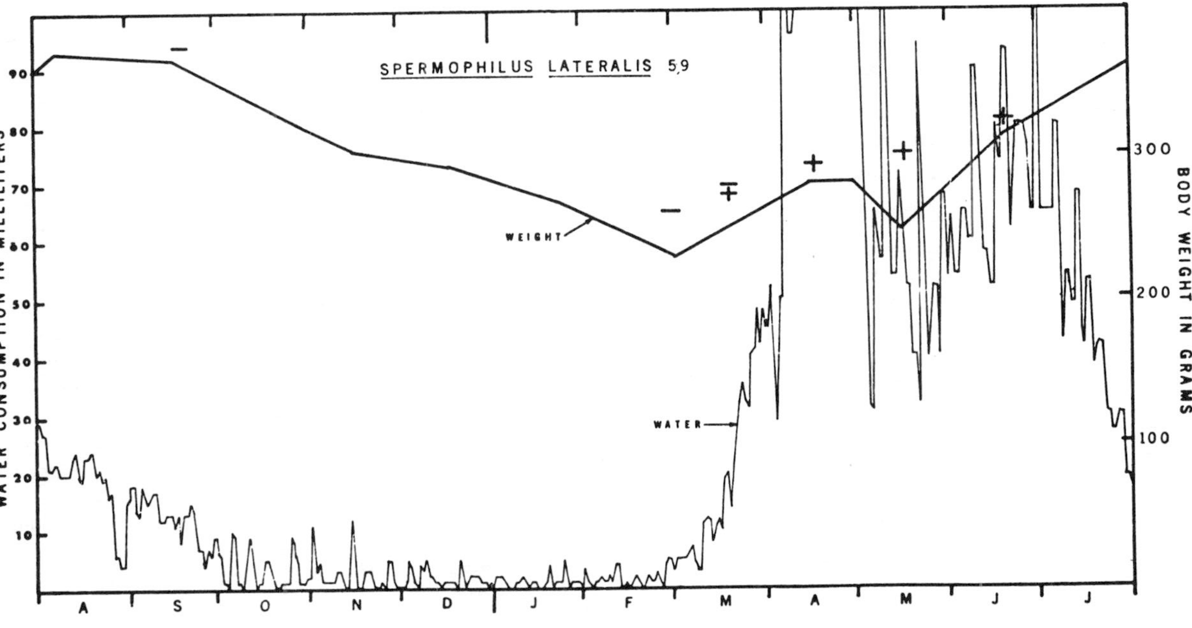

Fig. 6. Record of daily water consumption (fine line) and body weight (bold line) for *C. lateralis* male during its second year under constant laboratory conditions: + stands for reproductive competence: -stands for reproductive quiescence. Ambient temperature 16°C and photoperiod 12 hrs. (From Heller and Poulson, 1970.)

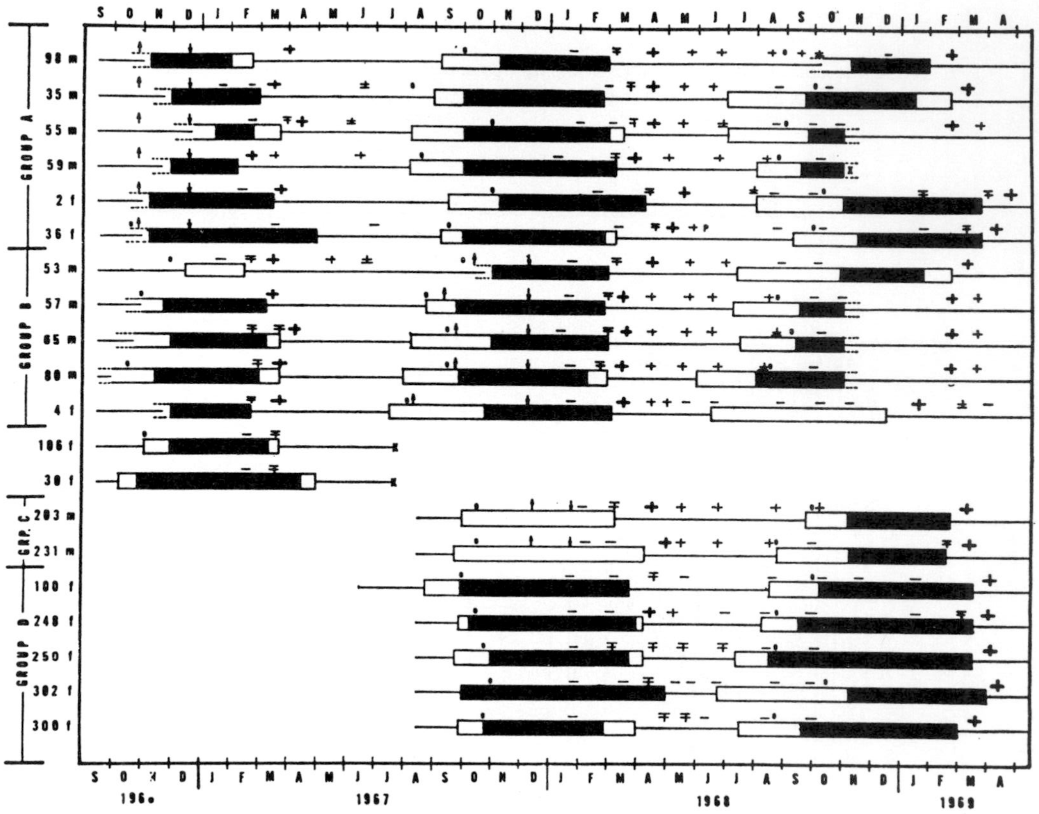

Fig. 7. Circannual rhythms of reproductive condition, hibernation and body weight in *C. lateralis* kept under constant laboratory conditions for 33 months, and the influence of prevention of torpor and dieting on these rhythms. Solid horizontal line is the life line and represents the time an animal was under observation under constant conditions. A solid black bar indicates the hibernation phase and a white bar indicates winter condition phase. A solid black circle above a record marks the occurrence of a body weight maximum. Full reproductive competence id denoted by +, reproductive quiescence by -, the transition into reproductive competency by ∓ and declining reproductive competence by ±. An arrow pointing up means dieting and prevention of torpor was begun and an arrow pointing down marks the return to *ad lib.* food and water consumption. The transition from either a solid bar or an open bar to a single life line is the date of terminal arousal. When the transition between active and inactive times was not clear, or when the record was incomplete, dashed lines are shown. In the record of animal No. 36, P (parturition) marks the birth date of a litter. An X indicates death. (From Heller and Poulson, 1970.)

considered reproductively incompetent, but this condition gradually changed at certain times to a competent condition where the testes became scrotal and the skin in that region became darkly pigmented. Similarly for most of the year the females had closed vaginae and unswollen vulvae, which was taken to be reproductive incompetence, but this gradually gave way to reproductive competence where the vaginae were open and the vulvae swollen. It is noteworthy that a reproductively competent pair did in fact mate in the laboratory and in due course the female produced a litter of six (end of May). From extensive observations in the wild where animals were live trapped and examined during early spring over many years, we (Pengelley and Asmundson - unpublished) can confirm that Heller and Poulson's (op cit.) method of determining reproductive competence is probably very accurate. Males at this time of year do in fact have scrotal testes and females open vaginae and swollen vulvae, and large numbers of females give birth to young some weeks later (Pengelley, 1966). In fig. 6 the declining part of the body weightrhythm corresponds temporally with the heterothermic part of the homothermic-heterothermic rhythm, which is here characterized by extremely low (often zero) water consumption and reproductive incompetence. Conversely the increasing part of the body weight rhythm corresponds with homothermy, high water consumption and reproductive competence. Heller and Poulson (op cit.) took the first sharp increase of water consumption, (which is obvious in fig. 6) as the time of terminal arousal from hibernation, (i.e. the end of the heterothermic period) and at this time they also observed a marked increase in activity, which confirms the observation of Pengelley and Fisher (1966). Using the first sharp increase of water consumption as the terminal arousal date, and the free-running circannual rhythm as terminal arousal to terminal arousal, the data for all *C. lateralis* in Heller and Poulson's work are plotted in fig. 7, and they indicate a remarkably constant rhythm. Using this method of determination, they were able to measure a total of 17 free-running circannual periods, the extremes of which were 44 and 59 weeks (308 and 413 days), with a mean of 51 weeks (358 days). From their carefully collected data Heller and Poulson conclude "regardless of the experimental manipulations we imposed upon the animals, the free-running period length measured from terminal arousal to terminal arousal is generally only slightly less than a year and shows remarkably little variation." When the data in figs. 6 and 7 are compared with those of Pengelley *et al.* in figs. 1, 2, 4 and 5 and tables 1 and 2, the similarities in the free-running periods are striking. Although Heller and Poulson's free-running periods more closely approximate a year's duration, this is probably due to their different method of determining the free-running period, though it is possible that it might in part also be due to the effect of their 16^{0}C ambient temperative acting as a zeitgeber.

The data shown in fig. 8 are also from the work of Heller and Poulson on *C. lateralis*. From this figure it may be seen that upon reaching its normal weight peak, along with high water consumption and the approximate beginning of the heterothermic period, the experimentors manipulated the food and water supply of the animal in such a way

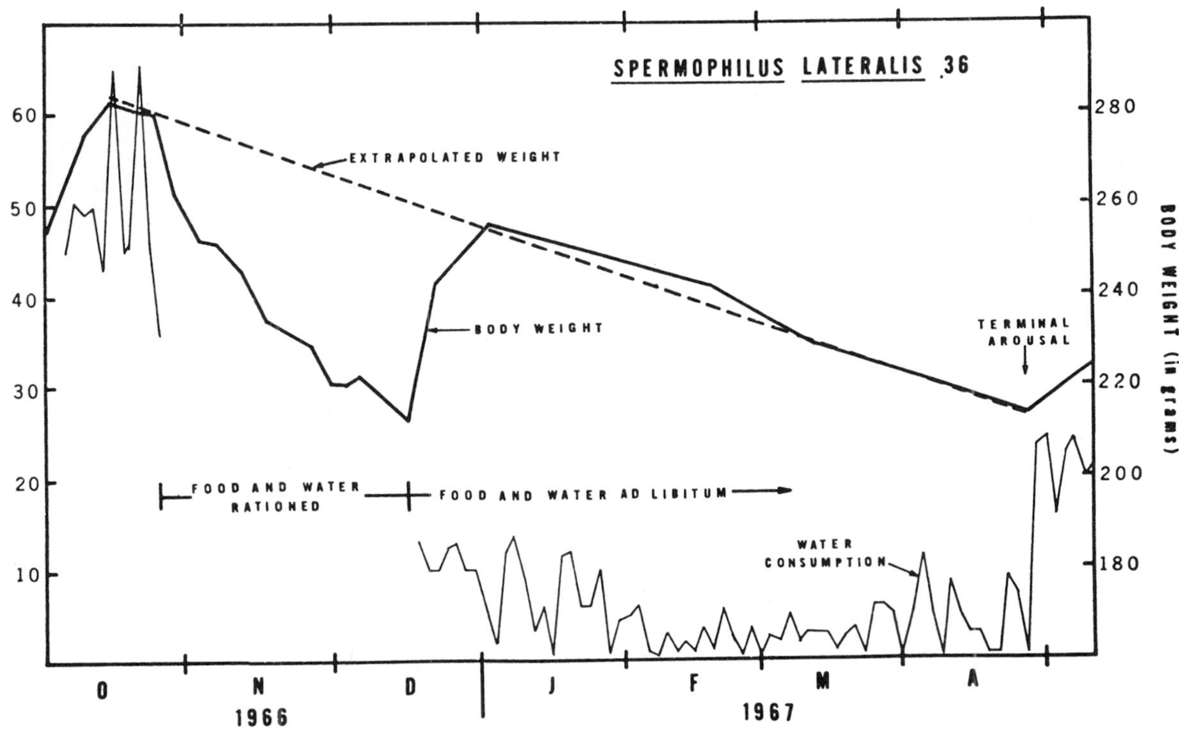

Fig. 8. Daily water consumption and body weight of a *C. lateralis* before, during and after dieting and prevention of torpor. This demonstrates the postdiet return of body weight to a programmed weight determined by a line extrapolated from weight peak to minimum weight at terminal arousal. (From Heller and Poulson, 1970.)

that a rapid decline in body weight was inevitable. About 2 months later when the food and water were restored *ad libitum*, it was found that the animal regained weight only to the point expected on the basis of a line extrapolated between weight peak and minimum weight at the end of the heterothermic period (i.e. terminal arousal). Of seven animals in this experiment, six behaved in the predicted manner. This same phenomenon has recently been demonstrated in another species, *C. tridecemlineatus*, by Mrosovsky and Fisher (1970), and is clear evidence of an internal programming mechanism for body weight. Heller and Poulson's data also demonstrate that excessive depletion of weight due to food and water rationing at the beginning of the heterothermic period has no effect on the time of terminal arousal, nor on the onset of the next heterothermic period, nor the length of the free-running circannual rhythm.

Other pertinent experiments on food manipulation are those of Pengelley (1968), the results of which are summarized in figures 9 and 10. As pointed out previously (Pengelley and Fisher, 1963)--see figs. 1 and 2, the attainment of a maximum body weight appeared to be a prerequisite and perhaps a trigger to the onset of the heterothermic period and hibernation. This, in turn, was followed by a loss of weight accompanying hibernation, until a minimum was reached, hibernation ceased, and the weight rise would commence again during the homothermic period. It was therefore decided to restrict the food supply during the homothermic period in such a way that the animals in question could not appreciably gain weight and would have to remain at approximately the trough of the body weight rhythm.

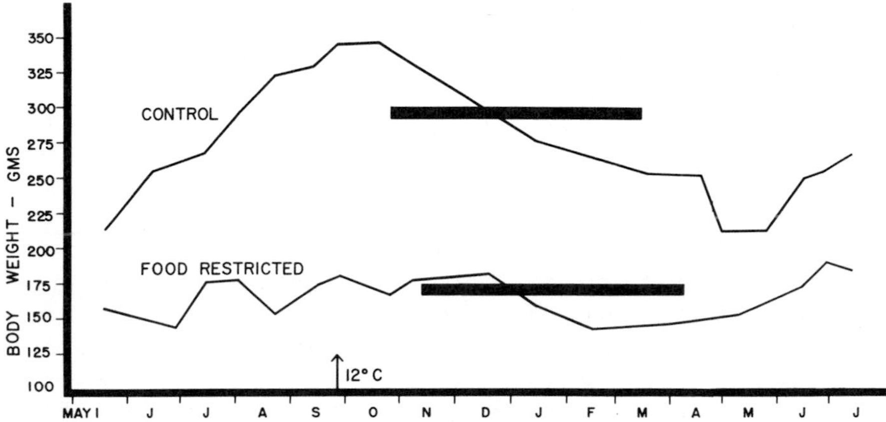

Fig. 9. Body weights and hibernation periods (black horizontal bars) for representative animals *C. lateralis* from control and food-restricted groups. Time of transfer from 23°C ambient to 12°C is indicated. Photoperiod 12 hr. (From Pengelley, 1968.)

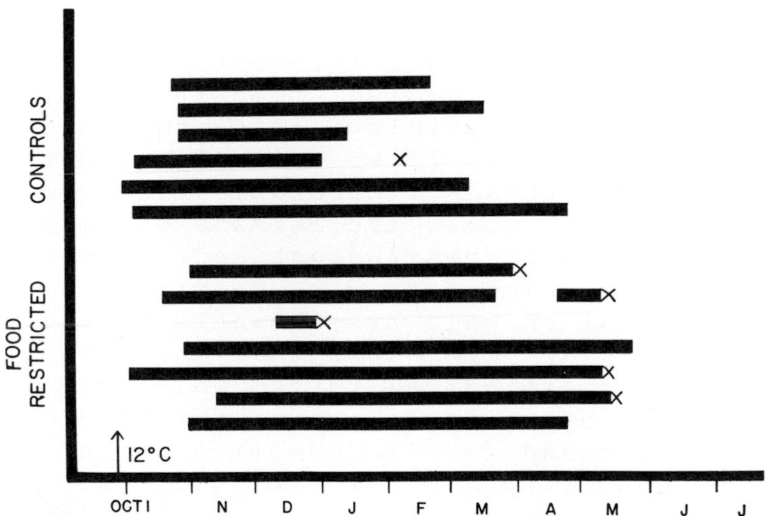

Fig. 10. Hibernation periods for all animals (*C. lateralis*) in the experiment. Time of transfer from 23°C ambient to 12°C is indicated. Photoperiod 12 hr. Death is indicated by a cross. (From Pengelley, 1968.)

The object of this was to see if the food restricted animals would hibernate at all and if so at what time in comparison to the control animals. Fig. 9 shows clearly the relation of the body weight in the two groups, and fig. 10 summarized the time of onset of hibernation in all the animals in the control and food restricted group. It is quite clear that there appears to be no appreciable difference in the time of onset of hibernation in the two groups, and that the attainment of a weight peak is certainly not a prerequisite for the onset of hibernation. Undoubtedly however, in the natural condition the body weight rhythm and the homothermic-heterothermic rhythm are phase locked with obvious biological advantages.

Attempts to separate the body weigth rhythm and the homothermic-heterothermic rhythm by temperature manipulation have also been undertaken by Pengelley and Fisher (1963). In their experiments animals were kept at an environmental temperature of 35^{0}C. At such an ambient temperature hibernation is impossible because the animal cannot appreciably lower its body temperature. Under these circumstances the question was asked would the animals display an annual rhythm of body weight and/or food consumption. The data for one of these experimental animals are shown in fig. 11 and it is plain from this that there is an internal physiological rhythm which gives rise to an approximate annual rhythm of body weight and food consumption which is not dependent upon overt expression of the hibernation (heterothermic) activity (homothermic)rhythm. This experiment also demonstrated that ambient temperature is a potential zeitgeber for these annual rhythms, for animals held at the high 35^{0}C temperature

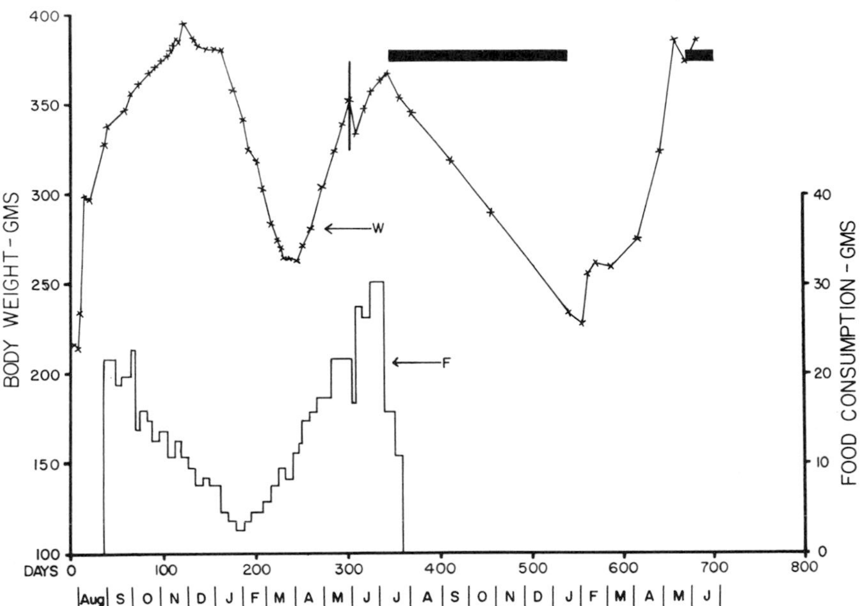

Fig. 11. Body weight (W), mean daily food consumption (F), and hibernating periods (solid horizontal bars) of an individual *C*. lateralis observed daily for two years. For first 302 days environmental temperature 35°C, photoperiod 12 hours. Transferred on day 302 (vertical line) to 0°C, same photoperiod. Food and water *ad lib.,* ample bedding. Rhythmic arousals not indicated. Food record stopped after 302 days. (From Pengelley and Fisher, 1963.)

have a shorter time span of peak to peak in body weight than those held at lower temperatures

In another important experiment, Pengelley and Fisher (1963) were able to demonstrate that by manipulating the ambient temperature to a high degree, it was possible to alter the heterothermic-homothermic rhythm such that it became 180^{0} out of phase with the presumptive seasons more rapidly than when the animals were held at lower and more natural ambient temperatures. In this experiment a number of animals were held at 35^{0}C for different lengths of time, and then transferred to an ambient temperature of 0^{0}C. The results are shown in fig. 12, and it may be seen that as a result of such temperature manipulation, the animals in group 5 are almost exactly 180^{0} out of phase with the presumptive seasons, and this after only one year of such exposure.

All the foregoing data seem to us to indicate clearly that in *C. lateralis*, which has been so extensively studied, that there is indeed an endogenous physiological circannual rhythm or rhythms with many overt expressions, with both ambient temperature and light as potential zeitgebers. Nevertheless, this theory has recently been questioned by Mrosovsky (1970), and the recent work of Dawe and Spurrier (1969); Dawe, Spurrier and Armour (1970); Dawe and Spurrier (in this symposium); is most pertinent. Mrosovsky's criticisms of the endogenous circannual rhythm concept are (a) that the phase of a relatively constant weight found in the spring in natural populations is missing under laboratory conditions, (b) there are instances where the free-running periods are too short (251 days) to be considered circannual, (c) that

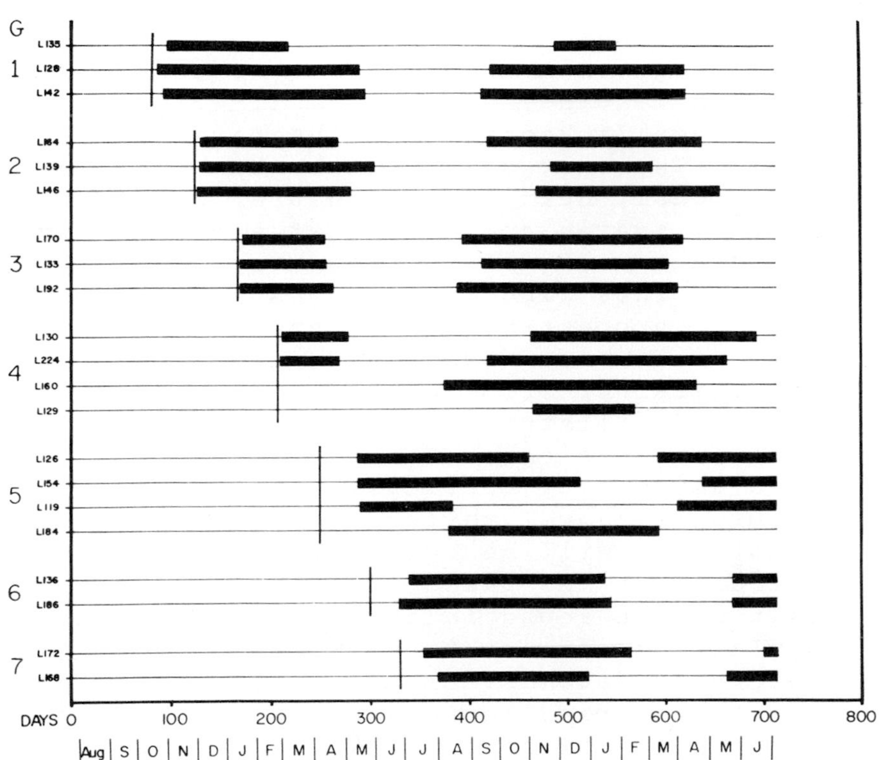

Fig. 12. Hibernating periods (solid horizontal bars) of 21 *C. lateralis* observed daily for two years. All seven groups (G) initially at 35°C and 12 hour photoperiod. Each group transferred to 0°C, same photoperiod, at time indicated by vertical lines. Food and water *ad lib.*, ample bedding. Rhythmic arousals not indicated. (From Pengelley and Fisher, 1963.)

the temperature independence is not as conclusive as appears at first sight, and (d) no one has yet discovered what the zeitgeber is. Mrosovsky's alternate theory is best summarized in his own words. "An alternate system that could give rise to annual cycles is a sequence of linked stages, each one taking a given amount of time to complete and then leading into the next, with the last stage linked back to the first one again. These stages might be based largely on hormonal, nutritional, neural or other internal states and each last a relatively standard time, just as pregnancy does. A cycle of this kind would be capable of persisting in constant conditions and in this sense would be endogenous, but it would be very different from the endogenous free-running oscillator mechanisms that seem to be involved in circadian rhythms." Mrosovsky's four citicisms of the endogenous circannual rhythm concept are worthy of note and will be considered briefly. It is true that the phase of a relatively constant weight found in the spring in natural populations appears to be missing under laboratory conditions, or to the animals' inability to breed in the laboratory, or to the presence of food and water *ad libitum* which is certainly not the reality in the natural state. It is true also that some free-running periods are as low as 251 days, but it should be noted that there are a few animals that never exhibit in the laboratory a heterothermic-homothermic rhythm at all. This in our opinion can be classed as simply biological variation (Darwin, 1859), and does not really detract from the basic circannual rhythm concept. We disagree that the temperature independence is not as conclusive as appears at first sight. Golden-mantled ground squirrels have been kept at 0°C, 3°C, 5°C, 12°C, 16°C, 22°C, and 35°C; and while there seem to be certain differences in the free-running periods, they nevertheless approximate a year, and it is far more likely that at certain times within the free-running period the animals are differentially sensitive to ambient temperature which is in fact acting as a zeitgeber. With regard to our lack of knowledge of specific zeitgebers, we have in fact implicated several and these will be discussed again. Considering also the appalling lengths of time and difficulties involved in studying circannual rhythms, it is not surprising that our lack of knowledge on entraining agents is so great. Taking all aspects of Mrosovsky's views into account, it does not seem to us that there is any basic disagreement with our circannual rhythm theory, and it is pertinent to point out here that there is no guarantee that the criteria which have been found valid for circadian rhythms will necessarily hold for circannual rhythms. As previously pointed out (Dawe and Spurrier op cit.) have reported that in *C. tridecemlineatus* they can induce hibernation during the normally homothermic period when recipient animals are transfused with blood from those in hibernation. By this method they have even induced hibernation in very young animals (Dawe and Spurrier - this symposium). This is a very important piece of work, because it clearly indicates that there is some chemical substance or substances which are controlling the various physiological states within the circannual rhythm, and more forthcoming data on this will be of extreme interest and value. However this adds new dimensions to the endogenous rhythm concept rather than changing its basic nature.

Another very important piece of experimental evidence which adds more information to the zeitgeber problem has recently been discovered by Pengelley and Asmundson (1974). In unpublished work, it has been noticed that in *C. lateralis* there were subtle but persistent differences in the behavioral responses of these animals to the environment. In general males hibernated less regularly than females, and since the one major event which females in the wild undergo which does not occur in males, is gestation and lactation, it appeared that the latter two events might be involved as circannual zeitgebers in the wild. An experiment was designed using two groups of female animals, one of which raised a litter of yound in the spring, while the second did not. Then on July 31st, which is long before any hibernation would normally be expected, both groups which had been kept at 22°C were placed in environmental chambers with a constant photoperiod of 12 hrs and a constant ambient temperature of 3°C ± 1°C. Food and water were *ad libitum*. The results are expressed graphically in fig. 13. It is clear from this that the group which had litters in the spring of 1970 hibernated for the first time considerably later than the no litter group ($P < .001$). In this respect there is no overlap between the two groups. By September 3 all animals but one in the no litter group had entered hibernation, while in the group with litters the onset of hibernation was spread over the months of September and October. From field observations it is known that the latter situation is the one which is much closer to the wild state, i.e. the group with litters hibernated more closely in synchrony with the

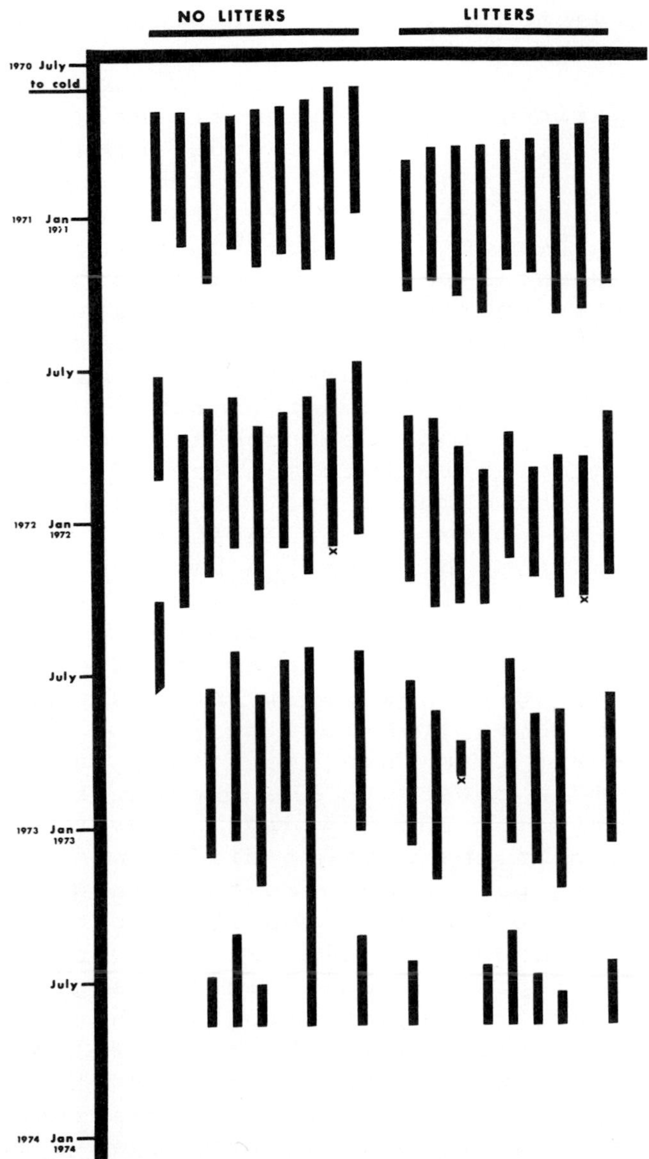

Fig. 13. Graphical representation of free-running circannual periods of 18 female animals (*C. lateralis*) for over 3 years. Upper group of 9 had a litter in the spring, 1970, lower group had no litters. Black bars indicate heterothermic period, clear space homothermic period, and the beginning to beginning of each black bar indicates one free-running circannual period. X = death. "to cold" indicates transfer of animals from 22°C to 3°C. Artificial photoperiod throughout of 12 hr. (From Pengelley and Asmundson, 1974.)

presumptive autumn. The time of onset of the second period of hibernation, i.e. the beginning of the second free-running circannual period, is also significantly different ($P < .001$). Thus the difference in the phase relationship of the two groups is quite clear for at least two free-running circannual periods. By the beginning of the third circannual period the difference, while still apparent, is not statistically significant. The data for this experiment is summarized more accurately in table 3, in which the actual number of days in each free-running period (measured from onset to onset of hibernation) is measured for all animals in the experiment. From this it seems that despite the differences in the time of onset of hibernation between the two groups, there is in fact no significant difference between the two groups in the length of the first or subsequent free-running periods. However in both groups the second and third free-running period are significantly less than the first, and the first is much closer to the presumptive annual period of 365 1/4 days. Now it is true also that there is no difference in the free-running period between the two groups, however, if it had been possible to get the group which had had a litter in the spring of 1970 to reproduce again in the spring of 1971, it appears highly probable that the onset of hibernation in the fall of 1971 would have been delayed again, and it may therefore be inferred that the free-running period of this group would have been altered in such a way that it would have been closer to the first free-running period, and the animals would have remained better in phase with the presumptive seasons, while the group without litters would have gotten

Table 3--Comparison of first, second and third free-running circannuan periods (onset to onset of hibernation) in days of *C. lateralis* under constant environmental conditions of 3°C and 12 hr photoperiod. Animals 1-9 had litters in the spring of 1970, 10-18 no litters. (From Pengelley and Asmundson, 1974.)

	Free-running circannual periods (days)		
Animals with litters in Spring 1970	1st (1970-1971)	2nd (1971-1972)	3rd (1972-1973)
1	353	337	310
2	397		
3	392	306	334
4	392	295	311
5	346	274	313
6	361	314	277
7	356	350	
8	317	349	
9	301	320	336
Animals with no litters			
10	355	348	337
11	345		
12	363	304	
13	362	297	318
14	376	323	345
15	343	302	336
16	340	337	341
17	383		
18	281	300	

more and more out of phase. Thus having a litter in the spring, not only assures the survival of the species, but also of the individual female who will be better programmed to the seasons of the year in her heterothermic-homothermic circannual rhythm. The importance of this experiment is simply that in these mammals the reproductive cycle is clearly linked, for the first time, to the observable circannual rhythm, that for females a zeitgeber is gestation and lactation, and it follows inevitably from this that for females also a sexually active male at the right place and the right time is in fact a circannual zeitgeber. Males are another problem which require further investigation.

2. Other species of hibernating mammals.

The knowledge collected for *C. lateralis* over many years of field observations and laboratory experiments, should be applied with great caution to other species of hibernating mammals, and to assume a circannual rhythm without experimental evidence would be folly. However, some discussion of evidence from other species is in order. Much work has been done on hibernation in the golden hamster, *Mesocricetus auratus*, and the evidence seems to be that a precise circannual rhythm is not one of the mechanisms evolved by these animals (Fawcett and Lyman, 1954; Hoffman and Reiter, 1965; Hoffman, Herter and Towns, 1965; Lyman, 1948, 1954; Mogler, 1958). However the recent and excellent study by Smit-Vis and Smit (1970) indicates quite clearly that hibernation in the hamster is dependent partly on age, and possibly on an endogenous rhythm as well. It should be noted however that the golden hamster has become so domesticated and so interbred that its physiology and behavior probably have little relevance to the wild animal. A good ecological study of this animal in the wild would be of great value.

Another species of hibernator with a very different phylogeny from *C. lateralis* is the European dormouse, *Glis glis*. The behavioral responses of this animal to the environment have been studied by Pengelley and Fisher (1963); Scott and Fisher (1970); Morris and Morrison (1964) and Morrison (1964). The latter workers performed a crucial experiment in which the responses of *Glis glis* and *C. tridecemlineatus* were accurately compared under controlled environmental conditions. Their experiment consisted of exposing the two species to boreal (normal) and austral (reversed) light cycles (9-19 hours day length) and a constant ambient temperature of 25°C for a year and a half in Alaska. Periodically the animals were moved to a 7°C environmental temperature for four days to test the degree to which the animals were capable of hibernating. The results are summarized in fig. 14, in which it is clear that the dormice quickly adjusted to the boreal or austral light schedules as the case might be. This was true not only for the test of ability to hibernate but also for testes growth and involution, as well as for growth and shedding. On the other hand, *C. tridecemlineatus*, showed an identical, synchronous cyclic behavior under both boreal and austral schedules. The day length did not regulate any of the aforementioned variables. It seems clear from this experiment that *Glis glis* does not have what we may term a "dominating" endogenous circannual rhythm, a fact confirmed by Pengelley and Fisher

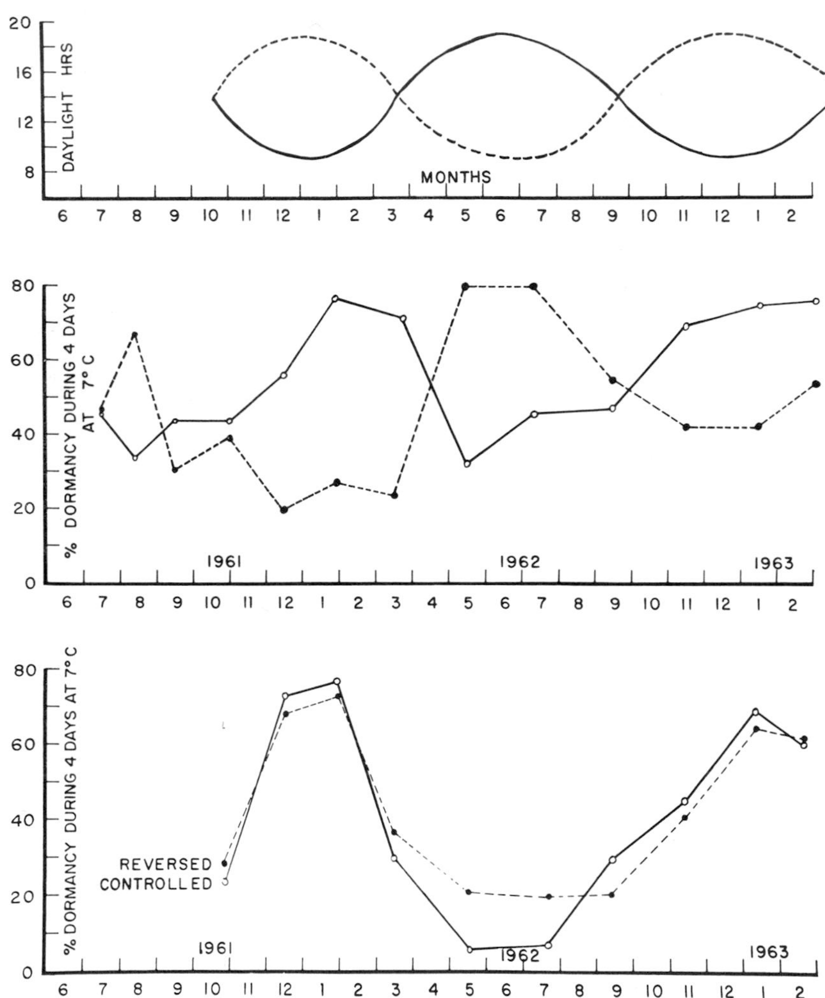

Fig. 14. Cyclic response in tendency toward torpor in relation to yearly light cycle in Alaska. Animals maintained in normal (solid lines) and reversed (broken lines) cycles at 25°C. Upper figure, light based on civil twilight (a.m.) to civil twilight (p.m.) for 54° N latitude. Middle figure, response of *Glis glis.* Bottom figure, response of *Citellus tridecemlineatus.* (Redrawn from Morrison, 1964.)

(1963); Mrosovsky (1964, 1968); and Scott and Fisher (1970). The latter authors suggest that there may be a rather imprecise rhythm in this species of from 5-9 months, but more experiments are needed. *C. tridecemlineatus* however, does exhibit a rather precise circannual rhythm, a fact amply verified by Pengelley and Fisher (1963); Mrosovsky (1970); Richter (1965, 1967); and others.

Good experimental evidence for the existence of circannual rhythms in other species of hibernators is very sparce, however what evidence there is indicates that all the following probably have some form of endogenous "annual" rhythm; the european hedgehog, *Erinaceus europaeus*, (Kristoffersson and Soivio, 1964; Kristoffersson and Suomalainen, 1964; Johansson and Senturia, 1972); the Vespertilionid bat, *Myotis lucifugus*, and probably other bats, (Menaker, 1961, 1962, 1964); many species of western chipmunks, *Eutamias sp.* (Heller and Poulson, 1970; Cade, 1963; Jameson, 1964; Jameson and Mead, 1964); also the eastern chipmunk, *Tamias striatus*, (Panuska, 1959; Scott and Fisher, 1970); several species of ground squirrels in addition to those already mentioned, i.e. *Citellus columbianus*, (Pengelley and Fisher, 1963; Scott and Fisher, 1970); *Citellus undulatus*, (Drescher, 1967); *Citellus variegatus*, *Citellus beecheyi*, *Citellus tereticaudus*, *Citellus mohavensis*, (Pengelley and Kelly, 1966); while *Citellus richardsonii* exhibits a behavior pattern difficult to interpret (Scott and Fisher, 1970).

There is one other major species of hibernator which is known to exhibit a circannual rhythm and that is the woodchuck, *Marmota monax*, (Davis, 1965a, 1965b, 1967a,

1967b). A very important experiment has recently been performed on this animal (Davis - personal communication). It was known that woodchucks in Pennsylvania exhibited an annual rhythm of deposition of fat which was independent of the environmental conditions in the field, and was not experimentally altered in the laboratory. The animals were normally at a minimum in April and a maximum in September, and under constant conditions this rhythm had a free-running period of about 10 months. Yearling animals were captured in April 1969, kept under constant conditions until nine months later (Jan. 1970) when they were flown to Sydney, Australia and housed in the Taronga Zoo. In australia they were exposed to external light and temperature conditions but with food and water *ad libitum*. Within two years these animals shifted their peak in body weight to April and maintained it there for an additional two years. It is quite clear that the endogenous annual rhythm of this northern hemisphere species entrained to the environmental conditions of the southern hemisphere. Dr. Davis speculates that the length of the day is the zeitgeber, though experimental evidence for this is at present lacking. However, the important thing is that for the first time complete entrainment of an endogenous circannual rhythm has been demonstrated.

3. Adaptive value of circannual rhythms.

The adaptive value of a circannual rhythm in such an animal as a hibernator is that it enables the animal to prepare well in advance for a future environmental condition, such as a combination of cold and food scarcity or optimal breeding conditions. Thus for example if the animal waited until the onset of cold weather, or worse--snow, in the fall as a stimulus to increase its fat stores for hibernation in the winter, it would be too late since the food supply would already be declining. But with the programming of an endogenous rhythm, the fat supply is automatically increased during the summer when the food supply is greatest. Similarly the animal must be adapted in such a way that reproduction takes place very early in spring, otherwise the young are not likely to have time to grow to sufficient size to survive the onset of winter. The evidence is in fact that reproductive competence is governed by a circannual rhythm which brings the females into estrus and makes the males sexually active (with scrotal testes) almost at once after the termination of hibernation. An obvious advantage also, is that arousal in the spring from hibernation should be from endogenous cues, since an animal deep in a winter burrow, and hibernating as well, is not likely to receive normal environmental cues. There is solid evidence that the spring arousal is triggered by an endogenous mechanism, for Pengelley and Fisher (1961) showed that in *C. lateralis* and all other species so far studied (Pengelley and Kelly, 1966) the frequency of periodic arousal steadily increased from about mid winter to spring when terminal arousal took place. On the other hand a circannual rhythm allows the hibernating mammal some flexibility in an environment that changes from year to year. This enables it to integrate a large number of environmental cues, and Heller and Poulson (1970) believe that environmental stimuli act to either shorten or lengthen some portion of the rhythm, i.e. to phase it without affecting its circannual period. It is clear also from the work of Pengelley and Kelly (1966) that the balance between phasing by environmental stimuli and annual endogenous control depends on the species ecology. Thus Heller and Poulson (1970) argue that *C. lateralis* and *C. beldingi*, which are sympatric and experience the same physical weather conditions, nevertheless have very different food habits. The animal's adaptations are that *C. lateralis* is mainly under the control of a circannual rhythm, whereas *C. beldingi* is much more susceptible to phasing by environmental stimuli. They were able to confirm this in the laboratory. Similarly they argue that temperate zone hibernators are more susceptible to phasing than are Arctic Zone animals, mainly because in the latter case the environment demands that the spring breeding be very accurately timed. Their speculations seem entirely justified, but in an animal such as a ground squirrel which has a potential life span of 5-10 years, it seems obvious that in addition to phasing there must be zeitgebers which entrain the animal to the exogenous annual rhythm of 365 1/4 days. All evidence points to the fact that in hibernating mammals with a circannual rhythm, the free-running period is some 300+ days or 50-60 less than the exogenous rhythm. Thus without entraining zeitgebers the animals would become 180^{0} out of phase in about 3 years. There are some 28 species of *Citellus* in North America, ranging in habitat from the low deserts to the Arctic tundra, and they are ideally suited to comparative study for determination of the relative relationships of endogenous and exogenous factors.

4. Problems for future investigation.

The problems future investigators face in studying circannual rhythms involve the whole complex interaction of the exogenous with the endogenous. Presumably much of this is mediated by the central nervous system but we are pitifully ignorant of how this takes place. In our opinion the whole concept must be related to the general ecology of the species, otherwise it has little meaning. Of course, another overriding problem is the actual biochemical and physiological nature of the endogenous oscillator. Hopefully investigators of circadian rhythms will shortly be able to give clues to this, but the difficulties are very great. Closely associated with this is the problem of whether there is one master annual oscillator which is driving all the other overt oscillators, or whether there are many oscillators which are normally phase locked but can on occasions get out of phase, a problem first raised by Bunning (1936). In circadian rhythms the evidence seems quite clear that the rhythms of different variables can run with different speeds (Wever, 1973), which means that the temporal order of the different rhythms is controlled by several clocks. There is little evidence one way or the other on circannual oscillators, but some evidence is available from Pengelley (1967), who felt that there was more than one oscillator. However this is an area where more sophisticated studies are much needed. It has been suggested by Pengelley, Bartholomew and Licht (1972) that attempts could be made to get key physiological events out of phase with others which it normally accompanies. For example, by endocrine injections it might be possible to induce experimental animals into reproductive competence and to breed out of season, and then to determine whether this affects the programming of other physiological events, such as the homothermic-heterothermic and body weight rhythms for both experimental animals and their offspring. In this kind of manipulative experiment, the value of comparative studies on different species is of the utmost importance.

Another area of investigation with great promise is the study of internal parameters rather than purely overt ones which have been studied up until now. Obviously the overt rhythms such as weight, activity, heterothermy-homothermy and reproductive competence must be dependent on subtle internal neurological, biochemical or physiological rhythms which are determining them. An attempt to uncover the neurological mechanisms has recently been made by Heller and Hammel (1972). A method of getting at some of these without sacrificing the animals has been described by Pengelley, Asmundson and Uhlman (1971), and further improvements in this are entirely possible. Yet other techniques have been used by Johansson and Senturia (1972), and their work on internal parameters is reported elsewhere in this symposium.

Considering the complexity of organisms, and the time necessary to study circannual rhythms under controlled conditions, our progress is certainly better than we have a right to expect. Nevertheless it seems imperative to us that even longer experiments (perhaps up to 10 years) should be attempted with the hope that more precise information can be obtained about the zeitgebers. In this regard it is important to note that the zeitgebers so common for circadian rhythms may not be the same for circannual rhythms. It is in fact far more likely that the zeitgebers are a complex of many factors and different for different species. Furthermore as Pengelley, Bartholomew and Licht (1972) have pointed out, some parts of the rhythm may be refractory to various environmental stimuli while other parts are not, though there may be discrete short periods of time, "windows", when the central nervous system responds to special stimuli (Licht, 1972). For example, an environmental factor such as photoperiod may produce an effect at one environmental temperature but not at another, at one season and not another, or on one sex and not another--which is highly probable in hibernating mammals. Indeed, any environmental factor may interact with another in such a way that both may be required for the animal to respond, or on the other hand one environmental factor may elicit a crucial response at one time in the circannual rhythm, while another environmental factor is necessary at another time. It is imperative also to realize that the response to a zeitgeber may be considerably delayed, and may become overt only after a passage of many months. The zeitgeber problem for circannual rhythms will not be easily solved.

CONCLUSION

In conclusion, the importance of circannual rhythms in general, and in hibernating mammals in particular, can scarcely be overestimated for the behaviorist, physiologist, and biochemist and indeed virtually all biologists in their attempts to understand and interpret living phenomena. In these mammals with circannual rhythms, the animal goes from one internal physiological state to another very different one and back again in a perfectly normal manner. These are in turn reflected in the animal's overt responses. Thus it is important to realize that experimental conclusions drawn at one period in the circannual rhythm may be utterly false at another period. In the light of this new knowledge many biological "dogmas" and theories may have to be modified if not wholly abandoned.

SUMMARY

1. The historical scientific background of endogenous rhythms is reviewed, and the criteria necessary to demonstrate them are discussed. The interaction of the driving and driven oscillators is also explained and the relationship of endogenous rhythms to zeitgebers are pointed out.

2. The discovery of circannual rhythms in their historical context is discussed, and particular attention is given to those in hibernating mammals. Circadian and circannual rhythms are contrasted and the phenomenon of hibernation is explained.

3. There is a thorough treatment of circannual rhythms in the golden-mantled ground squirrel, *Citellus lateralis*. It is pointed out that in this animal there are many overt circannual rhythms, such as body weight, homothermy-heterothermy, locomotor activity, food consumption, water consumption, and reproductive competence.

4. The problems of determining circannual

zeitgebers are explained in relation to circadian zeitgebers, and evidence is presented that in hibernators ambient temperature, light, access to locomotor activity and opportunity to breed are all potential zeitgebers. However it is argued that the zeitgebers for circannual rhythms are probably a complex of many environmental factors. Criticisms of the concept of circannual rhythms in hibernators are given, and an alternate theory in the form of a sequence of linked stages is discussed.

5. Overt evidences of circannual rhythms in several other hibernating mammals are described and experimental evidence is presented that in the woodchuck, *Marmota monax*, complete entrainment of the circannual body weight rhythm has been accomplished by transferring this northern hemisphere species to the southern hemisphere.

6. The adaptive value of circannual rhythms are explained in relation to the animals' total ecology.

7. The great importance of circannual rhythms for behaviorists, physiologists, and biochemists are explained, and various problems awaiting investigation are thoroughly discussed.

ACKNOWLEDGEMENTS

The preparation of this paper was supported by Grant GB-40827 from the National Science Foundation, and secondarily by an Intramural Grant from the University of California. We also acknowledge the able technical assistance of Mr. Brian Barnes and Mrs. Dorothy Whitson in the conduct of many of the experiments reported here.

REFERENCES

ALOIA, Roland C., PENGELLEY, Eric T., BOLEN, James and George ROUSER, (1974). Phospholipid analysis of cardiac muscle of the hibernating ground squirrel, *Citellus lateralis*, in the active and hibernating states. Lipids--in press.

ASCHOFF, J. (1960). Exogenous and endogenous components in circadian rhythms. Cold Spring Harbor Symp. 25: 11-28.

ASCHOFF, J. (1963). Comparative physiology: diurnal rhythms. Ann. Rev. Physiol. 25: 581-600.

ASCHOFF, J. (1964). Survival value of diurnal rhythms. Symp. Zool. Soc. London 13: 79-98.

ASCHOFF, J. (1965). Circadian Clocks. North Holland Pub. Co. pp 479.

BARTHOLOMEW, G. A. (1972). Aspects of timing and periodicity of heterothermy. Hibernation and hypothermia, perspectives and challenges. Elsevier/Excerpta/Medica/North Holland Pub. Co., Amsterdam, pp 663-680.

BARTHOLOMEW, G. A. and J. W. HUDSON (1960). Aestivation in the Mohave ground squirrel, *Citellus mohavensis*. Bull. Mus. Comp. Zool. Harvard Coll. 124: 193-208.

BENOIT, J., ASSENMACHER, I., and E. BRARD (1955). Evolution testiculaire du canard domestique maintenu a l'obscurite totale pendant une longue duree. C. R. Acad. Sci., 241: 251-253.

BENOIT, J., ASSENMACHER, I. and E. BRARD (1956a). Etude de l'evolution testiculaire du canard domestique soumis tres jeune a un elairement artificiel permanent pendant deux ans. C. R. Acad. Sci. 242: 3113-3115.

BENOIT, J. ASSENMACHER, I. and E. BRARD (1956b). Apparition et maintien de cycles sexuels non saisonniers chez le canard domestique place pendent plus de trois ans a l'obscurite totale. J. Physiol., Paris 48: 388-391.

BISSONETTE, T. H. (1935). Modification of mammalian sexual cycles. J. Expt. Zool. 71: 341-373.

BUNNING, E. (1936). Die endonome tages-rhythmik als grundlage de photoperiodischen reaktion. Ber. Deut. Botan. Ges. 54: 590-607.

BUNNING, E. (1964). The physiological clock. Academic Press, N.Y. pp 145.

BUNNING, ERWIN (1967). The physiological clock. Springer-Verlag, New York Inc. pp 167.

CADE, T. J. (1963). Observations on torpidity in captive chipmunks of the genus *Eutamias*. Ecology 44(2): 255-261.

CADE, T. J. (1964). The evolution of torpidity in rodents. Ann acad. Scientiarum Fennicae, Ser. A., IV 71: 77-112.

COLD SPRING HARBOR SYMPOSIUM on Quantitative Biology (1960). Biological clocks. Biological Laboratory, L.I. New York 25: 524 pp.

DARWIN, C. (1859). On the origin of species by means of natural selection. John Murray, London pp 502.

DARWIN, C. (1880). The power of movement in plants. John Murray, London.

DAVIS, D. E. (1965a). Behavioural aspects of torpor in woodchucks. Amer. Zool. 5(2): 196.

DAVIS, D. E. (1965b). Factors initiating

torpor in woodchucks. Amer. Zool. 5(2): 208.

DAVIS, D. E. (1967a). The annual rhythm of fat deposition in woodchucks *(Marmota monax)*. Physiol. Zool. 40: 391-402.

DAVIS, D. E. (1967b). The role of environmental factors in hibernation of woodchucks *(Marmota monax)*. Ecology 48: 683-689.

DAWE, A. R. and WILMA A. SPURRIER (1969). Hibernation induced in ground squirrels by blood transfusion. Science 163: 298-299.

DAWE, A. R., W. A. SPURRIER and J. A. ARMOUR (1970). Summer hibernation induced by cryogenically preserved blood "trigger." Science 168: 497-498.

DE MAIRAN, M. (1729). Observation botanique p. 35. Hist. de l'Academie Royale des Sciences, Paris.

DRESCHER, J. W. (1967). Environmental influences on initiation and maintenance of hibernation in the arctic ground squirrel, *Citellus undulatus*. Ecology 48: 962-966.

DUBOIS, R. (1896). Etude sur le mechanisme de la thermogenese et du sommeil chez les mammiferes. Physiologic comparee de la marmotte. Ann. Univ. Lyon 25: 1-268.

EISENTRAUT, M. (1956). Der winterschlaf mit seinen okologischen und physiologischen begleiterscheinungen. Veb. Gustav Fischer Verlag, Jena 160 pp.

FAWCETT, D. W. and C. P. LYMAN (1954). The effect of low environmental temperatures on the composition of depot fat in relation to hibernation. Jour. Physiol. Lond. 126: 235-247.

FOLK, G. E. (1960). Day-night rhythms and hibernation. Mammalian hibernation I.

Proc. First Intl. Symp. Nat. Mammal. Hib. Bull. Mus. Comp. Zool. Harvard Coll. 124: 209-232.

GWINNER, E. (1968). Circannuale periodik als grundlage des jahreszeitlichen funktionswandels bei Zugvogeln. Untersuchungen am fitis (*Phylloscopus trochilus*) und am Waldlaubsanger (*P. sibilatrix*). J. fur Ornithologie 109, Heft. 1, 70-95.

HALBERG, E., HALBERG, F., BARNUM, C. P. and J. J. BITTNER (1959). Physiologic 24-hour periodicity in human beings and mice, the lighting regimen and daily routine. pp. 803-78 in: Photoperiodism and related phenomena in plants and animals. Amer. Assoc. Adv. Sci. Washington. Pub. #55. 903 pp.

HELLER, H. C. and HAROLD T. HAMMEL (1972). CNS control of body temperature during hibernation. Comp. Biochem. Physiol. 41A: 349-359.

HELLER, H. C. and THOMAS L. POULSON (1970). Circannian rhythms-II. Endogenous and exogenous factors controlling reproduction and hibernation in chipmunks (*Eutamias*) and ground squirrels (*Spermophilus*). Comp. Biochem. Physiol. 33: 357-383.

HIBERNATION-HYPOTHERMIA, PERSPECTIVES and CHALLENGES (1972). Ed. South, Frank E., Hannon, John P., Willis, John R., Pengelley, Eric T., and Norman R. Alpert. Elsevier Pub. Co.

HOCK, R. J. (1955). Photoperiod as stimulus for onset of hibernation. Fe. Proc. 14: 73-74.

HOCK, R. J. (1958). Cold Injury. Trans. 5th Conf. Josiah Macy, Jr. Foundation: 61-133.

HOCK, R. J. (1960). Seasonal variations in

physiologic functions of arctic ground squirrels and black bears. Bull. Mus. Comp. Zool. Harvard Coll. 124: 155-171.

HOFFMAN, R. A. (1964). Speculations on the regulation of hibernation. Ann. Acad. Scientiarum Fennicae Ser A., IV 71: 199-216.

HOFFMAN, R. A., HESTER, R. J., and C. TOWNS (1965). Effect of light and temperature on the endocrine system of the golden hamster, *Mesocricetus auratus* Waterhouse. Comp. Biochem. and Physiol. 15: 525-533.

HOFFMAN, R. A. and R. J. REITER (1965). Pineal gland: Influence on gonade of male hamster. Science 148(3677): 1609-1610.

HUDSON, J. W. and G. A. BARTHOLOMEW (1964). Terrestrial animals in dry heat: estivators. Handbook of Physiology, Sec. 4, Adaptation to the environment. Amer. Physiol. Soc. 541-550.

JAMESON, E. W. (1964). Patterns of hibernation of captive *Citellus lateralis* and *Eutamias speciosus*. J. Mammal. 45(3): 455-460.

JAMESON, E. W. and R. A. MEAD (1964). Seasonal changes in body fat, water and basic weight in *Citellus lateralis*, *Eutamias speciosus* and *E. amoenus*. J. Mammal. 45(3): 359-365.

JOHANSSON, BENGT W. and JEROME B. SENTURIA (1972). Seasonal variations in the physiology and biochemistry of the European hedgehog *(Erinaceus europaeus)* including comparisons non-hibernators, guinea-pig and man. Acta Physiol. Scand. Supp. 380. pp 159.

KALABUKHOV, N. I. (1960). Comparative ecology of hibernating mammals. Bull. Mus. Comp. Zool. Harvard Coll. 124:

45-74.

KAYSER, C. (1940). Essai d'analyse du mechanisme du sommeil hibernal. Ann. Physiol. 16: 313-372.

KAYSER, C. (1953). L'Hibernation des mammiferes. Annee Biologique 29: 109.

KAYSER, C. (1957). Le sommeil hivernal, probleme de thermoregulation. Rev. Can. Biol. 16(3): 303-389.

KAYSER, C. (1960). Hibernation versus hypothermia. Bull. Mus. Comp. Zool. Harvard Coll. 124: 9-29.

KAYSER, C. (1961). The physiology of natural hibernation. Pergamon Press, New York. 325 pp.

KAYSER, C., G. NINCENDON, R. FRANK and A. PORTE (1964). Some external (climatic) and internal (endocrine) factors in relation to the production of hibernation. Ann. Acad. Scientiarum Fennicae, Ser. A., IV 71: 269-282.

KRISTOFFERSSON, R. and A. SOIVIO (1964). Hibernation of the hedgehog, *Erinaceus europaeus*. The periodicity of hibernation of undisturbed animals during the winter in a constant ambient temperature. Ann. Acad. Scientiarum Fennicae, Ser. A., N, 80: 1-22.

KRISTOFFERSSON, R. and P. SUOMALAINEN (1964) Studies on the physiology of the hibernating hedgehog: changes of body weight of hibernating and non-hibernating animals. Ann. Acad. Scientiarum Fennicae, Ser. A, IV, 76: 1-11.

LIGHT, PAUL (1972). Problems in experimentation on timing mechanisms for annual physiological cycles in reptiles. Hibernation and hypothermia, perspectives and challenges. Elsevier/ Excerpta/Medica/North Holland Pub. Co. Amsterdam, P. 681-711.

LYMAN, C. P. (1948). The oxygen consumption and temperature regulation of hibernating hamsters. J. Exper. Zool. 109: 55-78.

LYMAN, C. P. (1954). Activity, food consumption and hoarding in hibernators. J. Mammal. 35: 545-552.

LYMAN, C. P. (1958). Oxygen consumption, body temperature and heart rate of woodchucks entering hibernation. Am. J. Physiol. 194: 83-91.

LYMAN, C. P. (1958). Oxygen consumption, body temperature and heart rate of woodchucks entering hibernation. Am. J. Physiol. 194: 83-91.

LYMAN, C. P. (1961). Hibernation in mammals. Circulation XXIV: 434-445.

LYMAN, C. P., and P. O. CHATFIELD (1955). Physiology of hibernation in mammals. Physiol. Rev. 35: 403-425.

LYMAN, C. P. and REGINA C. O'BRIEN (1960). Circulatory changes in the thirteen-lived ground squirrel during the hibernating cycle. Bull. Mus. Comp. Zool. Harvard Coll. 124: 353-372.

MAMMALIAN HIBERNATION I., Proc. First Intl. Symp. Nat. Mammal. Hib. (1960). Ed. LYMAN, CHARLES P. and ALBERT R. DAWE Bull Mus. Comp. Zool. Harvard Coll. Vol. 124.

MAMMALIAN HIBERNATION II., Proc. Second Intl. Symp. Nat. Mammal. Hib. (1964). Ed. PAAVO SUOMALAINEN. Ann. Acad. Sci. Fenn. Ser. A IV Tome 71.

MAMMALIAN HIBERNATION III., Proc. Third Intl. Symp. Nat. Mammal. Hib. (1967). Ed. FISHER, KENNETH C., DAWE, ALBERT R., LYMAN, CHARLES P., SCHONBAUM, EDWARD and FRANK E. SOUTH, JR. Oliver and Boyd, Edinburgh (Amer. Pub.-Amer. Elsevier Co., New York).

MENAKER, M. (1961). The free-running period of the bat clock; seasonal variations at low body temperatures. J. Cell and Comp. Physiol. 57: 81-86.

MENAKER, M. (1962). Hibernation-hypothermia: an annual cycle of response to low temperature in the bat *Myotis lucifugus*. J. Cell and Comp. Physiol. 59(2): 163-173.

MENAKER, M. (1964). Frequency of spontaneous arousal from hibernation in bats. Nature 203, No. 4994: 540-541.

MOGLER, R. K. (1958). Das endokrime system des Syrischen Goldhamsters, *Mesocricetus auratus auratus Waterhouse*, unter berucksichtigung des naturlichen und experimentellen Winterschlaps. Zeit. Zur. Morphol. und Oekol. der Tiere 47: 267-308.

MORRIS, L. and P. MORRISON (1964). Cyclic responses in dormice, *Glis glis* and ground squirrels; *Spermophillus tridecemlineatus*, exposed to normal and reversed yearly light schedules. Science in Alaska: Proc. 15th Alaskan Sci. Conference AAAS: 40-41.

MORRISON, P. (1960). Some interrelations between weight and hibernation function. Bull. Mus. Comp. Zool. Harvard Coll. 124: 75-91.

MORRISON, P. (1964). Adaptation of small mammals to the arctic. Fed. Proc. 23(6): 1202-1206.

MROSOVSKY, N. (1964). The performance of dormice and other hibernators on tests of hunger motivation. Animal Behavior 12: 454-469.

MROSOVSKY, N. (1968). The adjustable brain of hibernators. Sci. Amer. 218(3): 110-118.

MROSOVSKY, N. (1970). Mechanism of hibernation cycles in ground squirrels: circannian rhythm or sequence of stages. Penn. Acad. Sci. 44: 172-175.

MROSOVSKY, N. and K. C. FISHER (1970). Sliding set points for body weight in ground squirrels during hibernation season. Can. J. Zool. 48: 241-247.

MULLALLY, D. F. (1953). Hibernation in the golden-mantled ground squirrel, *Citellus lateralis bernardinus*. J. Mammal. 34(1): 65-73.

PANUSKA, J. A. (1959). Weight patterns and hibernation in *Tamias striatus*. J. Mammal. 40(4): 554-566.

PENGELLEY, E. T. (1964). Responses of a new hibernator, *Citellus variegatus* to controlled environments. Nature 203: no. 4947, 892.

PENGELLEY, E. T. (1966). Differential developmental patterns and their adaptive value in various species of the genus *Citellus*. Growth 30: 137-142.

PENGELLEY, E. T. (1967). The relation of the external conditions to the onset and termination of hibernation and estivation. Proc. III Intl. Symp. Mammal, Hib. Ed. Fisher, K. C., A. R. Dawe, C. P. Lyman, E. Schonbaum, and F. E. South, Jr. Oliver and Boyd, Edinburgh. p. 1-29.

PENGELLEY, E. T. (1968). Interrelationships of circannian rhythms in the ground squirrel, *Citellus lateralis*. Comp. Biochem. Physiol. 24: 915-919.

PENGELLEY, E. T. (1969). Influence of light on hibernation in the mohave ground squirrel, *Citellus mohavensis*. Physiological systems in semiarid environments - Ed. C. Clayton Hoff and Marvin L. Riedesel. Univ. of New Mexico Press pp 11-16.

PENGELLEY, E. T., G. A. BARTHOLOMEW and PAUL LIGHT (1972). Problems in the chronobiology of hibernating mammals. Hibernation and Hypothermia, Perspectives and Challenges. Elsevier/Excerpta/Medica/North Holland Pub. Co., Amsterdam pp 713-716.

PENGELLEY, E. T. and SALLY J. ASMUNDSON (1969). Free-running periods of endodenous circannian rhythms in the golden-mantled ground squirrel, *Citellus lateralis*. Comp. Biochem. Physiol. 30: 177-183.

PENGELLEY, E. T. and SALLY J. ASMUNDSON (1970a). The effect of light on the free-running circannual rhythm of the golden-mantled ground squirrel, *Citellus lateralis*. Comp. Biochem. Physiol. 32: 155-160.

PENGELLEY, E. T. and SALLY J. ASMUNDSON (1970b). Circannual rhythmicity - Evidence and theory. Life Sciences and Space Research VIII. Ed., F. G. Favorite and W. Vischniac. Proc. XII Plenary Meeting COSPAR, Prague, Czechoslovakia. North-Holland Pub. Co. Amsterdam. pp 235-239.

PENGELLEY, E. T. and SALLY J. ASMUNDSON (1971). Circannual rhythmicity in hibernating ground squirrels, *Citellus lateralis*, in relation to sex and opportunity to breed. Proc. XXV Intl. Union Physiol. Sci. Munich, Germany, Vol. IX.

PENGELLEY, E. T. and SALLY J. ASMUNDSON (1972). An analysis of the mechanisms by which mammalian hibernators synchronize their behavioral physiology with the environment. Hibernation and Hypothermia, Perspectives and Challenges. Elsevier/Excerpta/Medica/North

Holland Pub. Co., Amsterdam, pp 637-661.

PENGELLEY, E. T. and SALLY J. ASMUNDSON (1974). Female gestation and lactation as zeitgebers for circannual rhythmicity in the hibernating ground squirrel *Citellus lateralis*. Comp. Biochem. Physiol. In press.

PENGELLEY, E. T., ASMUNDSON, SALLY J. and CHRISTINE UHLMAN (1971). Homeostasis during hibernation in the golden-mantled ground squirrel, *Citellus lateralis*. Comp. Biochem. Physiol. 38A: 645-653.

PENGELLEY, E. T. and R. R. J. CHAFFEE (1966) Changes in plasma magnesium concentration during hibernation in the golden-mantled ground squirrel *(Citellus lateralis)*. J. Comp. Biochem. and Physiol. 17: 673-681.

PENGELLEY, E. T. and K. C. FISHER (1957). Onset and cessation of hibernation under constant temperature and light in the golden-mantled ground squirrel, *Citellus lateralis*. Nature 180: 1371-1372.

PENGELLEY, E. T. and K. C. Fisher (1961). Rhythmical arousal from hibernation in the golden-mantled ground squirrel, *Citellus lateralis tescorum*. Can. J. Zool. 39: 105-120.

PENGELLEY, E. T. and K. C. Fisher (1963). The effect of temperature and photoperiod on the yearly hibernating behavior of captive golden-mantled ground squirrels, *Citellus lateralis tescorum*. Can. J. Zool. 41: 1103-1120.

PENGELLEY, E. T. and K. C. Fisher (1966). Locomotor activity patterns and their relation to hibernation in the golden-mantled ground squirrel, *Citellus lateralis*. J. Mammal. 47(1): 63-73.

PENGELLEY, E. T. and K. H. KELLY (1966a). A "Circannian" rhythm in hibernating species of the genus *Citellus* with observations on their physiological evolution. Comp. Biochem. Physiol. 19(3): 603-617.

PENGELLEY, E. T. and K. H. KELLY (1966b). Plasma potassium and sodium concentrations in active and hibernating golden-mantled ground squirrels, *Citellus lateralis*. Comp. Biochem. and Physiol. 20: 299-305.

PFEFFER, W. (1875). Die periodischen Bewegungen der Blattorgane pp 176, Wilhelm Engelmann, Leipzig.

PITTENDRIGH, C. W. (1958). Perspectives in the study of biological clocks. pp 239-268. Perspectives in marine biology; University of California Press. 621 pp.

POPOVIC, V. (1960). Endocrines in hibernation. Bull. Mus. Comp. Zool. Harvard Coll. 124: 105-130.

POPOVIC, V. (1964). Cardiac output in hibernating ground squirrels. Am. J. Physiol. 207: 1345-1348.

RICHTER, C. P. (1965). Biological clocks in medicine and psychiatry. C. C. Thomas, Springfield, Ill.

RICHTER, C. P. (1967). Comment p 27-28. Mammalian Hibernation III., Proc. Third Intl. Symp. Nat. Mammal Hib. Ed. Fisher, Kenneth C., Dawe, Albert R., Lyman, Charles P., Schonbaum, Edward and Frank E. South, Jr. Oliver and Boyd, Edinburgh, (Amer. Pub.-Amer. Elsevier Co., N.Y.).

RIEDESEL, M. L. (1960). The internal environment during hibernation. Bull. Mus. Comp. Zool. Harvard Coll. 124: 421-435.

RIEDESEL, M. L., L. R. KLINESTIVER and NANCY R. BENALLY (1964), Tolerance of

Citellus lateralis and *C. spilosoma* for water deprivation. Ann. Acad. Scientiarum Fennicae Ser. A., IV, 71: 375-388.

ROWAN, W. (1925). Relation of light to bird migration and developmental changes. Nature 115(2892): 494-495.

ROWAN, W. (1926). On photoperiodism, reproductive periodicity and the annual migrations of birds and certain fishes. Proc. Boston Soc. Natl. Hist. 38: 147-189.

ROWAN, W. (1938). Light and seasonal reproduction in animals. Biol. Rev. 13: 374-402.

SCOTT, GRACE WORKMAN and KENNETH C. FISHER (1970). The lengths of hibernation cycles in mammalian hibernators living under controlled conditions. Penn. Acad. Sci. 44: 180-183.

SMIT-VIS, J. H. and G. J. SMIT (1970). Hibernation in the golden hamster in relation to age and to season. Netherlands J. Zool. 20: 141-147.

STRUMWASSER, F. (1959a). Factors in the pattern, timing and predictability of hibernation in the squirrel, *Citellus beecheyi*. Am. J. Physiol. 196: 8-14.

STRUMWASSER, F. (1959b). Thermoregulatory, brain and behavioral mechanisms during entrance into hibernation in the squirrel, *Citellus beecheyi*. Am. J. Physiol. 196: 15-22.

STRUMWASSER, F. (1959c). Regulatory mechanisms, brain activity and behavior during deep hibernation in the squirrel *Cittelus beecheyi*. Am. J. Physiol. 196: 23-30.

STRUMWASSER, F., GILLIAM, J. J. and J. L. SMITH (1964). Long term studies on individual hibernating animals. Ann.

Acad. Scientiarum Fennicae Ser. A. IV, 71: 399-414.

STRUMWASSER, F., SCHLECHTE F. R., and STREETER, J. (1967). The internal rhythms of hibernation. Proc. III Intl. Symp. Nat. Mammal. Hib. Oliver & Boyd, Edinburgh, 110-139.

SWEENEY, BEATRICE M. (1969). Rhythmic Phenomena in Plants. Academic Press, pp. 147.

TWENTE, J. W. and TWENTE, J. A. (1964). The duration of hibernation cycles as determined by body temperature. Amer. Zool. 4(3): 295.

TWENTE, J. W. and TWENTE, JANET A. (1967). Seasonal variations in the hibernating behavior of *Citellus lateralis* Proc. III Intl. Symp. Mammal. Hib. Oliver & Boyd, Edinburgh, 47-63.

WEIHAUPT, J. G. (1964). Geophysical biology. Bioscience 14(6): 18-24.

WEVER, R. (1973). Hat der Mensch nur eine "innere Uhr"? Umschau in Wissenschaft und Technik Heft 18: 551-558.

CREDITS

Figures 1, 2, 11 and 12. Reproduced by permission of the National Research Council of Canada from the Canadian Journal of Zoology, Volume 41, pp. 1103-1120, 1963.

Figures 4, 5, 6, 7, 8, 9, 10, and 13, tables 1 and 3. Reproduced by permission of Comparative Biochemistry and Physiology.

Figure 14. Reprinted by permission from Federation Proceedings 23: 1202-1206, 1964.

Table 2. Reproduced by permission of Elsevier Publishing Company.

Figure 3. Reproduced by permission of Dr. Jean Manery Fisher.

COMMENT

by

N. MROSOVSKY

Departments of Zoology and Psychology
University of Toronto, Canada

It can reasonably be argued that the term "sequence of stages" is not very suitable for describing a control system different from one depending on a circannual oscillator, since an oscillator may also be thought of as a sequence of stages, as has been pointed out elsewhere (Mrosovsky, 1971). It might be better to distinguish between pendulum type and relaxation oscillators. Even here there may not be an absolute distinction but a continuum. Nevertheless, however one chooses the labels, there are differences between self-sustaining systems with relatively inflexible periodicities and those that require input and have relatively flexible periodicities.

My disagreement with Pengelley is really a matter of emphasis. It seems that he gives inadequate attention to some of the properties of the long term rhythms in hibernators. In particular cycle lengths are sometimes less than eight months, suggesting something at the flexible end of the continuum, and cycles sometimes disappear in constant conditions suggesting a system that is only marginally self-sustaining. Disappearance of cycles in constant conditions is especially obvious in thirteen-lined ground squirrels but also occurs sometimes in golden-mantled ground squirrels (Mrosovsky and Lang, 1971 and unpublished).

A matter of emphasis may appear trivial, but it has far reaching implications in guiding the types of experiments undertaken. If one is guided by an analogy between circannual rhythms in hibernators and circadian rhythms with pendulum type properties, one is led to experiment with photoperiods, provide further details of the analogy by studying aspects like phase angles, trying to come up with rules that describe the properties of the cycles in different conditions, and all the types of experiment that have been done so successfully with circadian rhythms. It becomes a matter of extending and articulating the paradigm, to use Thomas Kuhn's formulation. With circadian rhythms we have a beautifully worked out paradigm with all sorts of sophisticated manipulations and measurements and we can take it over and apply it over greater time scales.

If, on the other hand, one emphasizes the apparent differences in annual rhythms in hibernators and circadian rhythms, one may be led into doing somewhat different types of experiment. Rather than studying the properties of the rhythm in different conditions, one may try to pull it apart, block it or phase shift it by physiological and other manipulations; one may see how much the cycles can be shortened even below eight months and also concentrate more on working with those individuals that show no cycles in constant conditions.

Premature adoption of models devised for circadian rhythms may be restrictive in terms of the experiments they stimulate, and may distract from the special advantages that long term cycles, despite all their difficulties, have for understanding some of the physiological aspects of rhythmicity.

MROSOVSKY, N. (1971). Hibernation and the Hypothalamus, Appleton-Century-Crofts, New York, p. 118.

MROSOVSKY, N. and LANG, K. (1971). Disturbances in the annual weight and hibernation cycles of thirteen-lined ground squirrels kept in constant conditions and the effects of temperature changes. J. Interdiscipl. Cycle Res. 2: 79-90.

EVIDENCES FOR BLOOD-BORNE SUBSTANCES WHICH TRIGGER OR IMPEDE NATURAL MAMMALIAN HIBERNATION

ALBERT R. DAWE

and

WILMA A. SPURRIER

Department of Physiology
Loyola University
Stritch School of Medicine
2160 South First Avenue
Maywood, Illinois, 60153

and

Office of Naval Research
Chicago, Illinois, 60605

INTRODUCTION

Without question, one of the most striking circannual rhythmicities to be seen in the mammals is the phenomenon "practiced" by a few of them - i.e. natural mammalian hibernation. Of the three orders of mammals

(Rodentia, Chiroptera, and Insectivora) having the largest number of species exhibiting this state there is not much doubt that the rodential ground squirrels and woodchucks show season-wise circannuality most precisely and most clearly. Thus, to experimentally cause a direct biochemical or immunological interruption in the proclivity for maintaining hibernation states during the hibernating season (winter), or conversely, to cause a direct biochemical induction of hibernation in one of those animals during its usual non-hibernating season (summer) is worthy of investigation. It is particularly worthy if one is concerned with mechanisms for circannual rhythmic physiology. The former experiment (i.e. impeding of natural hibernation) is central to derivation of data for this paper. The latter - i.e. the triggering of hibernation in a nonhibernating season - had been the subject of much of the research done in recent years in our laboratory, and reported upon elsewhere (Dawe and Spurrier, 1969; Dawe, Spurrier and Armour, 1970; Dawe and Spurrier, 1972; Dawe and Spurrier, 1974).

A. Work by others:

I. Endogenous substances which trigger, induce, or maintain mammalian hibernation:

a. Torpors: A brief summary of work giving various indications that a torporous or dormant state may be induced autopharmacologically and out of season in living animals and plants (including mammals), is contained in a paper by one of us titled "Autopharmacology of Hibernation" (Dawe, 1973).

b. Hypothermias: As far as body cooling is concerned, there are numerous summaries of methods and numerous listings of results following dropping of body temperature: hypothermizing mammals. Endogenous changes favoring hypothermia in mammals (particularly as it occurs on a seasonal basis or circannual basis) center about blood electrolyte studies. For example, Salem and Osman (1973) found elevated calcium in blood serum and elevated magnesium in semen of cattle in winter. However, of course, hibernation in a mammal differs profoundly from the states of natural or artificial hypothermia insofar as a hibernating mammal primarily does not require external reintroduction of heat into its body in order for it to arouse from hibernation. In arousal from hypothermia, contrarily, heat must be introduced. Though hypothermia *per se* is therefore not considered herein, it is significant to note that a report exists which claims that extractable substance from the brown fat of hibernating animals will induce hypothermia in non-hibernators (mice). Zirm (1956) made this contention but experimental confirmation was not obtained at the laboratory of P. R. Morrison and Allan (1962). Similarly, Swan *et al.* (1968) prepared an extractable substance from an aestivating non-mammal (lungfish) which was found to be capable of lowering body temperatures in mice.

c. Hibernations: The existence of blood-borne substances which may initiate or "trigger" hibernation in a mammal which normally hibernates has been universally studied. Very generally, these researches have devolved from a presumption that one of four "glands" secretes hormones or hormone-like substances into the blood, that such material then enters the tissues inducing the animal to hibernate. A brief summary of researches on these four "glands" is as follows:

1) The "hibernating gland" or brown fat has been implicated for a long time. See Rasmussen (1923-24); Johansson (1959); Goldman and Bigelow (1964); and Bigelow *et al.* (1964). Its presumed role as an inducer of hibernation has certainly switched following the efforts of Smith and Hock (1963) who showed that brown fat *(per se)* is (or contains) an arousal substance, but evidently cannot be shown to contain a hibernation-inducing substance! This discovery was based on earlier work by Hook and Guzman-Barron (1941) who pointed to the extraordinary amount of metabolic heat which can be produced by brown fat found in very high amount in animals which hibernate. Despite this, brown fat remains as a possible contender for the production of other blood-borne substance which may trigger hibernation.

2) The pancreas. Insulin has also had a long-standing reputation as a hibernation-inducing substance (if introduced into the blood stream). See Laufenberger (1924); Dworkin and Finney (1927); and Suomalainen (1950).

3) The adrenal gland. Following upon original research by Popovic and Vidovic (1951) in which hibernation could be induced in season in adrenalectomized animals (which otherwise would not hibernate) by a simple graft of a small piece of adrenal cortex into the anterior chamber of the eye of such

an animal, the adrenal has been implicated as a hormonal source for hibernation. Kayser and Petrovic's research (1958) extended this observation remarkably. There is now little doubt that a careful balance of known hormonal secretions is required in order to have any hibernator in a condition wherein possible induction of hibernation can occur.

4) The brain. The contention by Kroll (1952) and Axelrod (1964) exists which states that a brain substance (presumably not circulated) can be extracted which induces recipient animals to hibernate which are not normally hibernators.

Finally, a listing of hibernation-inducing substances (besides our own "SM: trigger") should not be ended without, of course, noting that four ions in the blood have at various times and places been studied as possible hibernation-affective agents. They are Mg^{+} (Riedesel and Folk, 1958); Na^{+} (Pengelley and Kelly 1967); Ca^{++} (Myers and Sharpe 1968), or K^{+} (Willis *et al.* 1972). Of particular interest is a recent observation by Bito and Roberts (1974) that both K^{+} and Mg^{+} increased in blood serum and intracerbral fluid of hibernating woodchucks, as compared to anesthetized and aroused animals. Concomitantly, K^{+} and Mg^{+} concentrations decreased within brain tissue during hibernation.

Experiments are meager which are designed to truly induce hibernation in the summer time, and hence break the circannual rhythm. Most experimentation in which the above-named substances had been analyzed for and were introduced to induce hibernation were performed during a hibernation season.

Thus propensities for hibernation were set against the circannual backdrop which favored hibernation anyway. Experimentation on anesthetized recipients (or utilization of materials from anesthetized donors) will also compound the kinds of errors which can occur here. Perhaps of all hibernation investigators, both Kayser and Suomalainen should be cited most particularly as using season itself as the experimental variable. Kayser actually manipulated this variable by transferring ground squirrels from Belgrade, Yugoslavia (which normally has a different hibernation season) to his laboratory at Strasbourg, France. He found he could show physiological differences in these animals when brought to Strasbourg, thus favoring a viewpoint that they had endocrinologically or biochemically been circannually changed in some way prior to the transfer.

II. Endogenous substances which impede or act as "anti-triggers" to hibernation:

Absence of appropriate cortical hormone (Kayser 1957) and absence of appropriate insulin (Suomalainen 1948) have both been correlated with a lack of predisposition to hibernate. Similarly, high thyroxine concentrations (causing hyperthermia) are contra-indicated when hibernation is intended (Kayser *et al.* 1959). Hoffman (1964) and Lyman and Dempsey (1951) showed that a presence of high male or high female hormone seems also to rule against hibernation. Foster *et al.* (1939) were witness to the fact that hibernation was never observed in hibernators when exposed to cold during the breeding season.

A summary of others' reports on all naturally blood-borne substances considered (possibly) to be effective as inducers or triggers, as well as all substances thought (possibly) to be effective as inhibitors or anti-triggers of hibernation is as follows:

Possible "Triggers"

Adrenal cortex: cortical hormones
Pancreas: insulin
Brown fat: "hibernating gland"
Brain: extract
Electrolytes: Mg^{+}, Ca^{++}, K^{+}

Possible "Anti-triggers"

Thyroid: thyroxine
Gonads: hormones
Brown fat: extract

B. Data from our laboratory

A "trigger" factor(s) designated as "SM" (for "small molecule") has been discovered to be present in a dialysate of blood serum from a ground squirrel or woodchuck in deep hibernation which, when transfused into the circulation of an active ground squirrel or woodchuck in the summer will "trigger" the recipient animal to hibernate in the cold within several days or weeks of the procedure. Evidences later appeared in several experimental instances which led to a belief that residue left behind ("inside the bag") after dialysis (especially from blood of a summer-active animal) contained a factor(s) of larger molecular weight with activity which worked in an opposite manner - i.e. rendered "trigger" ineffective, and, conceivably removed the capability for the recipient animal to hibernate. Theory argued that an "anti-trigger" material may exist. We called it "LM" (for large molecule) capable of forming a composite material ("CM") with "SM" (trigger) thereby rendering trigger physiologically ineffective. Also it could be simple inactivation or denaturation of one compound by the other. Theory also argued that SM might "escape" rather quickly through excretory mechanisms particularly during the profound metabolic changes of arousal. There is positive evidence for just such a mechanism, since Zimny (1973) has found anatomical differences (by ultramicroscopy) between the basement membrane of the glomerules in a hibernating animal vs. that membrane from the non-hibernating state. She interprets this anatomical difference as accountable (concievably) for changes in concentrations of blood constituents as if this changed membrane acted like an exchange resin. Also, by theory it seemed that LM could be in varying concentration circannually in dialysate residues, whereas smaller SM is noted only in hibernation blood (not in bloods of aroused nor active animals). It seemed that if there was such a material in residues it could be mixed *in vitro* with dialysates containing "trigger" and assessed as to whether or not such residues were capable of rendering ineffective those dialysates known to be able to trigger hibernation - i.e. would certain residue materials tie up dialysate trigger material (form "CM")?

RESULTS

Earlier experiments in our laboratory have demonstrated that there is little or no LM in the residue of hibernation blood, because when such residue was mixed with "trigger", there is no change in the efficacy of "trigger". See Table I. Results given later in this paper will show contrarily that when we mixed the residue of active summer blood with "trigger" we were able to negate trigger's effectiveness. An experiment, therefore, was designed (the results of which are reported here) in which an adequately large group of laboratory-born ground squirrels was utilized on which very exact information was to be collected as to age, littermates, and previous hibernatory conditions.

Bloods were drawn from these animals in June. It was stored cryogenically (-50°C) until August, thawed to 7°C, and pooled. Bloods had been taken from 10 "young of the year" animals (1-2 months old), 10 "1-year" animals (14 months old) and 10 "2 year" animals (25 months old). Each of the pooled samples was centrifuged. The serum was dialysed against normal saline at 5°C. Both the dialysate and the residue that was left "in the bag" was collected and used for transfusate material. If fig. 1 (top) is consulted, it will be apparent that dialysate of one-year summer animals and two-year winter hibernating animals (B and E) as well as residues of young-of-the-year, one-year old and two-year old summer animals (A, C, and D) became ingredients for thirteen of fourteen kinds of transfusates to be given via the saphenous vein in 1 cc increments. The fourteen kinds of transfusates prepared for transfusions

Collection of Blood Samples

	Young-of-Year	1-Year Animal	2-Year Animal
	Spring Summer Fall Winter	Spring Summer Fall Winter	Spring Summer Fall Winter
DIALYSATE		(B)	(E)
RESIDUE	(A)	(C)	(D) (F)

BIRTH

(A) = Summer residue prepared from young-of-year ground squirrels' blood.
(B) = Summer dialysate prepared from one year old ground squirrels' blood.
(C) = Summer residue prepared from one year old ground squirrels' blood.
(D) = Summer residue prepared from two year old ground squirrels' blood.
(E) = Winter dialysate ("trigger") prepared from two year old hibernating woodchuck blood.
(F) = Residue prepared from hibernating woodchuck blood.

Results

First to Hibernate: August		Second to Hibernate: September	
Received Transfusate No.	Consists of:	Received Transfusate No.	Consists of:
#7	(E)	#5 (partial)	(D) + (E)
#8	(E dil.)	#10 (partial)	(E i p)
#1	(A) + (E)		
#2	(A dil.) + (E)		
#6	(D dil.) + (E)		
Third to Hibernate: October		**Last to Hibernate: November-December**	
Received Transfusate No.	Consists of:	Received Transfusate No.	Consists of:
#12	(A)	#4	(C dil.) + (E)
#9	(Saline)	#5 (partial)	(D) + (E)
#14	(B)	#3	(C) + (E)
#10 (partial)	(E i.p)	#11	(C)
#13	(D)		

Fig. 1. Criteria for the fourteen donor blood derivative mixtures, plus a rank ordering of the results obtained using these mixtures.

TABLE I

Induction of Summer Hibernation in 13-Lined Ground Squirrels

Interpretation of Results:

Whether hibernation dialysate alone (trigger) was administered, or "trigger" plus hibernation residue was administered - all induced hibernation. Contrarily, hibernation residue alone was ineffective: i.e. it neither triggered, nor acted as an anti-trigger to hibernation.

Transfusates	Recipients Hibernating
Dialysate (SM) of blood from hibernating woodchuck. (Trigger) diluted 1:1 with saline.	10/10
Dialysate (SM) of blood from hibernating woodchuck. (Trigger) diluted 1:2 with saline.	9/10
Dialysate (SM) of blood from hibernating woodchuck. (Trigger) diluted 1:1 with residue (LM) of hibernating woodchuck.	10/10
Dialysate (SM) of blood from hibernating woodchuck. (Trigger) diluted 1:2 with residue (LM) of hibernating woodchuck.	9/10
Residue (LM) of hibernating woodchuck alone and undiluted.	0/10

were made up, thereby, of the following materials:

<u>Transfusate No.</u>	<u>Consisting of</u>
#1 <u>(A plus E)</u>:	Young-of-year summer residue added (1:1) to winter hibernation dialysate ("trigger").
#2 <u>(A dil. plus E)</u>:	Young-of-year summer residue diluted 1:1 with saline. This mixture was added (1:1) to winter hibernation dialysate ("trigger").
#3 <u>(C plus E)</u>:	One-year olds' summer residue added (1:1) to winter hibernation dialysate ("trigger").
#4 <u>(C dil. plus E)</u>:	One-year-olds' summer residue diluted 1:1 with saline. This mixture was added (1:1) to winter hibernation dialysate ("trigger").
#5 <u>(D plus E)</u>:	Two-year-olds' summer residue added (1:1) to winter hibernation dialysate ("trigger").
#6 <u>(D dil. plus E)</u>:	Two-year-olds' summer residue diluted 1:1 with saline. This mixture was added (1:1) to winter hibernation dialysate ("trigger").

#7 (E dil.): Winter hibernation dialysate ("trigger") diluted 1:1 with saline.

#8 (E): Winter hibernation dialysate ("trigger").

#9 (Saline): Normal saline.

#10 (Eip): Winter hibernation dialysate ("trigger") but administered intraperitoneally.

#11 (C): One-year-olds' summer residue.

#12 (A): Young-of-year summer residue.

#13 (D): Two-year-olds' summer residue.

#14 (B): One-year-olds' summer dialysate.

Each of these fourteen transfusates was administered to five littermates - a total of 70 animals - on August 6, 1973, in the first year of life (i.e. - all recipients were three months old). In other words, each "transfusate number" in the succeeding text is also a "litter number." Under normal circumstances all of these recipient animals could have been expected to begin their first winter of hibernation within a month or two of the time of their transfusion anyway. The transfusates, it was presumed, would hasten or deter hibernation entry in recipients dependent upon the efficacy of each transfusate (high or low in "trigger" or high or low in "anti-trigger").

The first animals to begin to hibernate were three animals which had received transfusate #7, which was simply diluted "trigger." They fell into a bout of hibernation three days after transfusion, on August 9, 1973. Subsequently, all animals were observed and careful accounting was made of the <u>first day</u> that each of 70 recipients fell into hibernation. It is of interest that the last animal to begin a bout of hibernation was a male which received transfusate #11 (residue only) of a one-year-old active squirrel. It hibernated for the first time on December 14, 1973, 130 days post-transfusion.

Fig. 2 indicates (in blocks of 5 days) in each case the first day following transfusion on which a ground squirrel was observed to have entered a bout of hibernation. Several facts are strikingly apparent consequent upon tabulating the first dates of observable hibernations. First of all, in twelve of the fourteen litters the five animals within a litter always began hibernation within the same time frame, and in no instance more than 20 days apart. Second, the hierarchy of dates of first hibernations rather neatly subdivided into four groups of dates (I to IV). The Group I of dates is a segregation of twenty-four animals which hibernated very soon: that is, were very prone to hibernate following transfusion (they had received transfusates #1, #2, #6, #7, and #8 only). The Group II of dates represent animals somewhat less prone to begin hibernating (transfusates received were #5-partial and #10-partial only). The Group III of dates truly represents the time in mid-September to mid-October when animals

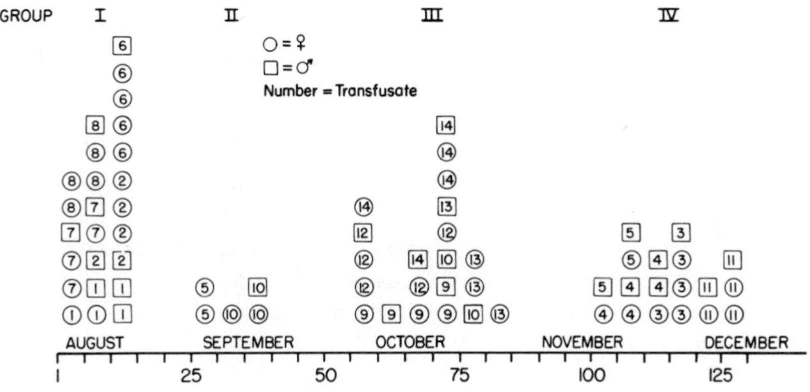

Fig. 2. Details of responses to the fourteen transfusates.

would have normally "fallen into" hibernation. These animals had received transfusates #9, #12, #13, #14 and #10 (partial only). Group IV of dates, contrarily, is 18 animals which, without doubt, seemed to have been very loath to hibernate - i.e. apparently they were retarded in starting their winter hibernation season; recipients had received transfusates #3, #4, #11 and #5 (partial only). We presume that an "anti-trigger" effect occurred in this group.

A closer examination of these results reveals the following relationships between transfusate compositions within each of the four groups, and propensity to hibernate (in rank order of quickness to hibernate):

Group I: This early group - this group of animals which hibernated soonest after receiving transfusates - had received (Rank 1) diluted "trigger" (#7); (Rank 2) "trigger" (#8); (Rank 3) young-of-year summer residue plus "trigger" (#1); (Rank 4) young-of-year summer residue diluted plus "trigger" (#2); and, finally (Rank 5) two-year-olds diluted summer residue plus "trigger" (#6).

Group II: This second group had in it (Rank 6) two animals which had received 2-year-olds' summer residue plus "trigger" (#5); and (Rank 7) three animals which had received "trigger" intraperitoneally rather than intravenously (#10).

Group III: This third group which began hibernation at a rather "normal" time of year - when these animals would be expected to begin hibernations anyway - consisted of recipients of: (Rank 8) the summer residue of young-of-the year (#12); (Rank 9) saline only (#9); (Rank 10) summer dialysate of 1-year-olds (#14); (Rank 10A) the other two animals which had received "trigger", but intraperitoneally rather than intravenously (#10); and (Rank 11) the summer residue of 2-year-old animals (#13).

Group IV: This last group to hibernate began to do so "out of season". They had quite obviously "hesitated" in their circannual rhythmicities. They may well have been deterred by introduction of "anti-trigger" by transfusion. These animals had received the following transfusates: (Rank 12) the diluted summer residue of one-year-old animals plus "trigger" (#4); (Rank 12A) the other three animals which had received summer residue of two-year-olds' blood, plus "trigger" (#5); (Rank 13) the summer residue of one-year-old animals plus trigger (#3); and, finally, (Rank 14) the summer residue of one-year-old animals (#11).

DISCUSSION

Recognition of the circannual quality of the natural phenomenon known as "hibernation" dates from the coining of that word, deriving (in the Latin form) from "hiber" or "winter". The Germanic rendering of the term into "winterschlaf" or "winter sleep" also reflects upon its circannualicity. Aristotle (several centuries B.C.) recognized the phenomenon (Creswell, 1862). Bartholomew and Hudson (1960) realized that in mammals, hibernation was perhaps an identity biochemically to aestivation and pleaded for a change of terminology to "facultative hypothermia". Recently, the tremendous surge of interest, research, and application in the entire field of hypothermia, essentially has stifled that plea to call natural hibernation and natural aestivation by a term which could be confused with another form of "hypothermia". The circannual elements inherent in both phenomena (hibernation and aestivation) permitted the etymological retention of the words, with the circumspect observation that perhaps aestivation was indeed a form of hibernation, with some kind of seasonal slippage due to a biochemical factor not knows. In any event, the circannual nature of the hibernating or aestivating events remain true to species. Thus, the thirteen-lined ground squirrel found in the northern part of our nation is a very seasonal winter hibernator. The Mohave ground squirrel found exclusively in the Mohave desert of California is also a very seasonal animal, but is a summer aestivator whose torpor blends into winter hibernation. The Antelope ground squirrel is active all year long, and has not been seen to hibernate or aestivate in nature. The Spotted ground squirrel (found in Texas) had not been seen to aestivate or hibernate in the wild, but was recently shown to be able to aestivate when placed in a cold, dark, quiet laboratory environment for 21 days in early summer (Smith, 1973). *Citellus citellus*, a European counterpart of our local American ground squirrels, performs very much like a hamster, insofar as it will exhibit a hibernation - aestivation physiological state by the simple procedure of subjecting it to the single expedient or parameter of cooling (Kayser, 1961). *Citellus perryi* as well as *Citellus columbianus* have much-lengthened seasons of hibernation, corresponding exactly to the longer winters of their extreme northern ranges in Alaska and Canada respectfully.

In sum, examination of the single genus *Citellus* (or *Spermophilus*) brings out a salient fact that (seemingly depending upon environmental constraints and upon life zones) each species of ground squirrel has a different time range (longer or shorter) of torpor, and each species also exhibits this torpor at circannually different seasons of the year. Of importance is the fact that none of these animals normally show hibernation or aestivation during a spring breeding season. If, therefore, the hibernation - aestivation states seen in various *Citellus* species is fundamentally the same from a biochemical-physiological point of view, but differ only in the times at which they are manifested, one is tempted to examine environmental as well as genetic parameters to see what might "trigger a trigger" to hibernate or aestivate in a given animal. Clearly the parameters of cooling, fattening, shorter days, lack of food supply, and nonreproductive activity all show positive correlations to entry into hibernation or aestivation. A fundamental start in a direction of discovery along this line was set by Pengelley (1967) who sought by all present knowledge (then) to uncover this most closely held "circannual secret" of the hibernators. Incidentally, the aestivation in a cool burrow below the ground and away from solar activity can indeed be a physical response to cooler temperature and shortened (ended) days, even though the situation above ground is one of longer and hotter days. The animal's response to certain environmental cues may start a secretion (later detectable by our experimental method) which appears in the blood of a hibernating animal. This is the triggering material so long sought after (Bartholomew and Hudson - 1960). Clearly, this substance (and its anti-trigger) should be identified and analyzed for.

As Hannon *et al* (1972) have nicely enunciated, any thermal-active substance, (such as a trigger for hibernation), must demonstrate its activity in one of three ways - i.e., by a) changes in its concentration, b) changes in chemical composition, or c) changes in the environment about it - particularly such changes as membrane permeabilities. We point out that there is a fourth way by which substances can act and that is in its biochemical harmony or dissonance with other substances. A "microecology" in other words, of substances favoring hibernation or not favoring hibernation is a conceivable *modus vivendi* within an organ, or a cell, or in the blood of a hibernator - aestivator.

Our approach to the problem has been identification. We presume that concentration of the trigger substance is an effective mode for modifying the capability of a recipient animal to hibernate. Very specifically, we note from the experiments reported upon here, as well as other experiments (Dawe *et al*, 1970) that the "lag time" between transfusion and the first sign of hibernation in a recipient correlates crudely with concentration gradients of "trigger". Second, we now believe that there is an "anti-trigger" which is capable of rendering this "trigger" ineffective. Therefore a "balance of power" ensues biochemically within an animal capable of hibernation. It is this balance, this homeostasis, which determines precisely whether a given animal at a given time of year will hibernate. We hypothesize that circannuality, in other words, in the hibernation of a Thirteen-lined ground squirrel reflects exactly the "trigger-anti-trigger" biochemical mechanism which in turn is modified by results from concentration differences found in bloods of these animals (taken at different times of the year). Effects of such concentration differences are noted by behavioral differences (hibernating or not hibernating) which we observe following transfusions of various blood derivatives and mixtures of blood derivatives. The differences we report upon are not taken from experiments showing analytical differences in chemical composition of blood substances.

CONCLUSIONS

The above results can be interpreted as follows:

Group I results tell us that "trigger" alone, even if diluted, is the most effective inducer (interms of days to first hibernation) for which we tested. Second, addition to "trigger" of summer residue from young-of-the-year animals, (full-strength or diluted) reduced "trigger's" efficacy only slightly. Similarly also, two-year olds' diluted summer residue, when added to "trigger", affected its potency only slightly.

Group II results tell us that several animals' performances indicated that two-year olds' summer residue added to trigger slowed down hibernation propensity somewhat, and that intraperitoneal transfusion of "trigger" was effective to some extent (in cases of three animals out of five).

Group III results can be interpreted as saying that time for hibernation, to begin (normal season) was relatively unaffected by giving summer residue of two-year old ground squirrels, by giving saline only, or by giving summer dialysate of one-year old ground squirrels, or summer residue of two-year old ground squirrels. In two cases two-year olds summer residue plus "trigger" was also ineffective.

Group IV results can be taken to mean that the following combinations will definitely deter hibernation from occurring: a) adding summer residue of one-year old animals' bloods (diluted or not) to "trigger"; b) giving summer residue of one-year old animals' blood; c) in three cases out of five -- giving summer residue of a two-year old animal's blood plus "trigger".

If it is posited, then, that "trigger" is in high concentration in dialysate of winter hibernation blood, and that "anti-trigger" is in high concentration in residue following dialysis of summer blood taken from an active hibernator (and that "anti-trigger" concentration is absent from young-of-the-year, high in one-year old animal's blood, and lower in a two-year old animal's blood) it is then possible to construct a circannual diagram which can be used to explain (in theory) the results. A scheme showing changes in relative concentrations circannually of both "trigger" and "anti-trigger" is given as fig. 3. In fact, from this diagram, explanations can be shown to "fall out" for all of our observations. Finally, what the diagram "says to us" is that in the first few months of life, the young squirrel has neither "trigger" nor "anti-trigger" in significant concentration in its blood. However, as the first fall approaches, "trigger" increases in concentration building up until hibernation

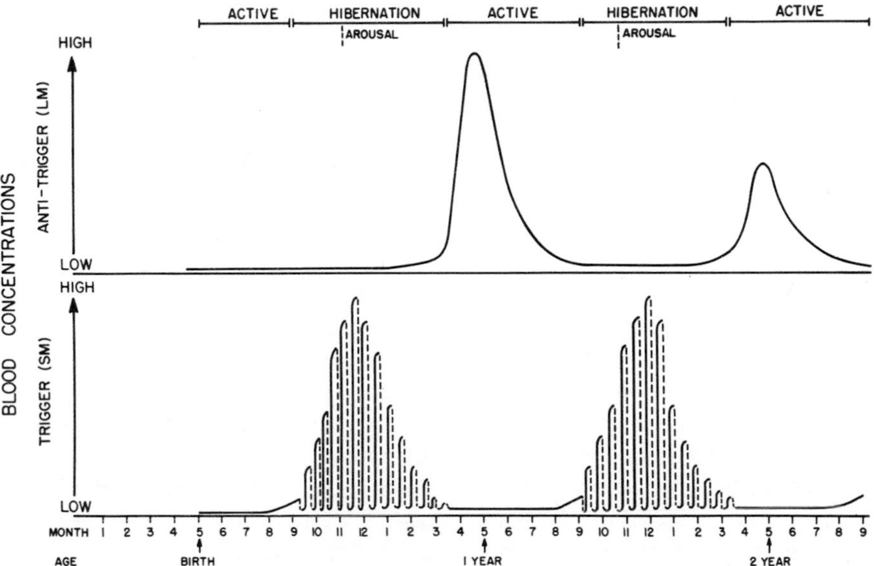

Fig. 3. A circannual "trigger - antitrigger" theory to account for these results.

occurs, followed by continuous increase in concentrations as the hibernation season progresses. At the second spring of life (a one-year old animal) there is a rapid fall in concentration of "trigger", but an increase in concentration of the "anti-trigger". (There is binding of LM to SM (forming CM) and finally a "spilling over" of LM into the blood, reaching its highest level of "anti-trigger" in the late spring blood of the one-year old animal.) This circannual cycle now repeats itself, but with the rather certain decrease in concentration of "anti-trigger" in the blood of the two-year old animal. What the circannual rhythmicities in "trigger" and in "anti-trigger" concentrations in blood may be later on in the animal's life can only be speculated upon. The observations of Coleman (cited by Kayser, 1961, p. 23) and other testifies to a weakening of capability for hibernation as the hibernator becomes older. This could be due to a reduction in "trigger" production or it could be a reflection on simultaneous reductions both of "trigger" and of "anti-trigger" concentrations. The latter - as a speculative probability - is more likely to be the case, particularly in view of the above evidences for reductions in effective LM concentration in bloods taken during the third summer of life.

Earlier experiments in our laboratory have demonstrated that when random-caught adult ground squirrels were induced to hibernate in the summer by transfusion of "trigger", they underwent bouts of hibernation and arousals throughout the summer and winter, but failed to terminally arouse and become fully active the following spring. Indeed, the majority of older animals continued bouts of hibernation and arousals continuously for several years. The circannual rhythm was altered dramatically, perhaps due to the impared ability of older animals to produce enough "anti-trigger" to override the effect of "trigger", except for short periods of time between bouts. When similar induction experiments were conducted on the young of the year, they hibernated the first summer after being transfused with hibernation "trigger", and continued to hibernate through the winter. However, the majority of transfused animals became fully active in the spring when they were one year old, at the time of the first breeding season. This may be correlated with a higher titer of "anti-trigger" in the one-year old's blood as reported here and/or the presence of increased amounts of circulating endocrines. We have been successful in inducing summer hibernation in six week old infant ground squirrels by transfusing them directly, or by having earlier transfused the pregnant mother with dialysate of hibernation blood (Dawe and Spurrier, 1974). Control infants born and reared under identical conditions in the coldroom did not hibernate until the winter months when they were six months of age. The fact that infant animals could be induced to hibernate a few weeks after their birth indicated to us that they were very susceptible to the "trigger", perhaps the reason was that they were producing little, if any, "trigger" or "anti-trigger". The data from this most recent experiment would suggest that that interpretation may be valid, since no "trigger" or "anti-trigger" could be demonstrated in the blood of one-two month old ground squirrels.

To recapitulate: by extractions of dialysates and residues from sera of hibernating as well as from active animals, and by various combinations of these dialysates and residues, the hibernation potentialities were studied (using as an "end point" the first day after transfusion when a hibernation bout was seen to begin) of 70 controlled-age ground squirrels. From this approach, a circannual pattern has been constructed in theory which accounts for these results on the basis of changes in concentrations of both "trigger" and "anti-trigger" in blood. An essentially satisfying explanation is thereby constructed for the results seen, (fig. 3). However, no indication emerges for an answer to the more fundamental question as to a possible "trigger for the trigger". We can only speculate that _that_ may be light-cycling changes during the first summer and fall in the life of the animal and that such an earlier response activates the formation of the blood-borne "trigger" we have described. Such photal responses might somehow set in motion the proper internal biochemistry to result in an elevation of "SM" in the blood, and hence result in the animal's first entrance into hibernation in its lifetime. (As a matter of fact, it is at this level at which the truly hormonal mechanisms may well be in effect, rendering "trigger" and/or "anti-trigger" to be in higher or lower absolute concentrations.) Thereafter, the animal undoubtedly becomes entrained to the effects of these two substances - the "trigger", and its contrary, the "anti-trigger". A full course theoretical approach is contained in the reference (Dawe and Spurrier, 1972) which accounts for one circannual cycle of hibernation (with the exception of a triggering mechanism for arousing) on a basis of relative concentrations of LM and

SM in the sera of an animal which normally hibernates - i.e., in the blood of the woodchuck or of the ground squirrel.

We can only be amazed at the tightly-correlated biological-clock-like results which we found by our own experiments and as explained in this paper. It gives substance indeed, to a "trigger-anti-trigger" circannual explanation for hibernation. We feel this explanation becomes far less of a "theory" and far more of "fact" consequent upon these researches.

SUMMARY

1. The circannual rhythms of natural seasonal hibernations in ground squirrels have been investigated by a.) inducing hibernations in the summertime or b.) impeding hibernations in the wintertime. This "manipulation" of seasonality was accomplished following intravenous transfusions of "trigger (SM) and/or "anti-trigger (LM)" blood derivatives from hibernators taken at various seasons.

2. By selected mixtures of dialysates and residues (of dialysis) of seasonal bloods of donor hibernators (*Citellus* and *Marmota*) induction times were accelerated or impeded in recipient ground squirrels.

3. The results of 1 and 2 (above) made possible a construction of a circannual diagram showing theoretical increases and decreases in concentrations of hibernation "trigger" and hibernation "anti-trigger" in the blood of a natural hibernator during a period from its birth to age 2 years.

ACKNOWLEDGEMENTS

This work was supported by NIH Grant HE 08682.

REFERENCES

AXELROD, L. R. (1964). Hibernation - a biological timeclock: Progress in Biomedical Research, Southwest Foundation for Research and Education 10.

BARTHOLOMEW, G. A. and HUDSON, J. W. (1960). Aestivation in the Mohave ground squirrel; *Citellus mohavensis*; Bul. Museum Comp. Zool., Harvard, 124 - Mammalian Hibernation 1959, Lyman and Dawe, eds., p. 193.

BIGELOW, W. G., TRIMBLE, A. S., SCHONBAUM, E. and KOVATS, L. (1964). A report on studies with *Marmota monax*. 1. Biochemical and pharmacological investigations of blood and brown fat; Annales Academiae Scientiarum Fennicae, Series A IV Biologica, 71, Mammalian Hibernation II, 1962, P. Suomalainen, ed. Helsinki. p. 37.

BITO, L. Z. and ROBERTS, J. A. (1974). The effects of hibernation on the chemical composition of cerebrospinal and intraocular fluids, blood plasma and brain tissue of the woodchuck (*Marmota monax*). Comp. Biochem. Physiol. 47(1), 173-193.

CRESSWELL, R. (1862). Aristotle: History of Animals; ch. 8, 215.

DAWE, A. R. and SPURRIER, W. A. (1969). Hibernation induced in ground squirrels by blood transfusion; Science 163, 298-299.

DAWE, A. R., SPURRIER, W. A. and ARMOUR, J. A. (1970). Summer hibernation induced by cryogenically preserved blood "trigger"; Science 168, 497-498.

DAWE, A. R. and SPURRIER, W. A. (1972). The blood-borne "trigger" for natural mammalian hibernation in the 13-lined ground squirrel and the woodchuck; Cryobiology 9, 163-172.

DAWE, A. R. (1973). Autopharmacology of hibernation; Proc. Pharmacology of Thermoregulation Symp., San Francisco, 1972, Karger, Basel. p. 359-363.

DAWE, A. R. and SPURRIER, W. A. (1974). Summer hibernation of infant (6 week old) 13-lined ground squirrels; Cryobiology, In Press.

DWORKIN, S. and FINNEY, W. H. (1927). Artificial hibernation in the woodchuck (*Arctomys monax*), Am. J. Physiol. 80, 75-81.

FOSTER, M. A., FOSTER, R. C. and MEYER, R.K. (1939). Hibernation and the endocrines, Endocrinol, 24, 603-612.

GOLDMAN, B. S. and BIGELOW, W. G. (1964). The transfer of increased tolerance to low body temperatures by the heterologous transplantation of brown fat and the infusion of plasma from hibernating ground hogs; Annales Academiae Scientiarum Fennicae, Series A, IV Biologica 71, Mammalian Hibernation II, 1962, P. Suomalainen, ed. Helsinki p. 175.

HANNON, J. P., BEYER, R. E., BURLINGTON, R. F., SOMERO, G. N. and BLAND, J. H. (1972). A guide for future studies of low temperature metabolic function; Proc. Symp. Hibernation and Hypothermia, 1971; South, Hannon, Willis, Pengelley and Alpert, eds. Elsivier Pub. Co. p. 99.

HOFFMAN, R. A. (1964). Speculations on the

regulation of hibernation, Annales Academiae Scientiarum Fennicae, Series A, IV, Biologica, 71, Mammalian Hibernation II, 1962, P. Suomalainen, ed. Helsinki, p. 199.

HOOK, W. E. and GUZMAN-BARRON, E. S. (1941). The respiration of brown adipose tissue and kidney of the hibernating and non-hibernating ground squirrel; Am. J. Physiol., 133(1), 56-63.

JOHANSSON, B. (1959). Brown fat: a review; Metabolism, 8, 221-240.

KAYSER, C. H. (1957). Le sommeil hibernal et les glandes surrenales. Etude faite sur le hamster ordinaire (*Cricetus cricetus*); C. R. Soc. Biol., 151, 982-985.

KAYSER, C. and PETROVIC, A. (1958). Role due cortex surrenalien dans le mecanisme du sommeil hivernal; C. R. Soc. Biol. 152, 519-522.

KAYSER, C. H., PETROVIC, A. and WERYHA, A. (1959). Effet de la thyrostimuline sur l'activite throidienne et l'hibernation du hamster ordinaire; C. R. Biol., 153, 469-472.

KAYSER, C. H. (1961). The physiology of natural hibernation; Pergamon Press, 1.240.

KROLL, F. W. (1952). Gibt es einen humoral schlaftoff im schlafhirn ? Dtsch. Med. Wschr. 77, 879-880.

LAUFENBERGER, V. (1924). Versuche uber die insulinwirkung; Ztschr. Ges. Exp. Med. 42, 570-613.

LYMAN, C. P. and DEMPSEY, E. W. (1951). The effect of testosterone on seminal vesicles of castrated hibernating hamsters. Endocrinology, 49: 647-651.

MORRISON, P. R. and ALLEN, W. T. (1962). Temperature response of white mice to implants of brown fat; J. Mammal 43 (No. 1), 13-17.

MYERS, R. D. and SHARPE, L. G. (1968). Temperature in the monkey, Transmitter factors released from the brain during thermoregulation; Science, 161, 572-573.

PENGELLEY, E. T. (1967). The relation of external condition to the onset and termination of hibernation and aestivation, Proc. Symp. Mammalian Hibernation III - 1965, Fisher, Dawe, Lyman, Schonbaum, South, eds. American Elsivier Co., p. 1.

PENGELLEY, E. T. and KELLY, K. H. (1967). Plasma Potassium and Sodium Concentrations in Active and Hibernating Golden -Mantled Ground Squirrels, *Citellus lateralis*; Comp. Biochem. Physiol. 20, 299-305.

POPOVIC, V. and VIDOVIC, V. (1951). Les glandes surrenales et le sommeil hibernal; Arch. Sci. Biol. (Belgrade) 3, 3-17.

RASMUSSEN, A. T. (1923-24). The so-called hibernating gland; J. Morph. 38, 147-205.

RIEDESEL, M. L. and FOLK, G. E., Jr. (1958). Serum electrolyte levels in hibernating mammals; Am. Nat. 92, 307-312.

SALEM, H. and OSMAN, S. A. (1973). Seasonal variations in the chemical constituents of blood and semen in buffalo cattle. Alexandria J. Agric. Res., 20-(1), 51-55.

SMITH, J. R. (1973). Hibernation in the Spotted ground squirrel, *Spermophilus spilosoma annecteus*; J. Mammal. 54(2), 499.

SMITH, R. E. and HICK, R. J. (1963). Brown fat: thermogenic effector of arousal in hibernators, Science, 149, 199-200.

SUOMALAINEN, P. (1948). Insulin and hibernation; Acta Physiol. Scand. 16 (Suppl. 53), 60-61.

SUOMALAINEN, P. (1950). Hibernation and insulin; Arch. Soc. Zool. Bot. Fenn., Vanamo, 5(1), 35-45.

SWAN, H., JENKINS, D. and KNOX, K. (1968). Antimetabolic extract from the brain of *Protopterus aethiopicus*; Nature 217, 67.

WILLIS, J. S., FANG, L. S. T. and FOSTER, R. F. (1972). The significance and analysis of membrane function in hibernation; Proc. Symp. Hibernation and Hypothermia, 1971, Elsevier, 123.

ZIMNY, M. L. (1973). Mucopolysaccharide in the glomerular basement membrane of a hibernator; Am. J. Anat., 138(1), 121-124.

ZIRM, K. L. (1956). Ein beitrag zur kenntnis des naturlichen winterschlafes und seines regulierenden wirkstoffes; I.Z. Naturf; 11b, 530-534; II, ibid. 11b, 535-538, III. ibid. 12b, 589-593.

EXTERNAL AND INTERNAL COMPONENTS OF THE MECHANISM CONTROLLING REPRODUCTIVE CYCLES IN DRAKES

IVAN ASSENMACHER

(Professor)

Department of Animal Physiology
University of Montpellier II
34060 Montpellier
FRANCE

INTRODUCTION

With such spectacular events as mating and territorial behavior, or the appearance of young specimens in a herd, the seasonal recurrence of reproduction undoubtedly belongs to the oldest biological cycles ever recognized by man, in his search for food and security. In fact, due to its adaptive importance, reproduction must occur during appropriate seasons, when the ecological conditions are most propitious for the survival of the young.

In the past decades, a number of recurring external stimuli have been recognized as ecological cues for reproductive activity in the manifold environmentsl situations of the globe (see Marshall, 1970). Even under middle and high latitudes, the natural environments usually exhibit rather precise cycles for several of their physical, chemical (including nutritional), and biotic parameters. However, among the fluctuating factors that surround the wild animal in these regions, the seasonal changes in day-length appear to act as a unique device for timing the reproductive periodicity.

Since the pioneer studies of Rowan, Bissonnette and Benoit, a great amount of research has been devoted to the problem of photoperiodic control of sexual cycles, more especially in birds and mammals. The extensive progress in our knowledge on so many aspects of photo-gonadal relationships can be followed through the recent reviews on this field. (See Benoit and Assenmacher, 1970; Van Tienhoven, 1968; Farner, 1973). However the basic question as to whether the photoperiodic control system acts on the timing of reproduction in a direct causal way, or rather as a synchronizer "zeitgeber") for an endogenous circannual periodicity, still awaits a decisive conclusion.

We shall try to draw on some original considerations of this rather complex problem from a number of studies, that were done on the male Pekin duck, which has been used for years by Benoit as an experimental model for ecophysiological research.

I. The sexual cycle of male ducks in "natural" cycling conditions.

Domestication seems not to have altered significantly the course of the sexual cycle in ducks kept under "natural" outdoor conditions. Like many temperate-zone birds, the Pekin duck is a "long-day" bird, with its reproductive season occurring in spring.

The classical profile of the testis cycle that was first described by Benoit (1936) is demonstrated in fig. 1. In this experiment 8 adult (3 years old) male Pekin ducks were reared under field conditions with free access to food (commercial granules Provimi, Paris) water (open-air pool), and shelters. At the beginning of each month, the testis size was checked on x-ray plates (Assenmacher *et al*, 1950), and the transverse diameter of the left testis in each bird was taken as an index for the actual volume of the gonad. The testis started their increasing phase after the winter solstice, in late December, to reach their maximal dimensions in April. Testicular regression set in as early as May-June, and was complete in August -September. This is a typical vernal reproductive cycle with a progressive phase of gonadal shape (and functions) during the days of steeply increasing day length, followed by a regressive phase that starts before the summer solstice, and a quiescent phase in fall.

However, this kind of monophasic cycle with a single vernal phase of testicular activity, is not the only pattern of the sexual cycle in this species. Alternatively, some individuals, even in a single flock from the same strain and of the same age, may exhibit a biphasic testicular cycle. In this case, reproduction occurs as usual in spring, but a second phase of testicular growth takes place from around August to October. The main characteristics of this peculiar biphasic mode of the testicular cycle have been very thoroughly studied by Garnier (1970, 1972, 1973). That a biphasic annual cycle in testicular activity should not be considered an exceptional curiosity in Pekin ducks, is evident from figure 2. This figure includes the testis size at autopsy, from 521 ducks that were taken at random from about 2000 outdoor control animals, belonging to various experiments over a 20 year period. The shape of the testicular cycle shown here definitely has a biphasic profile, that may therefore be considered as a more general pattern for Pekin ducks than the "classical" monophasic profile of fig. 1.

It is interesting to note that in a recent study, Cardinali *et al.* (1971) have described a similar profile for the testicular cycle of Pekin ducks reared in the southern hemisphere, with a 6 months phase-shift corresponding to the seasonal shift in the bioclimatic environment.

In an attempt to screen several external and internal factors that may be involved in the control of the testicular cycle in ducks, the two possible phases of the cycle will be studied successively, keeping in mind that the vernal phase is a constant feature, while the autumnal phase occurs eventually in addition to the former one.

II. The vernal testicular cycle.

a) Progressive phase.

The very close correlation between the onset of testicular growth and the seasonal increase in daylength, has for 40 years lead investigators in this field, (i.e. Rowan (1925) in the Junco; Bissonnette (1930) in the Starling; and Benoit (1934) in the Duck), to infer that this external factor could act as a stimulator to the gonadal cycle. Since their epochal observations, a great deal of experimental evidence has strengthened the view that the seasonally augmented light to darkness (L/D) ratio in

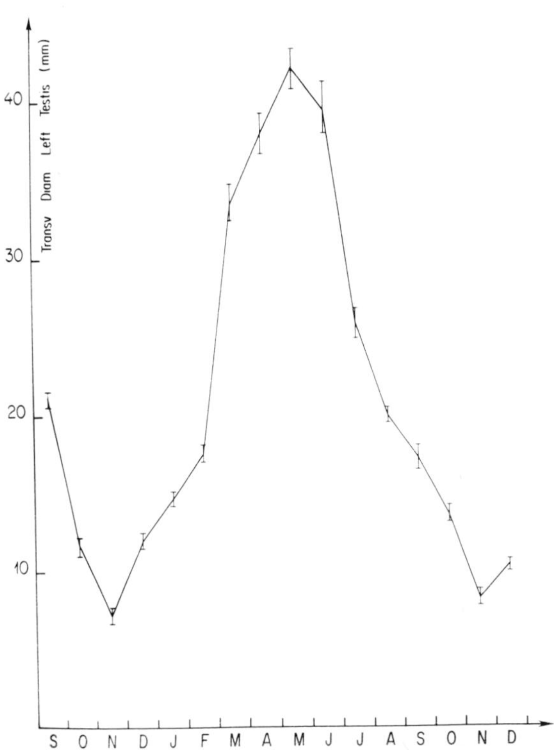

Fig. 1. "Classical" monophasic profile of the annual testicular cycle in Pekin ducks. Mean ± SE for 8 specimens reared outdoors.

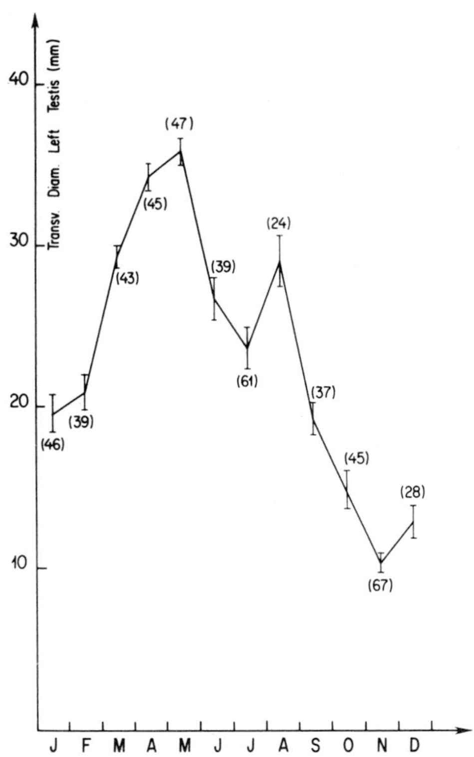

Fig. 2. Biphasic pattern of the annual testicular cycle in Pekin ducks. ( ) Number of control specimens sacrificed over a 20 year period. Mean ± SE.

the photoperiod provides the major clue for the annual stimulation of the sexual machinery in most temperate zone birds. An adequate experimental photoperiodic regime imposed on the birds during their autumnal resting phase can, either mimic the vernal progressive phase, or even induce an almost explosive gonadal growth. Fig. 3 illustrates the effect of 4 weeks of an experimental "long day" (18L-6D) device on the testis weight and plasma testosterone content in winter-ducks (December). It can be seen that three weeks of this lighting regime brings the testis to the fully active stage, which would be attained in about three months under the long-lasting photoperiodic increase in a natural environment.

In his first experiments, Benoit (1936) showed that the environmental cycles in temperature did not interfere with the light -induced stimulation of the gonads, since a hot environment did not stimulate winter-ducks kept under short-days, and persistent cold could not inhibit the photo-gonadal increment either. The absence of any stimulating (heat) or inhibiting (cold) effect of temperature on the testis activity is obvious in the experiment described in Table I.

The problem of the mechanism by which light, or better the photoperiod, exerts its stimulating effect on the testis of photo-sensitive birds, has been approached in recent years by very diversified techniques. They have lead to the certainty that this mechanism actually covers a rather complex chain of various structures and messages, extending between the photo-receptors and the target tissues in the testis, i.e. the seminiferous tubules and the interstitial tissue. Although many important links

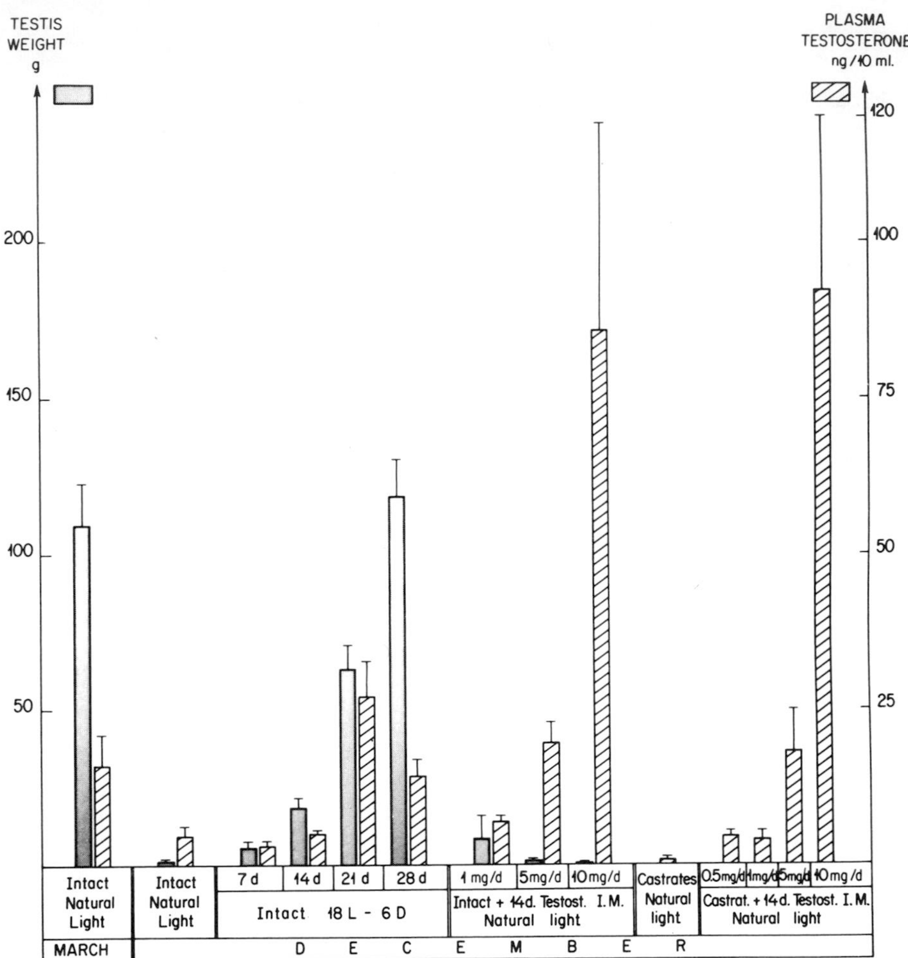

Fig. 3. Effect of long-days and of testosterone treatment on testis weight and plasma testosterone content in intact and castrated ducks. (after Jallageas and Assenmacher, 1973).

Table I

Effects Of Photoperiod vs Temperature On Testis Growth And Hormonogenesis In The Duck.

TREATMENT		N	Testis Weight (g)	Plasma Testosterone (ng/10ml)
Photoperiod	Temperature (^{o} C)		m ± SE	m ± SE
Natural	+ 12° C	3	3.8 ± 1.3	5.9 ± 2.5
Natural	21 days + 25°C	3	3.3 ± 1.5	2.3 ± 0.8
21 days: 18L-6D	Natural (+ 12° C)	3	66.8 ± 12.2	32.8 ± 11.1
21 days: 18L-6D	21 days + 4°C	4	105 ± 9	25.0 ± 3.8

from Jallageas and Assenmacher, 1974 (December ; Indoors)

of the chain are unknown, a good deal of evidence has accumulated on the general scheme of this control mechanism.

1. Benoit (1935a, 1935b) was the first to show that the eyes, which are essential for vision, do not play a unique role in the photo-gonadal response, since ducks which had been blinded by, either enucleation or transection of the optic nerve still reacted to long days by gonadal stimulation. And indeed, if conducted directly on the hypothalamic region of the brain by quartz rods (Benoit, 1938; Benoit *et al*, 1950) or more focally, be stereotaxically implanted "optic fibers" (Benoit, 1970), light retains its stimulating effect on the testis. In birds, that photoperiodic information proceeds from so-called extra-retinal photoreceptors (ERP) to the CNS lies beyond doubt. Open questions concern however (1) the site and the structure of the ERP, and (2) their relative importance vs the eye in this control mechanism. Regarding the location of the ERP, experiments on hypothalamic islands, isolated by stereotaxic lesions with "Halasz -type" knifes, have pointed to the infundibular nuclear complex (INC) as a possible site for ERP, since in quails with their INC isolated from other brain regions, the photo -gonadal response occurred, although at a subnormal level (Oliver *et al*, 1973). On the other hand, the fact that blinded ducks displayed a lower response than intact controls toward low intensities (Benoit *et al*, 1953) may indicate that both sites of photoreception, i.e. the eyes together with the ERP, could provide photic information to the CNS, while alternatively, a series of ingenious experiments on blinded house sparrows have lead Menaker (1971) to assign to ERP an almost exclusive role in nonvisual photic

adaptive regulations. However, the recent demonstration in the duck with both optic and electron microscopes (Bons and Assenmacher, 1969, 1973; Bons, 1974) and in the house sparrow (Oksche, 1970; Hartwig, 1973) of a nervous pathway leaving the optic tract close to the chiasma to terminate within the suprachiasmatic region of the hypothalamus, brings strong neuroanatomical support to the thesis of an involvement of both retinal and extra-retinal sources of photoperiodic information to the CNS.

2. It is also to Benoit's merit (1935c, 1937) that he pointed to the pituitary as an essential mediator in photo-gonadal regulation. Table II illustrates how three weeks of long-days (18L-6D) stimulate the plasma LH titer together with testis function.

On the other hand, it was found that the integrity of the neurovascular connections between the hypothalamus and pituitary was a further prerequisite for the photo-testicular response to occur, since the response was blocked by the transection of the hypophysial portal veins (Assenmacher and Benoit, 1953; Assenmacher, 1958), or by grafting the pituitary on the kidney (Assenmacher, 1958; Assenmacher and Bayle, 1964). The location in the hypothalamus of the main integrative centers to environmental stimuli, and of the site of neurosecretory biogenesis of the hypothalamic hormones that trigger the secretion of the pituitary gonadotropic hormones is still mysterious. There is, however a good piece of evidence that such functions could be assigned to at least two hypothalamic regions, e.g. the infundibular nuclear complex, and the preoptic and suprachiasmatic area (see reviews by Assenmacher, 1973 and Follett, 1973). Furthermore, central aminergic systems could play an important role in the light-induced stimulation of the gonads. Pharmacological depletion of the CNS amine stores (reserpine) or destruction of nor-adrenergic axons in the median eminence reduced markedly the photo-gonadal response in ducks and quail (Assenmacher, 1973).

Although increasing day-length may be considered as the primary environmental starter to both testicular functions, i.e. spermatogenesis and testosterone secretion, both responses, e.g. testis weight and plasma testosterone may affect quite different modalities. In figures 4 and 5, two vernal cycles of testis functions in drakes reared outdoors have been reproduced. During the first spring the testis weight that is correlated to the development of the seminiferous tubules, reached a maximum before that of the plasma testosterone, whereas the following year, the peak in circulating testosterone occurred long before the testes had attained their highest weight. The latter cycle depicts an extreme situation where both testis functions exhibit complete dissociation. However, it must be emphasized that the alternative dissociation with a testis weight cycle anticipating completely the endocrine cycle has never been observed, a fact that fits well to the stimulating role of testosterone on spermatogenesis. The origin of the internal desynchronization between testicular functions remains obscure.

b) Regressive phase.

The testes initiate their seasonal decline between late May and early June (fig. 1 and 2). The most acute regression of the gross parameters of spermatogenesis function, i.e. testis weight or size, usually takes place in June-July. On the other hand the level of circulating testosterone

Table II

Photo-sexual Response In Ducks

TREATMENT	Testis weight (g)*	Plasma Testosterone (ng/10 ml)*	M.C.R. Testosterone (ml/min/kg)*	Plasma LH (ng/ml)*
Control (December Outdoors)	6.4 ± 4.8 (4)	1.53 ± 0.55 (4)	104 ± 3 (6)	1.28 ± 0.29 (4)
21 days: 18L-6D	64.2 ± 6.4 [a] (9)	28.9 ± 4.9 [a] (9)	97 ± 7 (6)	4.09 ± 0.47 [b] (9)

After Jallageas, Follett and Assenmacher, 1974

* Mean ± SE ( ) N° of animals (a) $p<0.01$ vs controls

M.C.R.: Metabolic Clearance Rate (b) $p<0.02$ vs controls

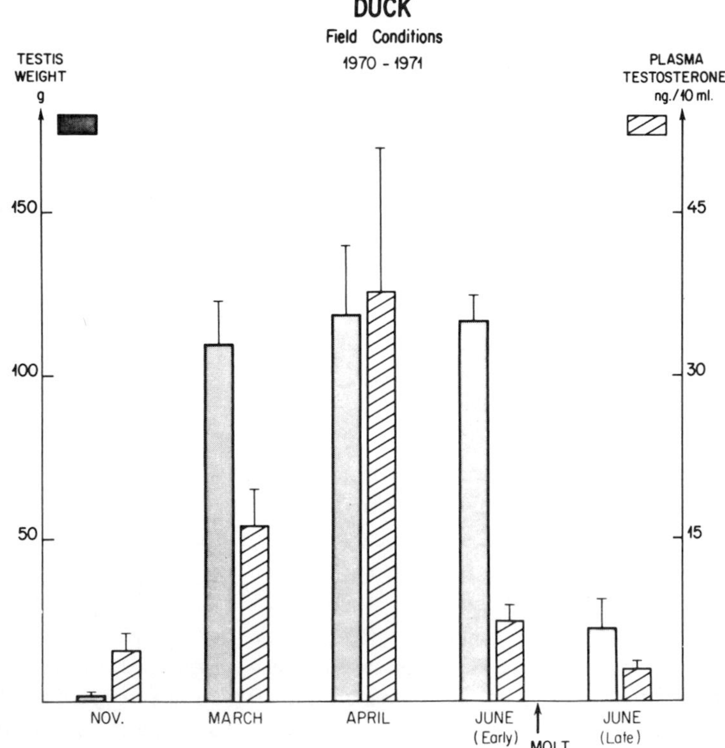

Fig. 4. "Classical" vernal gonadal cycle. Mean ± SE for 4 ducks (after Jallageas and Assenmacher, 1974).

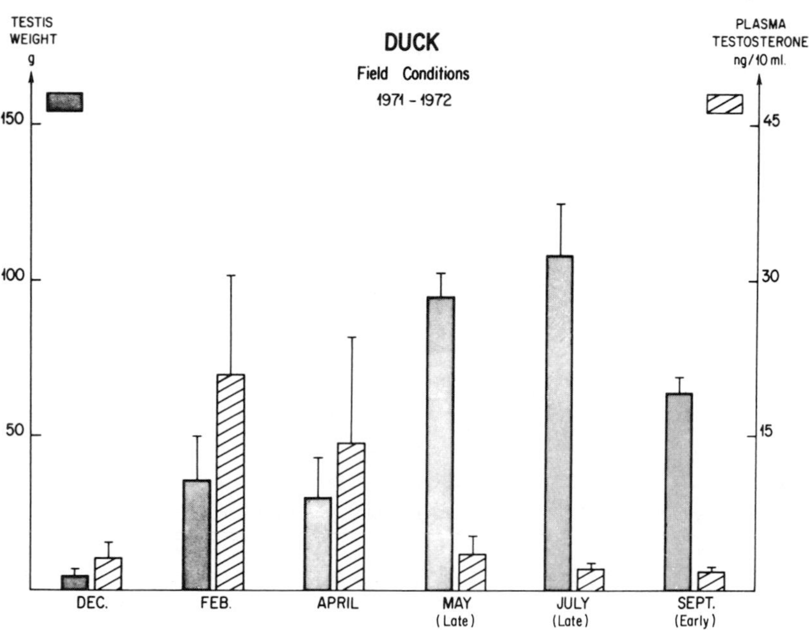

Fig. 5. Phase-dissociation between the annual cycles of testis weight and plasma testosterone respectively. Mean ± SE for 4 ducks. (after Jallageas and Assenmacher, 1974).

slopes down in May, and minimal annual titers may already be measured in late May or June (fig. 4 and 5). The histological and ultra structural pictures of both components of the testis correlatively point to an arrested spermatogenesis in June, while the interstitial tissue, and the Sertolli cells, lose their smooth endoplasmic reticulum in May, so that the former are reduced to an almost fibroblast-like appearance in June (Garnier, 1972, 1973).

1. External factors.

The most striking feature in this respect lies in the fact that the regressive phase occurs, and may even be completed before the summer solstice, i.e. before the onset of the seasonal decrease in daylength. As a matter of fact, if an increasing L/D ratio appears as the main stimulus to an active testis function, it must be kept in mind that the L/D ratio ceases any appreciable increase before the solstice. And indeed, a very strong photic stimulus, e.g. constant illumination imposed in June on ducks, which had already initiated seasonal regression, could induce the testes to grow back to their maximal size (Assenmacher and Tixier-Vidal, 1962). In the duck the natural photoperiodic conditions prevailing during the early phase of testicular regression might therefore be considered as a lack of adequate stimulus rather than as an actual depressing factor, as in species that are truly photorefractory (Farner, 1970).

Among the environmental clues that may act as inhibitors on testicular function, reduced food availability or food intake undoubtedly may play an important role in a number of ecophysiological situations. The deleterious effect of malnutrition on both components of the testis has been shown in ducks (Assenmacher *et al*, 1965; Jallageas and Assenmacher, 1973). On the other hand sexually active ducks reduce their food intake by about 8% (Assenmacher *et al*, 1965). The seasonal decrease in food consumption however, cannot be held responsible for the onset of testicular regression, since even a 50% food restriction had no effect on the testis in this species (Assenmacher *et al*, 1965). The situation might be different in the house sparrow, where Vaugien (1960) claimed a seasonal fall of 26% in food intake to occur, a decrease that would impede spermatogenesis in any control birds exposed to a stimulatory photoperiod.

2. Internal factors.

2.1. Negative feed-back of circulating androgens.

The possible implication of a negative feed-back mechanism in the annual termination of the sexual cycle has been postulated (Farner, 1970). That high doses of exogenous testosterone actually exert an inhibiting action on testicular activity in ducks has been demonstrated. Figure 3 shows that, even during the sexual quiescent phase (December) i.m. injections of testosterone propionate cause complete testicular involution. Gogan's work (1968-1970) has thrown some light on further details of this effect. First, it was shown that chronical administration of 800 ug/kg/d testosterone blocked the photogonadal response to constant illumination (LL) in 50% of the experimental ducks, while inhibition was complete with 1600 ug/kg/d. In the latter case the gonadotropic potency of the pituitary was reduced to 1/5 vs untreated light-stimulated ducks. The site of action of testosterone seems to lie primarily in the hypothalamus, since in a parallel experiment the stimulatory effect of light was either depressed, by stereotaxic microimplantations of the steroid within the preoptic or suprachiasmatic area, or even suppressed if implantation occurred within the ventro-medial tuberal area of the hypothalamus. On the other hand, the sensitivity of the gonadotropic machinery to the inhibiting action of testosterone seems to cycle during the year (Table III).

Whereas the gonads exhibited an increasing resistance toward the depressing action of the androgen during the progressive phase of the sexual cycle, the inhibiting effect of testosterone increased abruptly in late April, i.e. just before the onset of the natural regressive phase of the cycle. It is therefore tempting to speculate that a seasonal supervening hypersensitivity of the central machinery to the negative feed-back of the high circulating androgen level could be, at least part of the mechanism that brings about annual testicular regression. Whether this seasonal cycle in the negative feed-back effect of testosterone should have been determined by photoperiodic stimuli (i.e. persistent long days), rather than by some further internal factor lacks clarification.

2.2. Testicular sensitivity to circulating gonadotropic hormone.

According to another attractive hypothesis raised by Vaugien (1955) in the house sparrow, the regressive phase of the testis could be induced by seasonally occurring refractoriness of the testis to circulating gonadotropins. Vaugien's hypothesis has been ruled out in the ducks, where sexual activity could be resumed in regressing ducks by injecting high doses of gonadotropic extracts (Benoit *et al*, 1950b).

Table III

Seasonal Variations Of The Depressive Action Of Testosterone In Ducks

Month of year	Threshold Dosage of Testosterone for Testis Regression (1)	Threshold Dosage of Testosterone for Pituitary Gonadotropic Potency (2)
November	50	50-100
February	400	400-800
March	600-1200	800-1200
April (early)	1600	1600
April (late)	400-1200	1200-1600
May	Natural Regression	————

From Gogan, 1970.

(1) Daily testosterone dosage (2 weeks) inducing testis regression in 50% of the birds (mg/kg/d).

(2) Daily testosterone dosage (2 weeks) inducing significant reduction in the gonadotropic potency of the pituitary (Breneman's test).

Table IV

Gonadotropic Potency Of Pituitary Implants From Ducks At Different Stages Of The Sexual Cycle

planted N° Mice	Sexual State of Donor Ducks	Gonadotropic Response in Mice : $\frac{\text{Genital Tract Weight}}{\text{Body Weight}} \times 10^3$
25 implanted ntrol Mice	————	1.06 ± 0.06
10	Prepuberal (Inactive Testis)	1.41 ± 0.02
15	Reproductive Phase	5.89 ± 0.45
9	Regressive Phase	12.3 ± 1.4

om Benoit, Assenmacher and Walter. 950b) (2 duck pituitaries implanted per mouse. Autoposy 4 days after).

Moreover the trend to an increased gonadotropic potency evoking storage in the duck pituitary during the regressive phase that was demonstrated at the same time (Benoit *et al*, 1950a-table IV) fitted best to a primary central origin (i.e. cut in pituitary gonadotropin secretion) of testis regression.

This view that has been prevailing during the past two decades could however, be challenged by recent findings, illustrated in figure 6. In a cooperative work with B. K. Follett (Bangor, U. K.) the plasma - LH content was measured radio-immunochemically (see Follett *et al*, 1971), together with the plasma testosterone content in ducks reared outdoors. Surprisingly the circulating LH level remained high by the time of the seasonal downfall of the plasma testosterone titer. (Jallageas, Follett and Assenmacher, 1974). This disconcerting result might point at least to a marked increase of the sensitivity threshold of the testis toward available gonadotropin. This situation could also happen, either from external influences that could be mediated through the gonadal innervation, or from an internal origin.

2.3. Thyroid-testis interactions

It has been clearly established that a number of bird species including the common Starling (Woitkewitch, 1940; Wieselthier and Van Tienhoven, 1972), and several Indian passerines (see Thapliyal, 1969; Chandola, 1972) the seasonal regression of the testis can be suppressed by removing the thyroid. Alternatively, in other birds like domestic fowls and ducks, emphasis has been laid on thyroid hormones as a prerequisite for the juvenile or seasonal testicular growth to occur (Benoit and Aron, 1934). However, more recent experiments on

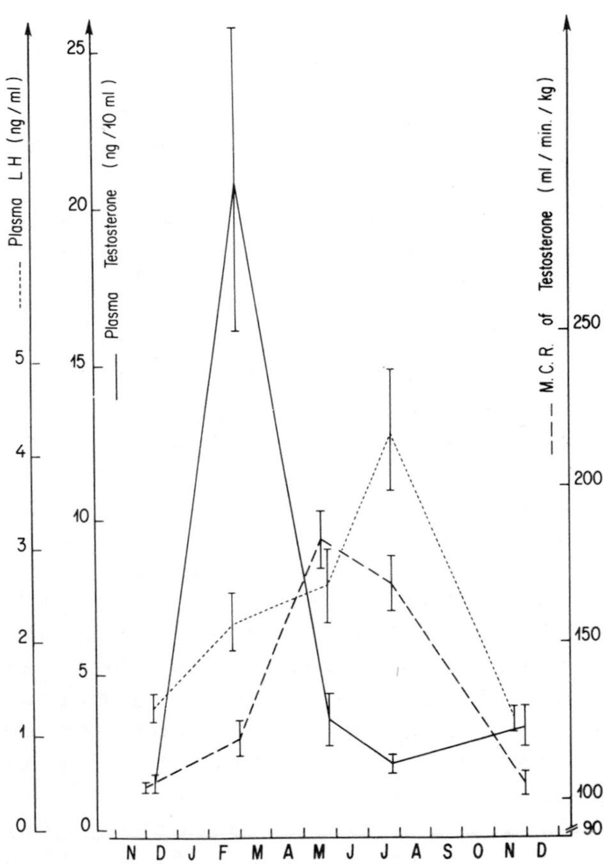

Fig. 6. Annual cycles in Plasma Level and Metabolic Clearance Rate (MCR) of testosterone, and in Plasma Concentrations of LH. (from Jallageas, Follett and Assenmacher, 1974).

domestic ducks have favored the thesis of a possible role of thyroid hormones in the induction of seasonal testicular regression, even in the latter species (Jallageas and Assenmacher, 1974). This assumption was based on the following data: (1) Facing the mild mediteranean climate prevailing at the University of Montpellier, the thyroid function of drakes living outdoors manifests an annual elevation in May-July (Astier 1970) (fig. 7); (2) At the same time the sexual function is characterized by the collapsing androgenic activity associated with an increased metabolic clearance rate of testosterone, together with a maintained high LH plasma titer (fig. 6); (3) The three parameters of the latter syndrome can be mimicked by thyroxine injections during the full reproductive period, while an experimental thyroxine load prevents the stimulating effect of artificial long days (18L-6D) on the gonads of winter-ducks (table V).

A seasonal increase in thyroid activity might therefore be considered as a prominent internal factor for the seasonal extinction of the androgenic function in the duck testis, although it is not yet demonstrated that the very initiation of this androgenic decline already originates in the plasma titer in thyroid hormones. Whether the seasonal hyperactivity of the thyroid itself is under exogenous or endogenous control is unknown.

One possible internal cause could lie in the decrease of testosterone secretion, since the depressing action of an active androgenic function on the duck thyroid has been fairly well established (Tixier-Vidal and Assenmacher, 1961a, 1962b; Jallageas and Assenmacher, 1972). Incidently it may be noted that the annual post-nuptial plumage

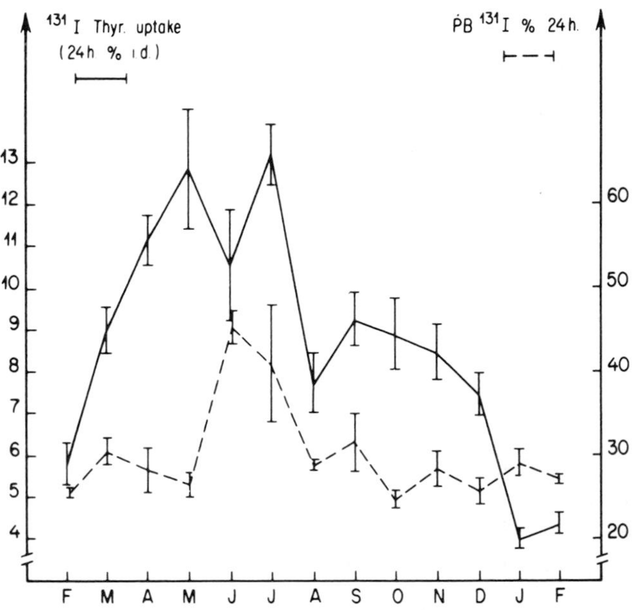

Fig. 7. Annual cycle in thyroid function in drakes reared outdoors at Montpellier (Southern France). (from Astier, Halberg and Assenmacher, 1970).

Table V

Thyroid Gonadal Interactions In Ducks.

TREATMENT	Testis Weight (g)*	Plasma Testosterone (ng/10ml)	MCR Testosterone (ml/min/kg)	Plasma LH (ng/ml)
Outdoor Controls. December	6.4 ± 4.8 (4)	1.53 ± 0.55 (4)	104 ± 3 (6)	1.28 ± 0.29 (4)
21 days : 18L-6D	66.8 ± 12.2 (3) (a)	32.8 ± 11.1 (3) (b)	97 ± 7 (6)	4.84 ± 0.18 (3) (b)
21 days : 18L-6D 9 days T_4 (1mg/d)	65.5 ± 6.2 (4) (a)	5.25 ± 1.05 (4) (bc)	128 ± 9 (4) (c)	2.92 ± 0.45 (4) (bc)
Outdoor Controls. March	27 ± 16 (4)	20.8 ± 9.6 (4)	119 ± 12 (4)	2.20 ± 0.67 (4)
9 days T_4 (0.1 mg/d)	9.4 ± 2.3 (4)	8.1 ± 4.0 (4)	141 ± 19 (4)	2.15 ± 0.31 (4)
9 days T_4 (1mg/d)	17.8 ± 5.6 (4)	4.2 ± 1.8 (4) (b)	205 ± 40 (4) (b)	2.29 ± 0.26 (4)
Outdoor Controls. April	146.5 ± 20.8 (3)	37.6 ± 13.6 (3)	n.m.	3.44 ± 0.74 (3)
9 days T_4 (1mg/d)	123.7 ± 2.3 (3)	2.8 ± 0.3 (3) (b)	n.m.	3.04 ± 0.88

After Jallageas, Follett and Assenmacher, 1974.

* Mean ± SE
( ) Number of animals

(a) $p<0.01$ vs outdoor controls
(b) $p<0.05$ vs outdoor controls
(c) $p<0.05$ vs 21 days 18L-6D.

molt occurs in late June, i.e. at the time when the circulating thyroxine to testosterone ratio attains an annual peak value. In the same line of thought it has been shown in ducks, that experimental molts, that always follow surgical disconnection of the pituitary from the hypothalamus, can be prevented either by testosterone supplementation (Assenmacher and Bayle, 1968), or by thyroid inhibition (Bayle, 1972) i.e., by lowering the thyroid/testosterone balance, which is increased after hypothalamic-pituitary disconnection.

Summarizing the preceding data, the following sequence of events could tentatively account for the regressive phase of the testis cycle: (1) The onset of a seasonal increase in the sensitivity to the negative feed-back of testosterone within the CNS centers controlling the gonadotropic function, may initiate a decreased testosterone secretion. The respective role of exogenous, environmental clues, and of endogenous factors, whether linked to unknown hormonal interactions or to an intervening circannual clock within the CNS, in inducing these alterations will need further clarification; (2) The decay of the androgenic function of the testis releases a brake on the thyroid function, and (3) the thyroidal hyperactivity could lead to further inhibition of testosterone secretion associated with an increased catabolism of the hormone; (4) The decreased testicular reactivity to circulating LH, that may be another cause of the suppression of androgenic function, could originate either in a thyroid-testis interaction, or in a transient inhibition mediated from the CNS by the direct innervation of the testis.

III. The autumnal testicular cycle.

In any flock of ducks a number of specimens display a secondary sexual cycle in early fall (see fig. 2). The pattern of this autumnal sexual cycle which seems to be restricted to the male, has been thoroughly investigated by Garnier (1970, 1972, 1973). The testes, that underwent marked regression in June-July proceed thereafter to a new but brief enlargement phase in August, that usually precedes the secondary augmentation in plasma testosterone. Compared with the vernal cycle, the autumnal secretion displays, however, remarkable pecularities: (1) Intratesticular binding of the hormone is very low, so that the synthetized hormone appears almost exclusively within the efferent venous blood of the testis. The autumnal testosterone content in the effluent blood samples therefore peaks at a higher level than during the spring cycle (fig. 8); (2) There is a notable phase-shift between the cycle of the testis weight that crests in August, and that of the plasma testosterone, which extends from September to November, with a nadir in October; (3) The Leydig cells do not seem to participate actively in the autumnal resumption of testosterone secretion, since ultrastructurally they keep an aspect of "undifferenciated" resting cells, that look almost like fibroblasts. On the other hand, the Sertoli cells reveal a moderate trend toward the reorganization of the smooth reticulum, that could point to them as the main contributors to testosterone secretion at this period (Garnier, 1973).

Although there is no formal evidence that the autumnal testicular cycle is under hypothalamic control, Tixier-Vidal (1965) has claimed that several clusters of fairly active, presumed LH (gamma) cells may

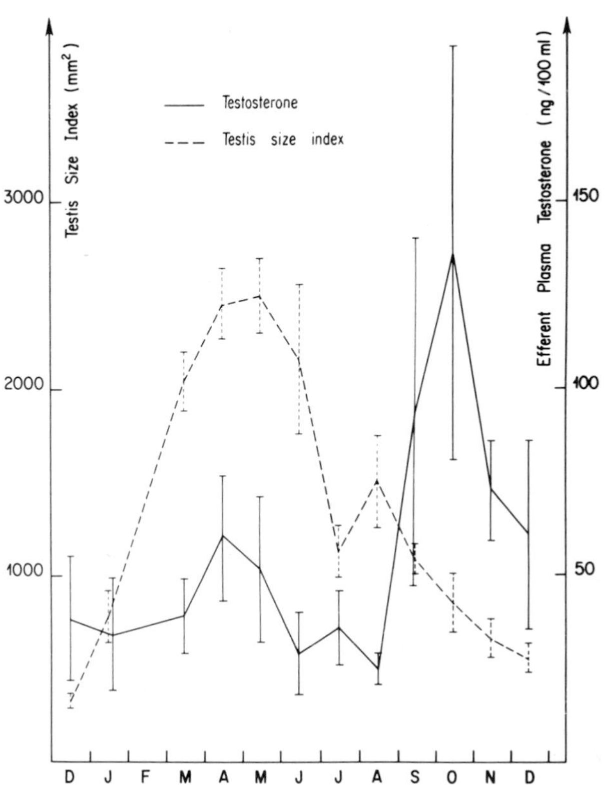

Fig. 8. Biphasic annual testicular cycle as measured by testis profile and testosterone content in effluent venous plasma samples. (after Garnier, 1970).

correlatively become apparent within the pituitary.

Regarding the possible causal factors of the secondary testicular cycle, no special environmental clue can be seen. More probably it must be of some endogenous origin. Among them, the termination of the seasonal phase of thyroidal hyperactivity may abandon the brake on testis activity that has been discussed earlier. Alternatively, the eventuality of a stimulus reaching the gonadotropic machinery from an endogenous master clock cannot be disregarded, as will be seen later on.

IV. The testis cycle in free-running conditions.

It is not an easy task to investigate free-running rhythms over several years. The literature, therefore indicates only scant contributions in this field, and it may be of interest to comment herein on a crucial experiment, that was performed on ducks between 1954 and 1959 (Benoit, Assenmacher and Brard, 1956; Benoit *et al*, 1970). In this experiment a colony of 28 male Pekin ducklings, from an inbred strain that were hatched on May 15th, 1954 were nursed all together until 18 days old. From this day on the animals were divided into 3 groups. The first one (LD : 10 specimens) was reared under natural light conditions. They were placed outdoors from age 4 weeks on, and they lived thereafter under field conditions. Group two (DD : 10 specimens) was housed in an insulated and completely darkened animal-house, that was kept at constant temperature (18° C). A third group (LL) of 8 specimens was grown in isothermic quarters provided with constant illumination (about 500 lux). All animals were given free access to water and solid food (standard mash). In the DD and LL groups, food supply and cleaning care were scheduled at various clock hours. Entrance to both special quarters occurred through locked chambers, and the only light source used in the DD room was an electric torch provided with a blue screen. Once a month, but at different days within the first two weeks, each animal underwent x-ray photography of the abdomen that permitted measurement of the diameter of the left testis, as an index of gonadal growth. During the manipulation, the DD ducks had their heads wrapped within opaque black cowls.

The present discussion will be restricted to the LD and DD groups, since the high mortality in the LL-group did not allow any quantitative rhythmometric analysis. As a matter of fact, two ducklings of the outdoor controls (LD) died within the first four months of the experiment. The 8 others were still alive 5 years later. Mortality was surprisingly low among the ducks living in constant darkness (DD), since the 10 animals survived for 3 years. One duck died at the age of 4 years and 7 others were still alive after 5 years. In contrast, in the constant light room (LL), one duckling died within the first month. After 2 years only 4 out of the 8 ducks survived, and this group was reduced to 2 animals after 3 years, and to one after 4 years. Five years after the beginning of the experiment, there were no survivors in this group. The development of the testis diameter for the control (LD) and (DD) ducks is indicated on figures 9 and 10. The sexual cycle of the outdoor controls (fig. 9) had a classical mainly monophasic profile throughout the experiment. The discontinuity of the graph for the first annual cycle was due to a technical failure in May

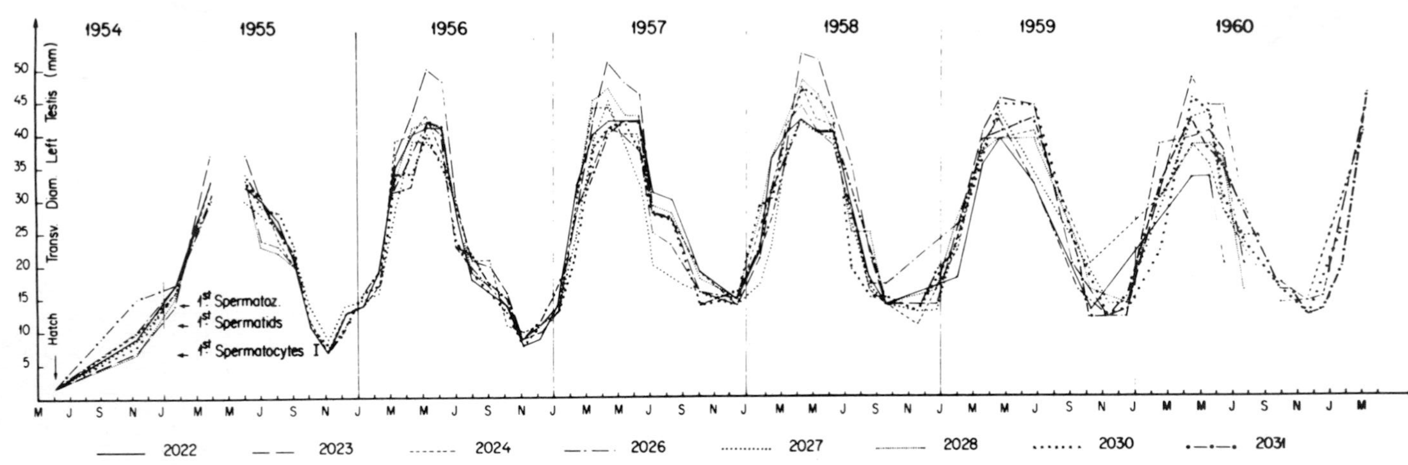

Fig. 9. Annual testicular cycles in 8 ducks reared outdoors. (These are the LD controls to fig. 10) (after Benoit, Assenmacher and Brard, 1956; Benoit *et al.*, 1970).

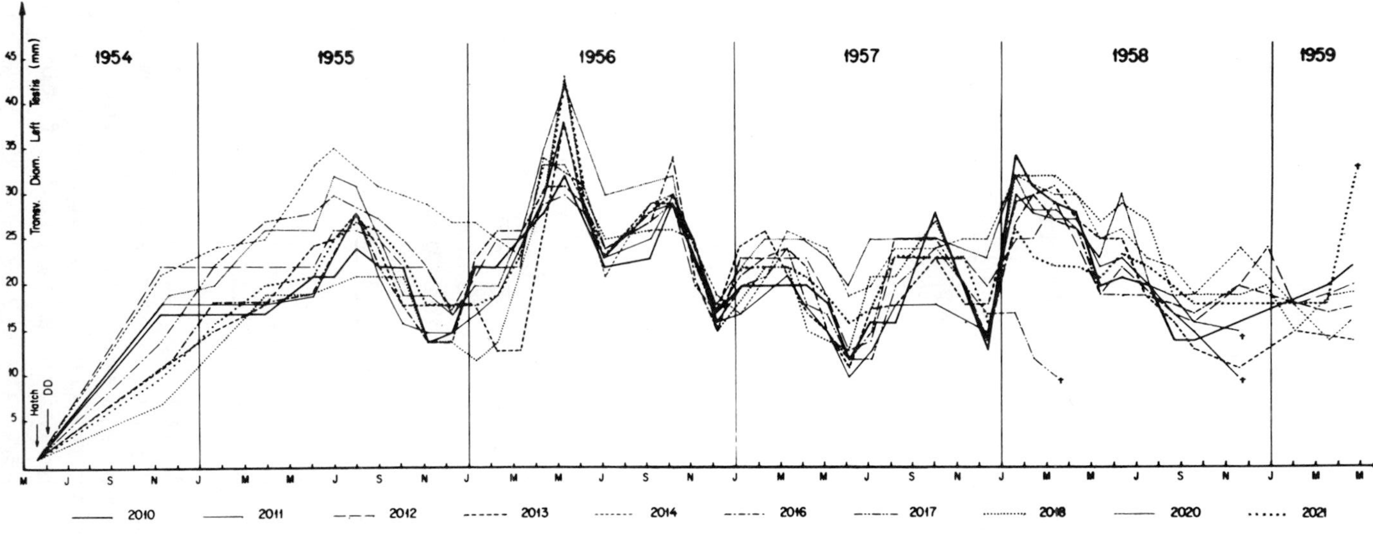

Fig. 10. Cyclic fluctuations in testis size of 10 ducks reared in constant darkness (DD) from age 18 days. (after Benoit, Assenmacher and Brard, 1956; Benoit *et al.* 1970).

1955. On the other hand the correlation between testis diameter and spermatogenetic maturation that is indicated on the graph was established on several hundreds of control ducks.

One remarkable feature in the darkened ducks (fig. 10) concerned the sexual maturation, that was not notably delayed. However, in 1955 the peak testis size occurred only in July-August. During the following three years unequivocal rhythmic fluctuations in the testis diameters were evident. Gross inspection of fig. 10 shows that the periodicity of the fluctuations was much shorter than that of the controls (fig. 9). Furthermore the occasionally recurring testicular regression did not always lead to complete arrest in spermatozoa production.

In an attempt to quantify the possible periodicities and related parameters of the testicular cycle in the LD and DD ducks, we have recently submitted the individual data from the 1954 experiment to a mathematical analysis based on the least square method (Halberg *et al.*, 1955), with the use of an IBM 360/65 computer programme. This method allows the isolation of "windows" corresponding to significant periodicities, by screening linearly a trial span of frequency (or period) with "steps" of various extent. In a second stage a finer screening leads to the determination of the most significant frequency (or period) within the preselected "windows". The results are given in table VI.

Several salient features are manifest on the table. First of all, the mathematical analysis demonstrates that in the LD Controls (group LD1 in table VI), there is a low amplitude rhythm with a period of about 147 days, which was hardly visible in figure 9,

Table VI

Rhythmic Fluctuations Of Testis In LD And DD Ducks.
Least Square Method Estimated Parameters.

TREATMENT	Period 2 (days)	Level (Co)		Amplitude (C)		Phase (1)	
LD 1 (Outdoor Controls to the DD Group)	147	25.33	2.0	2.32	0.68	-130	33
	365	25.36	0.77	15.44	0.15	101.5	1.1
DD	156	22.84	0.83	2.99	0.26	- 19.56	9.89
	319	23.14	0.84	2.89	0.35	-197.63	13.87
LD 2	140	23.77	2.29	5.89	1.37	-234.47	26.28 (Sept.)
(Outdoor Controls)	365	23.07	0.94	10.08	0.34	-116.54	3.75 (April)

LD 1 and DD were measured by x-rays (see fig. 9 and 10)

LD 2 were sacrificed over a 20 year span (see fig. 2)

(1) Origin : for LD 1 and DD : 1.01.1955; for LD 2 : January 1st.

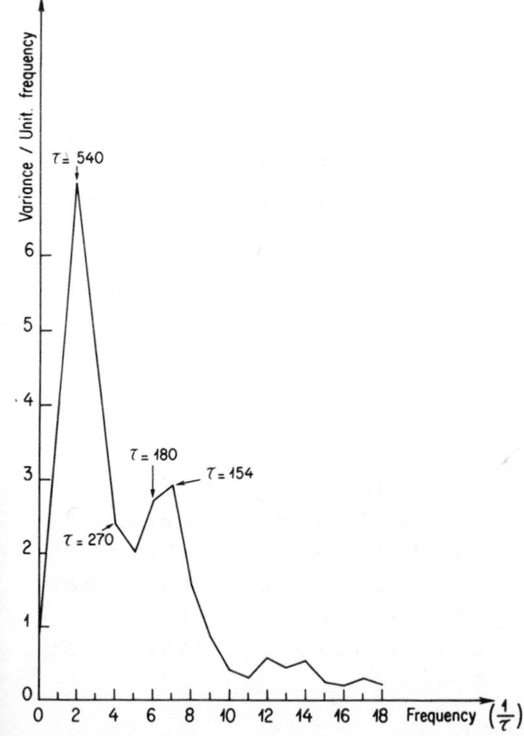

Fig. 11. Spectral analysis showing a circa-semestrial component of the cyclic fluctuations in testis size in the DD-ducks of fig. 10. Note: The resolution of the spectrum does not permit an identification of longer (*e.g.* circannual) periodicities.

and it was present together with the 365 day annual periodicity. If submitted to the same procedure, the data collected from the outdoor controls listed in figure 2 (LD2 group of table VI) similarly show a periodicity of 140 days together with the annual 365 day rhythm. It can therefore be concluded that even in synchronized LD conditions, the annual testis cycle definitely follows a biphasic pattern.

Turning to the DD ducks, two periodicities again emerge very clearly: (1) a circa-semestrial period, (156 days, i.e. between 5 and 5 1/2 months), and (2) a circannual period (319 days, i.e. between 10 and 11 months). From the analysis it cannot be inferred whether or not the latter periodicity is redundant to the former.

Another periodicity analysis procedure, e.g. spectral analysis (see Jenkins and Watts, 1969) was used correlatively to measure the shorter periodicity in the DD experiment. Figure 11 shows that it points to a 154 day period. The circannual periodicity actually lies beyond the discriminative power of the method, that would have necessitated more than 3 years of observations.

V. General Discussion.

Although the foregoing data provide some precise insights into the controlling mechanism of the sexual cycle in drakes, they cannot at this time be formulated into a coherent theory.

That the sexual cycle results from various interacting external and internal factors can be considered as established. Among the many environmental clues, the seasonal variations in photoperiodism play a dominating role in the annually synchronized repetition of the reproductive cycle. The photosensitivity of the neuroendocrine machinery that governs the testis functions is so high that increasing (or long) day lengths always induce a more or less explosive stimulation on the resting gonadal axis. Neither cold nor scarce food restrictions may hinder this sort of "photogonadal reflex". The vernal increase in day-length therefore, must be held as the main cause of the vernal maturation of the testis.

On the other hand, no definite environmental signal is concomitant with the initiation of testicular regression in ducks, whereas at least three internal endocrine or neuroendocrine factors were shown to interfere in the discontinuance of the active testicular phase: (1) The inhibiting effect of a seasonally increased output of thyroid hormones on testis function (associated with their stimulating effect on the catabolism of the androgen); (2) An abrupt enhancement in the reactivity of the neuroendocrine control of the gonads to the negative feedback of testosterone; and (3) A lowered sensitivity of the testis to the circulating level of gonadotropic hormone. The two latter events could clearly originate in the environment (e.g. the prolongation of "long-days", throughout the spring), or in some endogenous mechanism (and among them increased thyroxine secretion).

Now, the occurrence of a regular, free-running testicular cycle in DD-housed ducks, i.e. in the absence of photoperiodic fluctuations and of any known external "zeitgeber", clearly demonstrates the reality of an endogenous stimulating and timing mechanism (clock) to the gonadal function. As for other biological clocks, the site and nature of the master clock that controls endogenous testicular cycles remains hidden. Its location might be assigned to the CNS, since in ducks that underwent transection of the hypophysial portal veins, i.e. a chronic hypothalamic-hypophysial disconnection, the testes remained definitively regressed (Assenmacher, 1958). This endogenous oscillator could therefore consist of a population of neurons (or neuronal circuits), firing rhythmic outputs to the hypothalamic areas that synthetize gonadotropin releasing factors. As for other CNS activities, varying blood levels for several hormones naturally could interplay with this neuronal machinery. Alternatively, hormonal feed-backs and interactions could be a constitutive part of the clock, together with more or less complex neuronal circuits.

The speculations on where and what the clock may be can be extended to the problem of how it works. In other words, what is its basic periodicity? The data that have been previously analysed may be interpreted in different ways:

1. A circannual clock could be responsible for the endogenous testicular rhythm. As a matter of fact the mathematical computer analysis (table VI) indeed could be taken as evidence of a circannual periodicity (10-11 months) in the free-running fluctuations of testis diameter in the DD ducks. This hypothesis has already been raised by an earlier analysis with the cosinor method (see Halberg, 1970). In this eventuality, the annual testicular cycle in the natural environment would result from a classical entrainment of the endogenous circannual rhythm by the exogenous annual "zeitgeber". It should then be admitted that the autumnal testicular cycle, that frequently appears in natural conditions, as well as every second

testicular cycle in the free-running conditions (DD), would result exclusively from a temporary inhibition of the full testis activity by other interfering endogenous factors, viz. negative feed-backs or other endocrine interactions, that would not act on the clock, but on its output, the hypothalamic-pituitary-testis axis.

2. It is evident, however, that the occurrence of a circa-semestrial clock could equally control the testicular cycle in both "natural" and "constant" conditions. A circa-semestrial periodicity (5 to 5 1/2 months) in fact appears in the LD as well as in the DD ducks (table VI); when exposed to the annual synchronizers of the natural environment, the circa-semestrial clock either could be entrained by the exogenous "zeitgeber", and the overt testicular cycle would be strictly annual. This really seems to occur in the "classical" monophasic testicular cycle (fig. 1 and 10). Alternatively synchronization could be incomplete, and even in a natural environment the measured testis cycle would have a biphasic profile, a situation that indeed occurs frequently (fig. 2 and fig. 8). The occurrence of a circa-semestrial periodicity, though of low amplitude, in group LD1 of table VI that corresponds to the apparent monophasic pattern of figure 10, demonstrates that even in this group, some individuals might not have followed the classical pattern. Which possibility actually takes place, might depend on biometeorological conditions, or on internal conditions (e.g. multihormonal interactions on the neuroendocrine machinery, or on the testis reactivity to the gonadotropic signals). Let us for instance consider the data of table VI. In the synchronized ducks (LD1 and LD2) a 140 to 147 day periodicity

was measured. It was fairly close to the 156 day period in the free-running experiment (DD). Assuming that a nearly 5 month (140 to 156 day) endogenous periodicity competes with an annual "zeitgeber", it could be considered that the former can only be manifested under certain external and internal conditions. As a matter of fact, if one speculates that in a given annual cycle (N^0 1) the endogenous rhythm is in phase with a vernal cycle peaking in March, the next endogenous stimulus will occur in August (+5 months) leading to an overt secondary cycle. During the 2nd year a new endogenous output will occur in January, i.e. at the onset of the vernal cycle that can therefore be considered as an "open" season. On the other hand, the two following endogenous outputs (June and November) will fall into "closed" seasons, due to internal hormonal interactions (June), or to unfavorable external conditions. Following these phase jumps the two endogenous stimuli of the 3rd year will again induce two overt testicular rhythms (April and September).

3. The data at hand do not allow us to rule out a third hypothesis, i.e. that of two endogenous rhythms fluctuating respectively at a circa-semestrial and at a circannual periodicity. It muse however, be emphasized that the mathematical analytical procedure available, cannot isolate from the data of the DD experiment a circannual periodicity that would be independent from the circa-semestrial period.

Although no experimental evidence would at present favor one of the three hypotheses, it can however be speculated that the eventuality of a circa-semestrial clock (hypothesis 2), controlling the gonadal cycle could account for the occurrence of an almost complete reproductive cycle, that may be observed in fall in drakes hatched very early in the season (winter) without imparing the following "normal" vernal cycle.

In the same line of thought, it is tempting to envisage that the biological clock operates in connection with structures analogous to an electronic stage counter. The input to the stage counter would be connected to a basic self-sustained oscillator, i.e. the clock itself, and the stage outputs would drive the different biological rhythms fluctuating at various periodicities. In such a unitarian view, a number of medium and low frequency rhythms controlled by the CNS, e.g. circadian, circa-mensual, circannual, could originate in the same basic self-sustained oscillator with outputs firing at fixed intervals of the highest common frequency (here circadian), on frequency dividing structures that would be firing again at fixed but lower frequencies. These frequency dividing structures would finally be responsible for the overt rhythmicities of lower frequencies such as circa-mensual rhythms (i.e. 30 circadian cycles), circa-semestrial rhythms (i.e. 180 circadian cycles), and finally circannual rhythms (i.e. 360 circadian cycles).

CONCLUSION

The annual cycle of the testis function in Pekin ducks provides an interesting experimental model for the study of interfering exogenous and endogenous factors controlling annual biological rhythms.

Like most temperate-zone birds, ducks display their reproductive season in spring. However, in a number of specimens, a secondary peak for both the spermatogenetic and androgenic activity of the testis occurs in early fall. The trend toward a biphasic pattern of the annual testis cycle has been mathematically assessed by exploring it with a computer analysis program based on the least square method. This has been done on the month to month variations in testis size of some 520 outdoor control ducks belonging randomly to various experiments over the last 20 years.

A great variety of experimental designs has demonstrated several prominent aspects of the mechanism that controls the testis cycle in the "natural" outdoor environment.

The progressive phase of the annual gonadal cycle is mainly directed by the annual augmentation of day-length that follows the winter solstice. The seasonal stimulation of the testis can be mimicked or accelerated unseasonally by exposing the birds to appropriate experimental photoperiods. The basic knowledge of the site of action of the photoperiodic information (retinal and extraretinal photoreceptors), and of the neuroendocrine route whereby the resulting signals act on the testis, through neurosecreted hypothalamic hormones and pituitary gonadotropins, can be considered as established although a number of important details remain obscure.

No precise environmental clue can be correlated with the onset of the regressive phase of the testis cycle that usually precedes the summer solstice. On the other hand several internal factors have been shown to possibly play a role in inducing the gonadal regression, and among them are: (a) A seasonally increased sensitivity of the CNS centers controlling the gonadotropic function to the negative feed-back of circulating testosterone; (b) A decreased testicular reactivity to the plasma titer in LH; (c) A seasonal hyperactivity of the thyroid gland, for thyroid hormones were proved to depress testosterone secretion and to increase its metabolic clearance rate from the plasma. Although the direct interference of environmental stimuli in timing these events cannot be ruled out, a primary endogenous origin, whether due to multiendocrine interactions or to some clock machinery, could alternatively be postulated.

The determining factors for the secondary autumnal testis cycle, with its lowered amplitude and some salient peculiarities regarding intra-testicular binding of androgen are essentially unknown. It could be interpreted as a result of the suppression of the gonadal inhibition that brings on the regressive phase of the vernal cycle, or as a response to a new stimulating output of central origin. In the former hypothesis the autumnal cycle should be regarded as the last part of a single annual cycle, with a temporary inhibition of internal origin separating both parts of the cycle. In the second hypothesis, the autumnal cycle could be considered as an aborted semestrial cycle following a first, completely fulfilled semestrial cycle (reproductive season). The latter thesis may find some support in the pattern of the endogenous rhythmicity of the testis cycle that could be observed in a "constant" environment.

In a recent periodicity analysis (least square method) of the free-running fluctuations in testis size that had been previously reported on ducks reared from age 18 days and over a 5 year span in constant darkness (DD), two endogenous periodicities of equal amplitude could in fact be measured. One of 156 days (5 to 5 1/2 months), and another of 319 days (10-11 months). With respect to the limited duration of the analysed data, the discriminative power of the analytical procedure made it impossible to decide whether or not the circannual periodicity was independent of the circa-semestrial periodicity. However, it must be emphasized that the occurrence of two endogenous testis cycle within a circannual period in the free-running (DD) conditions is reminiscent of the biphasic pattern of the annual cycle in the natural (LD) environment.

Two main hypotheses are therefore, offered for speculation: (1) A circannual clock may control the endogenous testicular rhythm. In natural conditions (LD) this circannual endogenous rhythm would be entrained to the strictly annual periodicity by the photoperiodic "zeitgeber". In this case, the vernal regressive phase (LD), as well as the testicular inflation between one out of two successive cycles of the endogenous overt rhythm (DD) should be related exclusively to some endocrine interaction unrelated to the clock. (2) Alternatively, the biological clock controlling the testis cycles in ducks could be basicly a circa-semestrial clock. The mathematically demonstrated circannual periodicity in the endogenous free-running rhythm of the testis would then actually appear as redundant to the circa-semestrial rhythmicity. In the natural annually synchronized environment this circa-semestrial clock could be fairly well entrained to the annual periodicity, leading at first appearance to the classical (mathematically proved) annual reproductive periodicity. However, the frequent appearance of a physiologically and mathematically

assessed secondary (autumnal) cycle in testis activity would then demonstrate the incompleteness in the entrainment of the circasemestrial clock, that would result from a margin which was too wide, between both the exogenous and endogenous periodicities.
With the present state of knowledge both hypotheses can be regarded as acceptable. The latter hypothesis would probably have some adaptive advantage for the bird, since ducklings hatched very early in the season would be "offered" the possibility of running through a "normal" reproductive cycle in fall, without imparing the succeeding vernal sexual cycle. A situation that has in fact been repeatedly observed.

SUMMARY

1. The annual cycle of the testis function in Pekin ducks has been explored as an experimental model for the study of interfering exogenous and endogenous factors controlling annual biological rhythms. Organ weight or transverse diameter of the testis were used as a gross index for the spermatogenetic function, while plasma testosterone levels were indicative of the endocrine testicular function. Rhythmometric measurements were performed on testis weight or diameters, with a computer analysis program based on the least square method.

2. In natural outdoor conditions the reproductive season occurs with an annual periodicity in spring. However in a number of specimens in any flock, a secondary peak of testicular activity supervenes in early fall. The biphasic pattern of the testicular rhythm could be mathematically assessed, by submitting to periodicity analysis, the month to month variations in testis weight

of 520 ducks taken at random among some 2000 outdoor controls for the past 20 years.

3. A great variety of experimental designs have demonstrated that the progressive phase of the annual gonadal cycle is basicly controlled by the annual augmentation of day-length that follows the winter solstice.

4. On the other hand no precise environmental clue could be correlated with the onset of the regressive phase of the testis cycle that usually precedes the summer solstice.

5. However, several internal factors were shown to play a possible role in inducing gonadal regression: (a) A seasonally increased sensitivity of the CNS centers controlling gonadotropic secretion, to the negative feed-back of testosterone; (b) A decreased testicular reactivity to the plasma titer in LH; (c) A seasonal hyperactivity of the thyroid gland, inhibiting testosterone secretion and stimulating its catabolism. These events may in turn be timed either by signals from exogenous or endogenous (hormonal interaction or biological clock) origin.

6. The frequently occurring secondary autumnal testis cycle can be regarded either as the second part of a single annual cycle, following a temporary inhibition of internal origin, or as an aborted second (semestrial) cycle, following a first, fulfilled semestrial (reproductive) cycle. The latter thesis has some support from the pattern of the endogenous rhythmicity of the testis cycle.

7. The free-running fluctuations in testis size that were previously reported on ducks reared from age 18 days, over a 5 year span in a dark environment (DD), were sub-

mitted to a periodicity analysis. Two endogenous periodicities of equal amplitudes could be measured. One of 156 days (5 to 5 1/2 months), and another of 319 days (10 to 11 months). The alternative hypotheses of a circannual vs circa-semestrial clock are discussed.

ACKNOWLEDGEMENTS

I am very indebted to Dr. Janine Nouguier-Soule, Informatics-Engineer at our Research Unit, who was in charge of the periodicity analyses of this work. Dr. T. K. Bahattacharyya's stylistic advise was also much appreciated.

This work was in part supported by a C.N.R.S. Grant (A.T.P. "Physiologie Ecologique").

REFERENCES

ASSENMACHER, I. (1958). Recherches sur le controle hypothalamique de la fonction gonadotrope prehypophysaire chez le Canard. Arch. Anat. Micr. Morphol. Exper. (Paris), 47, 447-572.

ASSENMACHER, I. (1973). Reproductive Endocrinology: The Hypothalamo-hypophysial Axis. In : Breeding Biology of Birds (D.S. Farner edt.). Nat. Ac. Sc. Publ. Washington, D.C. p. 158-191.

ASSENMACHER, I. and BAYLE, J. D. (1964). Repercussions endocriniennes de la greffe hypophysaire ectopique chez le Canard male. C.R. Ac. Sci. (Paris) 259, 3848-3850.

ASSENMACHER, I. and BAYLE, J. D. (1968). Balance endocrinienne et mue du plumage chez le Canard Pekin male adulte. Arch. Anat. Histol. Embryol. (Colmar), 51, 67-73.

ASSENMACHER, I. and BENOIT, J. (1953). Repercussions de la section du tractus porto-tuberal hypophysaire sur la gonado-stimulation par la lumiere chez le Canard domestique. C.R. Ac. Sci. (Paris), 236, 2002-2004.

ASSENMACHER, I. and TIXIER-VIDAL. A. (1962). La reactivite a divers agents de l'axe hypophyse-testicule du Canard traite a l'extreme debut de la "Periode Refractaire" du cycle sexuel. C.R. Soc. Biol. (Paris), 156, 267-272.

ASSENMACHER, I., GROS, CH., BENOIT, J. and WALTER, F. (1950). La radiographie appliquee a l'etude macroscopique de l'evolution des testicules d'Oiseaux. C.R. Soc. Biol. (Paris), 144, 1107-1111.

ASSENMACHER, I., TIXIER-VIDAL, A. and ASTIER,

H. (1965). Effets de la sous-alimentation et du jeune sur la gonadostimulation du Canard. Ann. Endocrinol. (Paris), 26, 1-26.

ASTIER, H., HALBERG, F. and ASSENMACHER I. (1970). Rythmes circannuels de l'activite thyroidienne chez le Canard Pekin J. Physiol. (Paris) 62, 219-230.

BAYLE, J.-D. (1972). Mue experimentale et equilibre endocrinien chez le Canard Pekin male a autogreffe hypophysaire ectopique. Gen. Comp. Endocr. 18, Abstr. 14.

BENOIT, J. (1934). Activation sexuelle obtenue chez le Canard par l'eclairement artificiel pendant la periode de repos genital. C. R. Ac. Sci. (Paris), 199, 1671-1673.

BENOIT, J. (1935a). Stimulation par la lumiere artificielle du developpement testiculaire chez des Canards aveugles par section du nerf optique C. R. Soc. Biol., 120, 133-135.

BENOIT, J. (1935b). Stimulation par la lumiere artificielle du developpement testiculaire chez des Canards aveugles par enucleation des globes oculaires. C. R. Soc. Biol. (Paris) 120, 136-138.

BENOIT, J. (1935c). Hypophysectomie et eclairement artificiel chez le Canard male. C. R. Soc. Biol. (Paris), 120, 1326.

BENOIT, J. (1936). Facteurs externes et internes de l'activite sexuelle. I. Stimulation par la lumiere de l'activite sexuelle chez le Canard et la Cane domestiques. Bull. Biol. Fr. et Belgique 70, 487-533.

BENOIT, J. (1937). Facteurs externes et internes de l'activite sexuelle. II. Etude du mecanisme de la stimulation par la lumiere de l'activite testiculare chez le Canard domestique. Role de l'hypophyse. Bull. Biol. Fr. et Belgique., 71, 393-437.

BENOIT, J. (1938). Action de divers eclairements localises dans la region orbitaire sur la gonadostimulation chez le Canard male impubere. Croissance testiculaire provoquee par l'eclairement direct de la region hypophysaire. C. R. Soc. Biol. (Paris), 127, 909-914.

BENOIT, J. (1970). Etude de l'action des radiations visibles sur la gonadostimulation, et de leur penetration intra-cranienne chez les Oiseaux et les Mammiferes (J. Benoit and I. Assenmacher edts). C.N.R.S. Editeur, Paris 1970, pp. 121-136.

BENOIT J. and ARON, M. (1934). Sur le conditionnement hormonique du developpement testiculaire chez les Oiseaux. Resultats de la thyroidectomie chez le Coq et le Canard. C. R. Soc. Biol. (Paris), 116, 221-223.

BENOIT, J. and ASSENMACHER, I. (1970). La Photoregulation de la Reproduction chez les Oiseaux et les Mammiferes. Congr. Internat. C.N.R.S. N° 172, Montpellier 1967, C.N.R.S. Editeur, Paris 1970, 588 p.

BENOIT, J., ASSENMACHER, I. and BRARD, E. (1956). Apparition et maintien de cycles sexuels non saisonniers chez le Canard domestique eleve pendant plus de trois ans a l'obscurite totale. J. Physiol. (Paris), 48, 388-391.

BENOIT, J., ASSENMACHER, I. and WALTER, F. X. (1950a). Activite gonadotrope de l'hypophyse du Canard domestique, au cors de la regression testiculaire saisonniere et de la prepuberte. C. R.

Soc. Biol. (Paris), 144, 1403-1408.

BENOIT, J., ASSENMACHER, I. and WALTER, F. X. (1953). Dissociation experimentale des roles des recepteurs superficiel et profond dans la gonado-stimulation hypophysaire par la lumiere chez le Canard. C. R. Soc. Biol. (Paris), 147, 186-191.

BENOIT, J. ASSENMACHER, I., BRARD, E. and KORDON, C. (1970). Influence des facteurs lumineux et sociaux sur le conditionnement testiculaire et le comportement sexuel du Canard male de race Pekin. Probl. Act. Endocrinol. Nutrit. Fr., 14, 109-125.

BENOIT, J., MANDEL, P., WALTER, F. X. and ASSENMACHER, I. (1950b). Sensibilite testiculaire a l'action gonadotrope de l'hypophyse chez le Canard domestique au cors de la regression testiculaire saisonniere. C. R. Soc. Biol. (Paris), 144, 1400-1403.

BENOIT, J., WALTER, F. X. and ASSENMACHER, I. (1950c). Contribution a l'etude du reflexe opto-hypophysaire gonadostimulant chez le Canard soumis a des radiations lumineuses de diverses longeurs d'onde. J. Physiol. (Paris), 42, 537-541.

BISSONNETTE, T. H. (1930). Sexual maturity, its modification and possible control in the European Starling *Sturnus vulgaris* : General statement. Amer. J. of Anat. 45, 289-305.

BONS, N. (1974). Mise en evidence, au microscope electronique, de terminaisons nerveuses d'origine retinienne dans l'hypothalamus anterieur du Canard. C. R. Ac. Sci., 278, 319-321.

BONS, N. and ASSENMACHER, I. (1969). Presence de fibres retiniennes degenerees dans la region hypothalamique supra-optique du Canard apres section d'un nerf optique. C. R. Ac. Sci. (Paris), 269, 1535-1538.

BONS, N. and ASSENMACHER, I. (1973). Nouvelles recherches sur la voie retino-hypothalamique chez les Oiseaux. C. R. Ac. Sci. (Paris), 277, 2529-2532.

CARDINALI, D. P., CUELLO, A. E., TRAMEZZANI, J. H. and ROSNER, J. M. (1971). Effects of pinealectomy on the testicular function of the adult male duck. Endocrinology, 89, 1082-1093.

CHANDOLA, A. (1972). Thyroid in Reproduction. Reproductive physiology of *Lonchura punctulata* in relation to iodine metabolism and hypothyroidism. Ph.D. Thesis N° 7774, Banaras Hindu University, Banaras (India), 99 p.

FARNER, D. S. (1970). Day length as environmental information in the control of reproduction of birds. In : La Photoregulation de la reproduction chez les Oiseaux et les Mammiferes (J. Benoit and I. Assenmacher edts). C.N.R.S. Editeur, Paris 1970, pp. 81-88.

FARNER, D. S. (1973). Breeding Biology of Birds. Nat. Acad. Sci. Publ. Washington, D.C. 515 p.

FOLLETT, B. K. (1973). The Neuroendocrine Regulation of Gonadotropin Secretion in Avian Reproduction. In : Breeding Biology of Birds (D.S. Farner edt.). Nat. Ac. Sci. Publ. Washington D.C. p. 209-243.

FOLLETT, B. K., SCANES, C. G. and CUNNINGHAM, F. J. (1971). A radioimmunoassay for avian luteinizing hormone. J. Endocrinol. (London), 51, v-vi.

GARNIER, D. H. (1972). Etude de la fonction endocrine du testicule chez le Canard

Pekin au cours du cycle saisonnier. Aspects biochimique et cytologique. These Doct. Sc. Nat. Univ. Paris, N° C.N.R.S.-A.O. 7338, 129 p.

GARNIER, D. H. and ATTAL, J. (1970). Variations de la testosterone du plasma testiculaire et des cellules interstitielles chez le Canard Pekin au cours du cycle annuel. C. R. Ac. Sci. (Paris), 270, 2472-2475.

GARNIER, D. H., TIXIER-VIDAL, A., GOURDJI, D. and PICART, R. (1973). Ultrastructure des cellules de Leydig et des cellules de Sertoli au cours du cycle testiculaire du Canard Pekin. Z. Zellforsch. 144, 369-394.

GOGAN, F. (1968). Sensibilite hypothalamique a la testosterone chez le Canard. Gen. Comp. Endocrinol., 11, 316-327.

GOGAN, F. (1970). Role de la retroation par la testosterone dans la regulation de la fonction gonadotrope du Canard male. These Doctorat Sc. Nat. Paris, N° C.N.R.S.-A.O. 4278.

HALBERG, F. (1970). Discussion in : La Photoregulation de la Reproduction chez les Oiseaux et les Mammiferes (J. Benoit and I. Assenmacher edts.) C.N.R.S. Editeur Paris, pp. 520-528.

HALBERG, F., ENGELI, M., HAMBURGER, C. and HILLMAN, D. (1965). Spectral resolution of low frequency small-amplitude rhythms in excreted 17-ketosteroids; probable androgen-induced circaseptan desynchronization. Acta Endocrinologica, Suppl. 103, 54 p.

HARTWIG, H. G. (1973). A possible retinohypothalamic connection in the House-Sparrow. Sympos. on "Annual Cycles, Photoperiodism, and its endocrine mechanisms in Vertebrates" : Dtsche

Forschungsgemeinschaft. Erling-Andechs Seewiesen 9-12 Oct. 1973.

JALLAGEAS, M. and ASSENMACHER, I. (1972). Effets de la photoperiode et du taux d'androgene circulant sur la fonction thyroidienne du Canard. Gen. Comp. Endocrinol., 19, 331-340.

JALLAGEAS, M. and ASSENMACHER, I. (1973). Effets du jeune et de la castration sur la cinetique du metabolisme de la testosterone. Gen. Comp. Endocrinol. 20, 401-406.

JALLAGEAS, M. and ASSENMACHER, I. (1974). Thyroid-gonadal interactions in the male domestic duck in relationship with the sexual cycle. Gen. Comp. Endocrinol. 22, 13-20.

JALLAGEAS, M., FOLLETT, B. K. and ASSENMACHER, I. (1974). Testosterone secretion and plasma luteinizing hormone concentration during a sexual cycle in the Pekin duck, and after thyroxine treatment. Gen. Comp. Endocrinol. (In press).

JENKINS, G. M. and WATTS, D. G. (1969). Spectral analysis and its applications. Holden-Day Edit., San Francisco, 525 p.

MARSHALL, A. J. (1970). Environmental factors other than light involved in the control of sexual cycles in birds and mammals. In : La Photoregulation de la Reproduction chez les Oiseaux et les Mammiferes. (J. Benoit and I. Assenmacher edts.). C.N.R.S. Editeur Paris, 1970, p. 53-64.

MENAKER, M. (1970). Rhythms, Reproduction and Photoreception. Biol. of Reprod. 4, 295-308.

OLIVER, J., HERBUTE, S. and BAYLE, J. D. (1973). Repercussions de la desafferentation hypothalamique combinee a la section des nerfs optiques sur le reflexe photosexuel chez la Caille (*Coturnix coturnix japonica*). C.R. Ac. Sc. (Paris), 277, 2021-2024.

OKSCHE, A. (1970). Retino-hypothalamic pathways in Mammals and Birds. In : La Photoregulation de la Reproduction chez les Oiseaux et les Mammiferes. (J. Benoit and I. Assenmacher edts.) C.N.R.S. Editeur, Paris, p. 151-161.

ROWAN, W. M. (1925). Relation of light to bird migration and developmental changes. Nature 115, 494.

THAPLIYAL, J. P. (1969). Thyroid in avian reproduction. Gen. Comp. Endocrinol. Suppl. 2, 11-122.

TIXIER-VIDAL, A. (1965). Caracteres ultrastructuraux des types cellulaires de l'adenohypophyse du Canard male. Arch. Anat. Micr. Morphol. Exper. (Paris), 54, 719-780.

TIXIER-VIDAL, A. and ASSENMACHER, I. (1961a). Etude comparee de l'activite thyroidienne chez le Canard male normal, castre ou maintenu a l'obscurite permanente. I Periode de l'activite sexuelle saisonniere. C.R. Soc. Biol. (Paris), 155, 215-220.

TIXIER-VIDAL, A. and ASSENMACHER, I. (1961b). Etude comparee de l'activite thyroidienne chez le Canard male normal, castre, ou maintenu a l'obscurite permanente. II Periode du repos sexuel saisonnier C. R. Soc. Biol. (Paris), 155, 286-290.

VAN TIENHOVEN, A. (1968). Reproductive Physiology of Vertebrates. W. B. Sanders Co., Philadelphia, 498 p.

VAUGIEN, L. (1960). L'appetit du Moineau domestique s'abaisse graduellement durant le cycle sexuel jusqu'a une valeur nefaste a la gonadostimulation experimentale par illumination, C.R. Ac. Sc. (Paris), 251, 1570-1572.

WIESELTHIER, A. S. and VAN TIENHOVEN, A. (1972). The effect of thyroidectomy on testicular size and on the photorefractory period in the Starling, *Sturnus vulgaris*. J. Ex. Zool. 179, 331-338.

WOITKEWITSCH, A. A. (1940). Dependence of seasonal periodicity in gonadal changes on the thyroid gland in *Sturnus vulgaris*. C.R. Ac. Sc. SSSR, 27, 741-745.

CREDITS

Figure 6, tables II and V. Reproduced by permission from General and Comparative Endocrinology, in press, 1974. Academic Press Inc., New York.

Figure 7. Reproduced by permission from Journal de Physiologie, 62, 219-230, 1970. Masson, Paris, France.

Figures 9 and 10. Reproduced by permission from Act. Endocrinol. Nutrit. France, 14, 109-125, 1970. Expansion Scientifique Francaise, Paris, France.

Figure 8. Courtesy of Dr. D. H. Garnier.

Table III. Courtesy of Dr. F. Gogan.

RELATIONSHIPS BETWEEN CIRCANNUAL RHYTHMS AND ENDOGENOUS LUNAR AND TIDAL RHYTHMS

J. T. ENRIGHT

Scripps Institution of Oceanography
La Jolla, California, 92037

ABSTRACT:

Endogenous annual rhythms have, to date, been documented in a very small number of species, primarily hibernating mammals and migratory birds. Endogenous rhythms which can be synchronized by lunar or tidal factors have also, to date, been conclusively documented in a very small number of species, primarily intertidal invertebrates. Hence, it is not surprising that no organism has been investigated which clearly shows both types of endogenous systems.

Many marine and intertidal species, under field conditions, show tidal and fortnightly or monthly changes in behavior and physiology, and annual changes as well; but it is entirely possible that the vast majority of these phenomena represent exogenous, environmentally-driven rhythms rather than endogenous ones. Of the species which show endogenous lunar and/or tidal rhythmicity most if not all show changes in behavior and physiology associated with season, but the data available have no bearing on the question of endogenous annual rhythms; the simpler alternative, at present, is that such seasonality is environmentally induced.

Even were a case to be documented, however, in which both lunar and annual endogenous systems occur in the same animal, this would not represent convincing evidence for a causal relationship between annual and lunar or tidal rhythms. One might postulate that an endogenous annual rhythm could be derived from a lunar rhythm by frequency "demultiplication", and in principle, this is so, but to me such an hypothesis, which would imply the equivalent of "counting" successive full moons, is no more plausible than the contention that an endogenous annual rhythm is based upon "counting" circadian cycles: far less complicated alternatives seem available.

The one case in which field evidence suggests that the moon might play a role as a synchronizing agent for circannual rhythms is the breeding of the sooty tern on Ascension Island. The evidence in this case is not compelling, since different colonies on the island begin nesting as much as 3 weeks out of phase with each other. Even if substantiated by more data, however, this case would seem to involve an exogenous influence of the moon. There is nothing in these data to suggest endogenous lunar rhythms nor any strong reason to suspect that the moon itself is a direct cause of the 9.3 month reproductive cycle.

Hence, I must conclude that there is at present no good empirical evidence that endogenous tidal and lunar rhythms are related in physiological mechanism with endogenous circannual rhythms, nor any good intuitive reason to suspect such a relationship.

PHYSIOLOGICAL AND BIOCHEMICAL REFLECTIONS OF CIRCANNUAL RHYTHMICITY IN THE EUROPEAN HEDGEHOG AND MAN

JEROME B. SENTURIA and BENGT W. JOHANSSON

Department of Biology and Health Sciences
The Cleveland State University
Cleveland, Ohio 44115

and

Heart Laboratory, General Hospital
S-214 01 Malmo, Sweden

INTRODUCTION

In this paper we have tried to visualize the "grand scheme" for the integration between external stimuli, the existence of an endogenous annual cycle and neuroendocrine control. It is the interaction of these controls that we believe to be responsible for the seasonal changes in the biochemistry and physiology which we have noted in one particular hibernator, the European hedgehog. We have studied the European

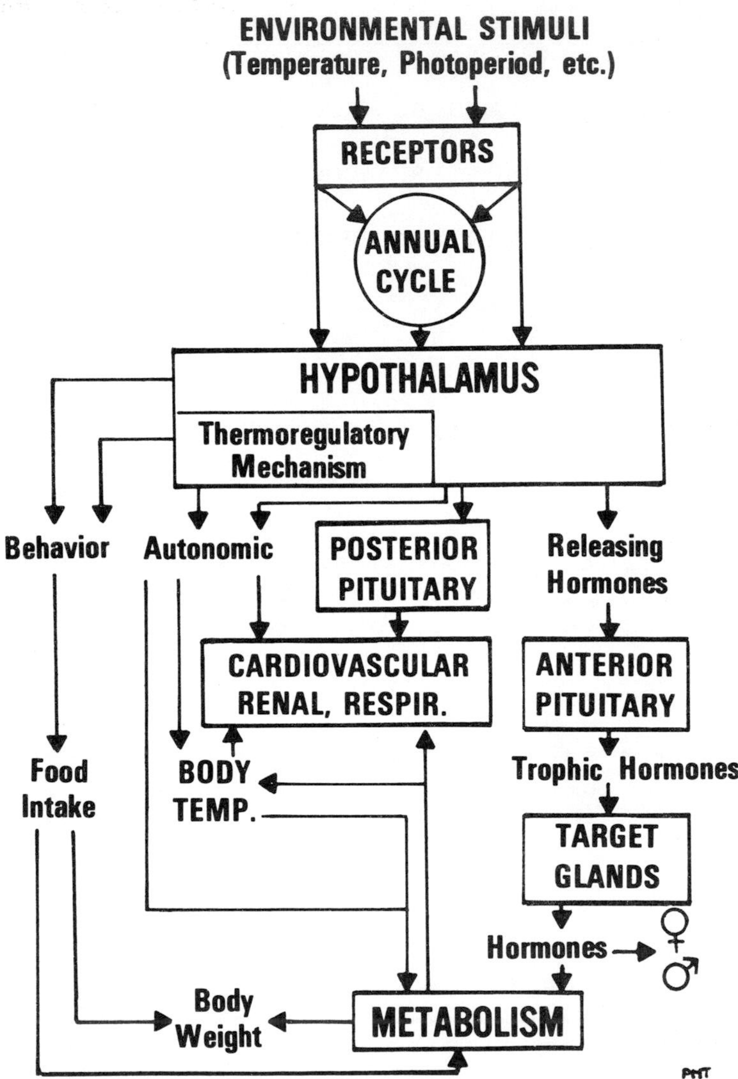

Fig. 1. A scheme showing how the cyclic physiological changes seen in hibernators might be related to hypothalamic control. Feedback loops are not included. Other details in text.

hedgehog's biochemistry and physiology in as much detail as one hibernating species has been studied in a short time in one laboratory. These studies also required the efforts of Bjorn Akesson, Anders Bjernstad, Barbro Eklund, Hans Ericsson, Kai C. Nielsen (now deceased), Sven-Olle R. Olsson, Christer Owman, Maurizio Pandolfi, Wilfried von Studnitz and Jan Thorell. Additional assistance was provided by Harriet-Carole Senturia, Catharine Sjoberg and Julianna E. Szilagyi.

Our studies parallel those of the workers in Finland, Inkovaara and Suomalainen (1973), Kristoffersson (1961, 1965a, 1965b), Kristoffersson, Ahlstrom and Suomalainen (1966), Kristoffersson and Soivio (1964a, 1964b), Kristoffersson, Soivio and Suomalainen (1965), Kristoffersson and Suomalainen (1964, 1969), Laukola and Suomalainen (1971), Saarikoski and Suomalainen (1970, 1971, 1972), Sarajas (1967), Soivio (1967), Suomalainen and Ahlstrom (1970), Suomalainen, Laukola and Seppa (1969), Suomalainen and Rosokivi (1973), Suomalainen and Suvanto (1953), Suomalainen and Saarkoski (1969, 1970, 1971), Suomalainen and Walin (1972), and Uuspaa and Suomalainen (1954). Both sets of studies, however, concentrate on periods less than two years in length. To the best of our knowledge, there exist no studies demonstrating a true endogenous circannual rhythm in the European hedgehog. The ideas leading to the syntheses expressed in this paper have their origins in our work (Johansson and Senturia, 1972), the work of Mrosovsky (1971), the work of Pengelley and Asmundson (1972) and others. Our speculation (fig. 1) is that seasonal cycles such as those shown in the European hedgehog are the results of synchronization of an endogenous cycle by environmental stimuli--primarily photoperiod and temperature, but not excluding others such as food availability, food type, etc. These stimuli would be sensed using receptors such as photoreceptors and thermal receptors. The stimulus could possibly be used to synchronize the phases of the annual cycle to the environment.

By currently accepted theory, thermal receptors act through the hypothalamus. At the level of the hypothalamus, integration of the immediate environment with the annual cycle might well take place. Of possible relevance here is that the hypothalamus is one of the areas of the brain active at low temperatures, at least during arousal (Kayser and Malan, 1963; Mihailovic, 1972). Our comments about the hypothalamus should not be interpreted as a statement that the driving functions originate there, but rather that it represents a common site for regulation of many of the variables associated with the hibernation cycle.

First - It is the seat of autonomic control of thermal regulation. The schemes set forth by Myers and others (Myers and Yaksh, 1972; Luecke and South, 1972: Heath *et al.*, 1972) clearly place the hypothalamus as the locus for these mechanisms of temperature regulation.

Second - The hypothalamus is also the site for integration of many hormonal activities. Releasing hormones from the hypothalamus are responsible for regulating the hormones of the anterior pituitary (Schally, Arimura and Kastin, 1973). Direct neural connections are responsible for release of hormones from the posterior pituitary.

Third - Certain behaviorial mechanisms, especially those relating to thermal regulation and food intake, have their seat in the hypothalamus (Mrosovsky, 1971).

It is the outputs from these mechanisms in the hypothalamus which we believe regulate in large part the seasonal variations in the biochemistry and physiology of the European hedgehog. Such outputs are apparent as changes in food intake, body temperature, body weight, changes produced by autonomic nervous system control, changes in metabolism as produced by changes in hormone levels and food intake, and changes leading to a seasonal reproductive cycle.

Clearly, for a hibernator to be able to synchronize its physiological events to changes in the environment, it must be able to "anticipate" environmental change. Thus, it is not good enough to begin preparation for hibernation when the weather becomes very cold. If the organism is a hibernator, it must begin changing long before the physiologically demanding events of winter arrive to test its preparation.

These preparation changes, which have been noted in many species, constitute a part of the annual cycle. They permit the hibernator to carry out the rest of his annual cycle.

There is no evidence for photoperiodic control of the seasonal physiological changes associated with the hibernation cycle in the European hedgehog. We are in the position of speculating the existence of an endogenous annual rhythm behind these seasonal changes and the role of environmental stimuli in synchronizing this hypothetical annual cycle because the experiments which ask the appropriate questions about the existence of an annual cycle and its "zeitgebers" have not been reported. What we will present here are some of the kinds of seasonal changes which we have observed in the European hedgehog by examining a group of animals for a period of one year.

THE HEDGEHOG

The European hedgehog (*Erinaceus europaeus*) is a very old mammal from an evolutionary point of view. The genus *Erinaceus* appeared in the miocene (about 30 million years ago). The hedgehogs are, together with the shrews, the best known contemporary members of the mammalian order *insectivora*. The primitive *insectivores* have been credited as the stock from which evolved the present placental mammals (Romer, 1959). Hedgehogs are currently found only in Europe, Asia and Africa, although their distribution included North America in earlier times. The European hedgehog now thrives in Norway, Sweden and Finland. Its natural northern distribution follows a latitude of 61°N, but man has been credited with introducing the hedgehog to more northerly areas, changing its distribution to $63^{\circ}30'$ N (Kristoffersson and Soivio, 1966). Hedgehogs probably reached Sweden during the time when there was a land connection between Denmark and Sweden (about 9,000 years ago). The oldest dated findings of hedgehogs in Sweden are from the Malmo area and are dated about 6,000 years ago(Landell, 1971).

The hedgehog's diet consists of insects snails, slugs, snakes, frogs and mice. Peculiar use of the quills and resistance to venom have been seen as adaptations to eating snakes (Bourliere, 1964). The term "*insectivora*" is thus not literal in the modern meaning of the word "insect". This diet, together with the hedgehog's strange quilled appearance, have given rise to some interesting stories. One is the ability to transport apples on its quills; another is that it carries pheasant eggs away on its quills --to feed its young. There is more folklore than truth to these stories (Landell, 1971). Fruits and vegetables are an inadequate diet for the hedgehog. In captivity it is very fond of leftovers from the human diet. Both herring and meatballs seem to be especially palatable to hedgehogs. It is well known in Sweden that the hedgehog likes milk. Placing a saucer of milk in the garden is a popular way of inviting a hedgehog for a visit on a mild summer evening. The hedgehog has much folklore surrounding it. In summary, the hedgehog is loved, awed and respected. When he is well cared for, it is said that he brings good luck; but when he is mistreated, he is believed to bring misfortune, especially to the livestock (Landell, 1971).

The hedgehog is a nocturnal animal. His activity goes from dusk until midnight, with another period of activity sometimes before sunrise. The hedgehog's eye is an all rod eye, being devoid of cones (Bourliere, 1964). The hibernation period is between November and March. Courtship and mating starts in May or June. The hedgehog has a peculiar mating dance with the male circling around the female (fig. 2). The gestation period is 40 days. The quills are very soft at birth--which is fortunate for the mother. The cubs are weaned one month after birth (Landell, 1971).

Scientists studying hibernation make only a small dent in the hedgehog population. It is the automobile that is the hedgehog's major enemy (Landell, 1971; Kristoffersson and Soivio, 1966). The European hedgehog is now a protected species in Sweden.

Fig. 2. Photograph of two hedgehogs in their natural environment. Male (right) is engaged in the "mating dance", circling the female (left) in alternating clockwise and counterclockwise direction (from Landell, 1971). Reproduced with permission.

METHODS AND MATERIALS

The following methods and materials were common to the studies reported in this paper. Details of the study can be found in Johansson and Senturia (1972).

Biological materials - Man was used to compare some of the blood chemistry findings seen in the European hedgehog.

Feeding and caging - The hedgehogs were kept in semisheltered enclosures in large groups (25-60) and provided with hay as bedding. No attempt was made to control temperature in these enclosures and their ambient temperature and light cycle was that of the natural environment.

The seasonal changes in temperature and other meteorological factors are shown in fig. 3.

The hedgehogs were fed the standard human diet from the central kitchen of the Malmo General Hospital. Table I shows the nutritional content of the diet as fed in six week cycles. They were not fed during the winter (hibernating period).

Time of sampling - To compare the annual seasonal phases of the hibernation cycle, samples were taken in four seasonal periods. In each day of a sampling period, no more than two samples of each species were taken. In order to minimize the effect of circadian changes the samples were taken five to six hours after sunrise for the hedgehogs and between 08.00 and 09.00 for man.

All samples were fasting samples. Food was not accessible to the animals after 17.00 the day before sampling. Samples from man were taken before breakfast.

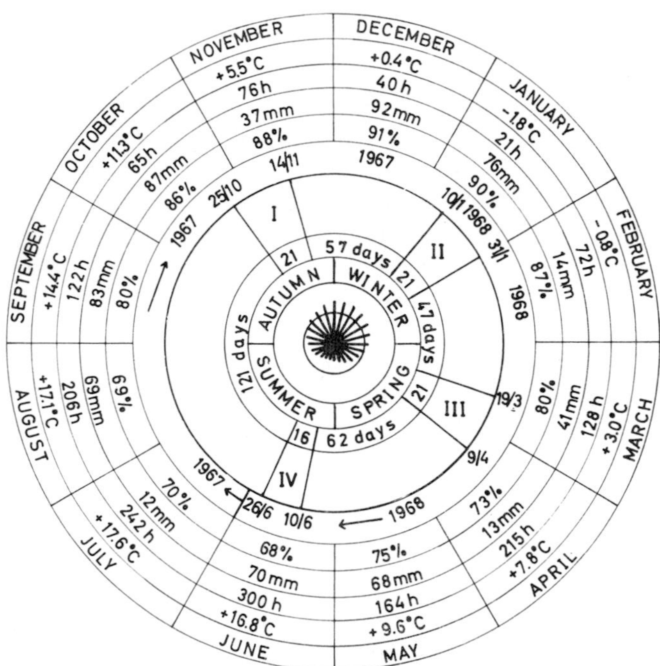

Fig. 3. The temporal relationship between the periods of sampling in this study. Also included are the meteorological conditions associated with the study. It starts July, 1967. In the outermost circle is the month. In the next circle is the average ambient temperature in degrees Celcius. The next circle, hours of sunlight; next precipitation in millimeters; next average percent relative humidity. The next two circles represent the time of sampling. The next circle represents the duration of the sampling period in days and the time between each period. In the next circle the four seasons are presented. In the very center, the time in hours between sunset and sunrise is graphically presented by the length of the radius from the center. Photoperiod data is taken from the 1st and 15th day of each month. The distance from the center to the innermost circle represents 12 hours. The meteorological data is from the Swedish weather burea (Statens meterorologiska och hydrologiska institut). (From Johansson and Senturia, 1972, reproduced with permission).

TABLE I: DAILY NUTRITIONAL CONTENT OF THE DIET FED TO HEDGEHOGS/100 CALORIES

DAILY AVERAGE:	Protein g.	Fat g.	Carbohydrate g.	Ca^{++} mg.	Iron mg.	Vitamin A IE	Thiamin mg.	Riboflavin mg.	Niacin mg.	Vitamin C mg.
Week A	5.02	3.55	11.28	60.63	0.925	141.9	0.0838	0.122	0.872	6.75
Week B	5.18	3.72	10.71	61.86	0.964	288.1	0.0814	0.130	0.945	8.44
Week C	5.09	3.44	11.51	61.85	0.940	303.9	0.0790	0.131	0.906	5.43
Week D	5.00	3.73	10.95	57.15	0.897	224.0	0.0950	0.123	0.882	7.12
Week E	5.19	3.49	11.31	59.41	0.921	326.1	0.0834	0.120	0.868	6.94
Week F	5.18	3.45	11.39	63.71	0.930	149.25	0.0887	0.112	0.838	6.35
Daily Average for 6-week Period	5.11	3.56	11.19	60.75	0.928	238.7	0.0850	0.123	0.884	6.86
Total Caloric Distribution (6-week Average)	Protein %	Fat %	Carbohydrate %							
	21	33	46							

In general, each sample consisted of six males and six females of each species. The samples were taken as follows:

Season I.

Fall period: Some of the hedgehogs are just beginning to enter hibernation.

October 25-November 14 between 12.00 and 13.00 hours.

Season II.

Winter period: All hedgehogs are hibernating.

January 10-31 between 13.30 and 14.30 hours.

Season III.

Spring period: Most hedgehogs have completed their final arousal from hibernation.

March 19-April 9 between 11.00 and 12.00 hours.

Season IV.

Summer period: All hedgehogs are active.

June 10-26 between 08.30 and 09.30 hours.

Figure 3 shows the temporal relationship between these periods and the changes in meteorological conditions. It starts January 7th, 1967 and ends June 30th, 1968. In the outermost circle is the month, in the next circle is the average ambient temperature in degrees Celcius, next number of hours of sunlight, next precipitation in millimeters, next the average percent relative humidity. The next two circles represent the time of sampling. The next circle represents the duration of the sampling period in days and the time between each period. In the next circle the four seasons are presented and in the very center the time in hours between sunset and sunrise is graphically presented by the length of the radius from the center. The data for photoperiod is taken from the 1st and 15th day of each month. The distance from the center to the innermost circle represents 12 hours. The meteorological data is from the Swedish weather bureau (Statens meteorologiska och hydrologiska institut).

Anesthesia - The hedgehogs, except during season II, were anesthetized with 3.3 ml/kg of chloral hydrate, sterile water qs to 100 ml). Hibernating animals were not anesthetized.

General measurements - Before anesthesia each animal was weighed to the nearest gram and superficial abdominal body temperature was measured to the nearest 0.1 C using a thermocouple (Ellab, Copenhagen). These data plus the date, time of day and sex of the animal were recorded.

Blood samples - Ten milliliters of venous blood was drawn from each unanesthetized human subject using a cannula without syringe, allowing the blood to flow directly into the sample tube. This technique was used to minimize hemolysis and the consequent changes in serum ion ratios. This was the only sample taken from the human subjects in this study.

After anesthesia the thoracic cavity of the hedgehog was opened and a hypodermic needle size no. 1.8 was placed in the aortic arch of the hedgehog.

1. Whole blood was taken by micropipette for enzymatic blood glucose determination.

2. As much as possible of the blood remaining in the animal was then removed by syringe. The blood was allowed to clot. This was then centrifuged at 1600 rpm for 10 minutes. The resulting supernatant was centrifuged at 1500 rpm for 5 minutes.

The serum so obtained was divided for the following analyses:

A. Urea.
B. Insulin.
C. Total fat.
D. Cholesterol.

Tissue samples - Tissue samples were taken from the heart, liver, adrenals, pancreas, testes, vas deferens, uterus, thyroid gland, pituitary, brown adipose tissue and white fat for weight, concentrations of certain enzymes and fluorescence histochemistry for sympathetic innervation and electron microscopy (for details see Johansson and Senturia, 1972).

Data handling and statistical methods - Wherever the results from the studies reported here yielded digital results amenable to statistical analysis, they were analyzed together with the results from other tests by means of a computer program, "Season". This program, modified from units of the IBM Scientific Subroutine Package was run on the Cleveland State University IBM 360/40 computer. The Program allowed selection of data on the basis of the other parameters of that sample, e.g. season, sex and non zero value. The output gave mean, standard deviation, standard error of the mean, maximum, minimum, sum, sum of squares, number of samples for each variable with each season and Students "t", "F" values and degrees of freedom for comparison of significance of difference between season, sex and species.

RESULTS AND DISCUSSION

Body Temperature

Not to belabor the obvious, body temperature was lowest in winter (fig. 4). The mean body temperature of the hedgehogs was lowest in the hibernating group (Season II) and slightly depressed in the pre- and post-hibernating season groups (I and III); this is accounted for by those animals which were just entering hibernation in Season I and those just leaving hibernation in Season III.

Body Weight

We noted a cycle of body weight, but lower in amplitude than that observed by Kristoffersson (1964). The body weight was highest in fall and lowest in spring (fig. 5).

Autonomic Nervous System

1. Liver and heart glycogen content were high in winter, low in summer. This might indicate a decreased sympathetic nervous system effect on metabolism (figs. 6 & 7).

2. Noradrenaline concentration was highest in summer in both heart and adrenal. Heart noradrenaline decreased in fall (figs. 8 & 9).

3. Adrenal adrenaline concentration was essentially constant (fig. 10).

4. Uterine sympathetic innervation, as measured by flourescence histochemistry, increased in spring and summer (figs. 11 & 12). It is possible that the increased level of neuronal noradrenaline in the uterus of the hedgehog during its active period

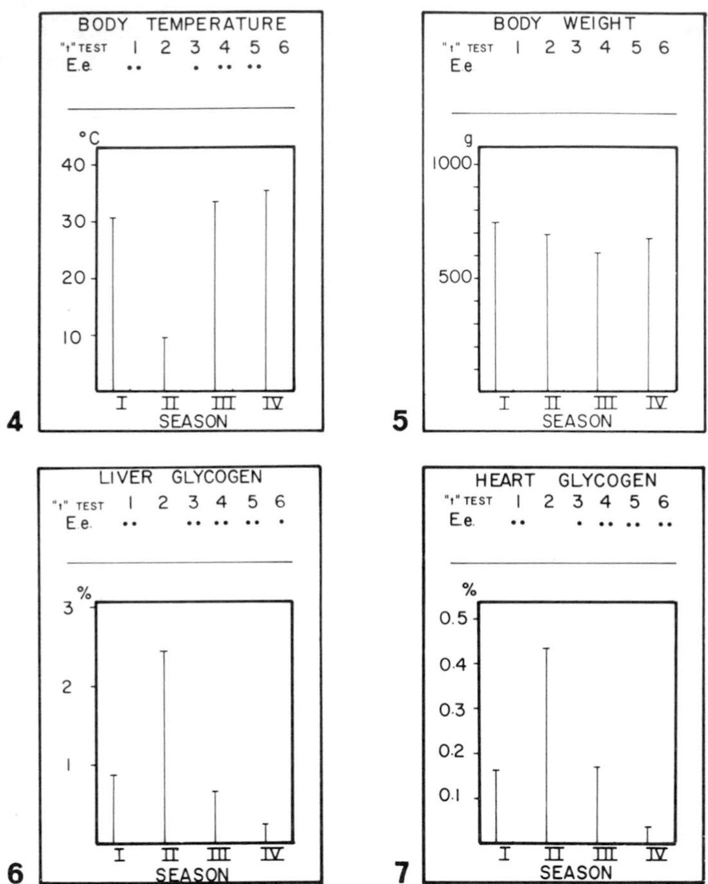

Fig. 4. Body temperature in degrees Celsius. Season refers to the prehibernating (I), hibernating (II), posthibernating (III) and active non-hibernating (IV) phase. See further fig. 3. The solid line refers to hedgehogs. The number of * denote the significance level according to the "t" test, one * means $0.05 > P > 0.01$ and two ** $P < 0.01$. E.e. refers to the hedgehog *(Erinaceus europaeus)*. 1 indicates the difference between seasons I and II, 2 between I and III, 3 between I and IV, 4 between II and III, 5 between II and IV and 6 between III and IV. (From Johansson and Senturia, 1972, reproduced with permission.)

Fig. 5. Body weight. See fig. 4 for legend (From Johansson and Senturia, 1972, reproduced with permission).

Fig. 6. Liver glycogen. See fig. 4 for legend (From Johansson and Senturia, 1972, reproduced with permission).

Fig. 7. Heart glycogen. See fig. 4 for legend (From Johansson and Senturia, 1972, reproduced with permission).

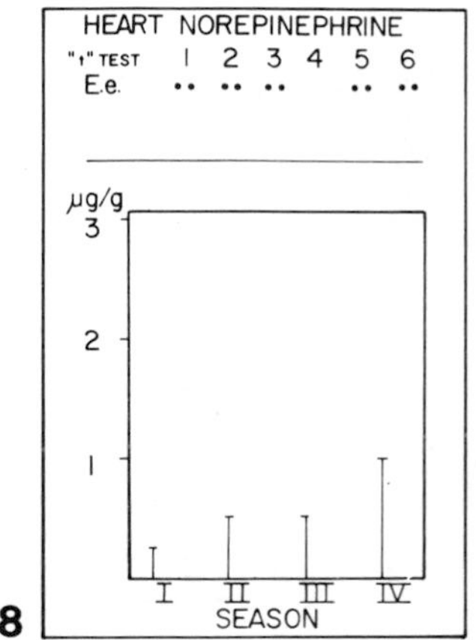

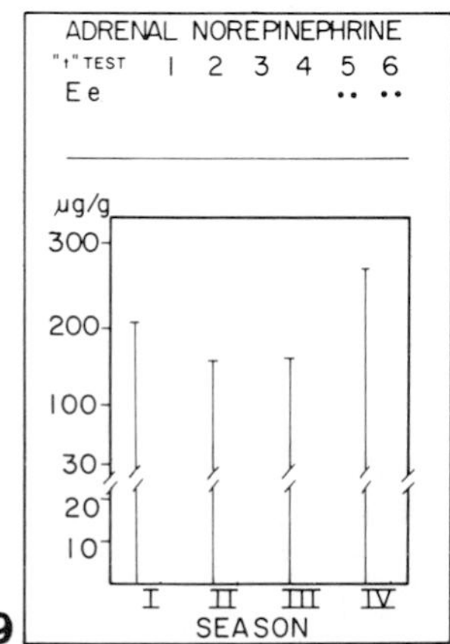

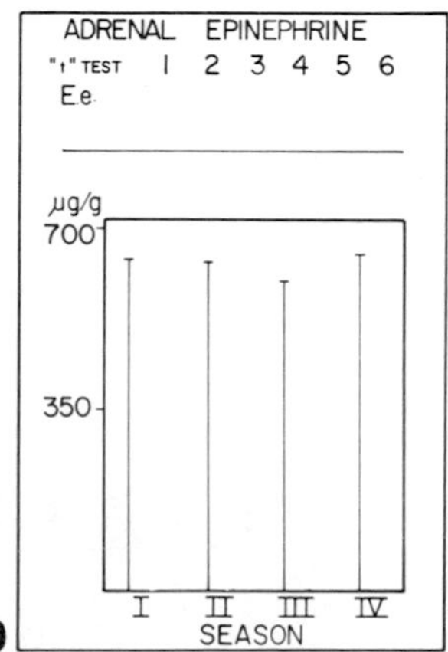

Fig. 8. Heart Norepinephrine (noradrenaline). See fig. 4 for legend (From Johansson and Senturia, 1972, reproduced with permission).

Fig. 9. Adrenal Norepinephrine (Noradrenaline). See fig. 4 for legend (From Johansson and Senturia, 1972, reproduced with permission).

Fig. 10. Adrenal Epinephrine (Adrenaline). See fig. 4 for legend (From Johansson and Senturia, 1972, reproduced with permission).

Fig. 11. Winter hedgehog, transverse section of uterus. The more peripheral part of the organ circumference is in the upper portion of the figure, more centrally located parts (in the direction of the endometrium) in the lower portion. Magnification indicator 100u. Adrenergic innervation in the myometrium is well distributed as indicated by the green fluorescence (light colored in this photograph). (From Johansson and Senturia, 1972, reproduced with permission).

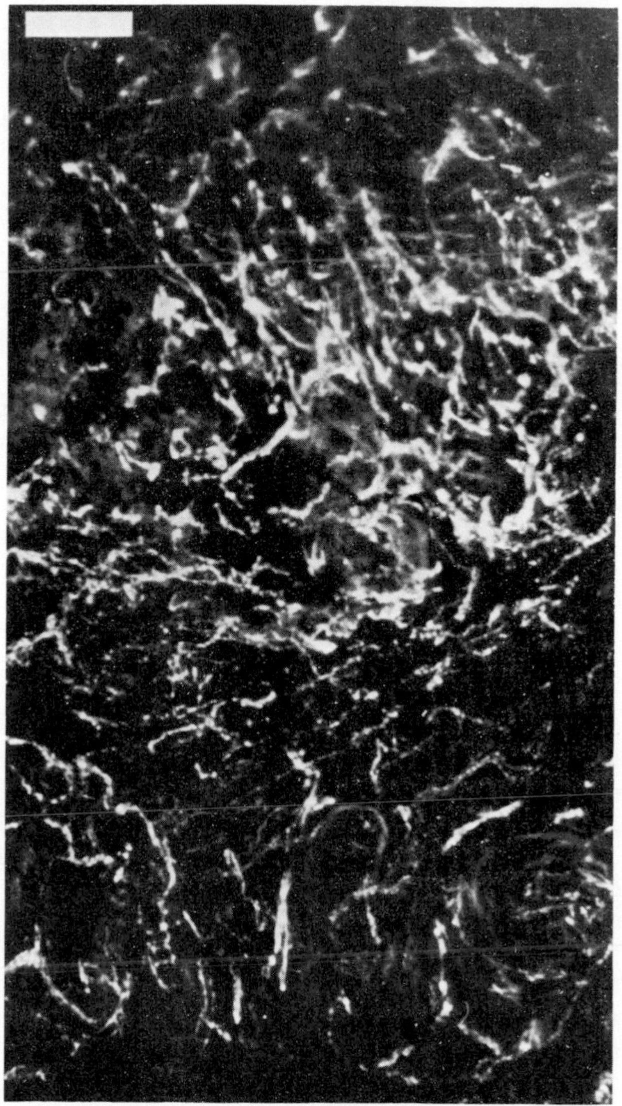

Fig. 12. Summer hedgehog, transverse section of uterus. Orientation and magnification same as fig. 11. The amount of adrenergic nerve terminals has increased markedly. This is well illustrated by the very dense adrenergic nerve supply in the outer parts of the myometrium. (From Johansson and Senturia, 1972, reproduced with permission.)

is a natural counterpart to the higher content of uterine noradrenaline that can be induced by estrogen treatment of the rabbit (Sjoberg, 1968b; Falck *et al*, 1969).

5. No seasonal changes were noted in the sympathetic innervation to the anterior pituitary, thyroid and vas deferens.

Endocrine Changes

1. Adrenal weight was lowest in the winter season (fig. 13). This finding is similar to that obtained in Mexican ground squirrels (fig. 14).

2. There is a marked reduction of the blood glucose value in the hedgehog during hibernation. Part may be accounted for by the decreased glycogenolysis implied by the glycogen data presented above. The lack of food intake during hibernation may also account for the low blood glucose. Variation in blood glucose has previously been shown in the hedgehog by Clausen (1965) and in great seasonal detail by Saarikoski and Suomalainen (1970). The non-hibernating values during summer are low in our study and this is also true for man (fig. 15). A similar result may be the increase in blood glucose seen in hedgehogs by Saarikoski and Suomalainen (1970) between May and September.

We were interested in following not only the blood glucose but also the insulin values since it has been claimed that insulin is the pharmacological agent that induces hibernation. Indeed, the low blood glucose values during hibernation fit this theory. To our surprise our studies failed to detect any immunoreactive insulin activity at all and this was so during all the periods. The method used was a modification of the radio-immuno assay method according

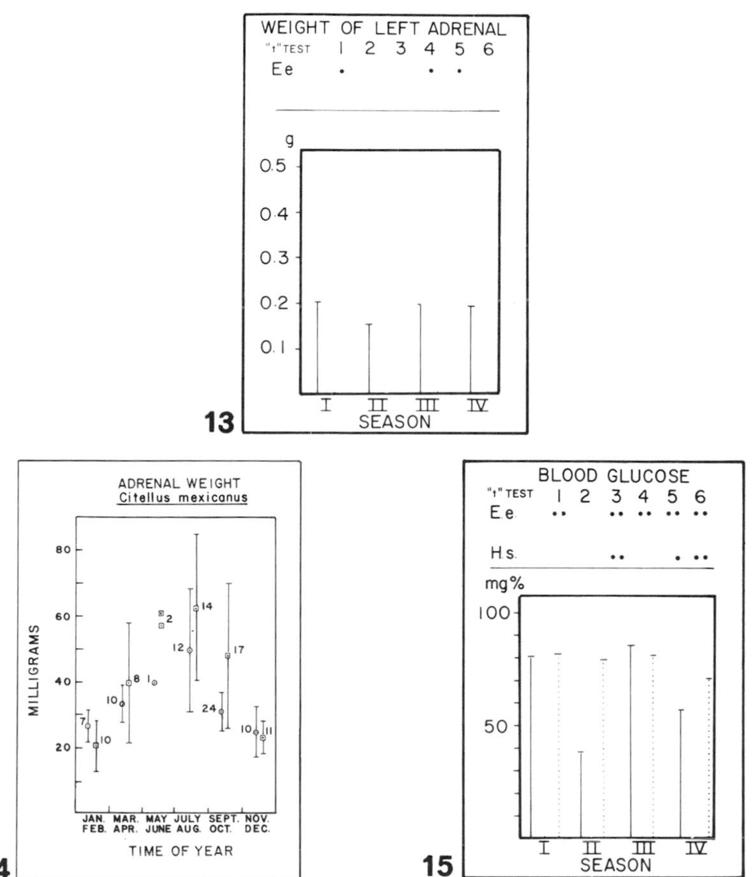

Fig. 13. Weight of left adrenal of the hedgehog. See fig. 4 for legend. (From Johansson and Senturia, 1972, reproduced with permission.)

Fig. 14. Weight of adrenals of *Citellus mexicanus.* Circles represent males, squares represent females. Vertical lines represent standard deviation. Numbers of animals sampled are shown next to each mean point. Conditions the squirrels were kept under are reported in Senturia, Stewart and Manaker (1970).

Fig. 15. Blood glucose in mg per 100 ml. Season refers to the prehibernating (I), hibernating (II), posthibernating (III) and active non-hibernating (IV) phase. See further fig. 3. The solid line refers to hedgehogs and the dotted line to man. The number of * denote the significance level according to the "t" test, one * means $0.05 > P > 0.01$ and two ** $P < 0.01$. E.e. refers to the hedgehog (*Erinaceus europaeus*), and H.s. to man (*Homo sapiens*). 1 indicates the difference between season I and II, 2 between I and III, 3 between I and IV, 4 between II and III, 5 between II and IV and 6 between III and IV. (From Johansson and Senturia, 1972, reproduced with permission.)

to Yalow and Berson (1960). Two possible explanations presented themselves.

a. The insulin activity in hedgehogs is so low that it cannot be assayed with this method, which in fact is highly sensitive.

b. The structure of the insulin molecule is so different from that found in other species that the immunological assay procedure is not applicable. This explanation, however, is not very probable because insulin activity can be assayed with this technique in many species, including fish.

To get an answer to these questions we continued with a further series of experiments. Extracts were prepared from the pancreas and these consistently displayed high concentrations of immunoreactive insulin. These values are shown in Table II. There was no difference between the values found in January during hibernation and those found in November just prior to hibernation. The values from both these seasons were much lower than the values obtained during summer.

These results do not support the insulin theory but they show that there is insulin in the hedgehog which can be assayed. A very low peripheral concentration might explain our negative results with blood plasma. Since a glucose load might be expected to increase the insulin activity, the hedgehogs were given 1 g/kg body weight i.v. of a 25% glucose solution. Blood glucose and plasma insulin were determined at different intervals after the load. It will be seen in Table III that the blood glucose levels increased after the glucose administration and it was possible to assay insulin activity in two of the three animals.

It still seemed rather surprising that the insulin activity should be so low

Table 2

Date of collection	No. of animals	Pancreatic weight gram	Insulin		Significance of differences in insulin/pancreatic weight in relation to seasonal group	
			mU/g of pancreas	mU in total pancreas		
5/11	10	1.79±0.65	156±20	250±71	0.05	16/1
					0.001	1/8
6/1	12	2.38±0.76	173±30	415±260	0.05	5/11
					0.001	1/7
6/7	6	5.18±1.22	762±425	1040±390	0.001	5/11
					0.001	16/1

Extractable insulin in hedgehog pancreas at various seasons. All figures given are mean SD.
(From Johansson and Senturia, 1972, reproduced with permission)

Table 3

Date of test	Type of test		Before	10 min.	20	40
27/7	G	Insulin uU/ml	3	3	3	3
		Glucose mg/100 ml	122	330	300	264
13/8	G	"	3	36	40	40
			64	230	193	164
13/8	G	"	3	3	3	25
			112	527	339	298
15/1	G+P	"	3	6	25	150
			93	790	636	613
15/1	G+P	"	3	22	42	65
			7	417	304	291

Plasma insulin and blood glucose at various intervals after a glucose load in non-hibernating hedgehogs. Two types of tests were performed: G = Glucose load, 1 g/kg body weight intravenously. G+P = the same glucose load, and continuous infusion of phentolamine, 0.03 mg/kg body weight and minute.
(From Johansson and Senturia, 1972, reproduced with permission.)

despite very high blood glucose values, e.g. 527 mg % and no insulin activity in one of the animals. The hedgehogs were anesthetized, and this might be an explanation since the anesthesia might induce a stress reaction which blocks the activity of the islets of Langerhans. The blocking is transmitted via the α-adrenergic receptors. The glucose load experiments were then repeated after we had given an α-adrenergic blocking agent, phentolamine, in a dose of 0.03 mg/kg body weight per minute intravenously. This study was performed in December in animals in which arousal from hibernation had been induced. The results show good insulin values after the pancreas block had been inhi bited.

Summarizing, we were able to assay insulin activity in hedgehogs but the results hardly support the theory that hibernation is induced by a high insulin activity, since the highest pancreas values were obtained in non-hibernating summer animals.

We have discussed the insulin results in some detail to illustrate the difficulties one can run into. Anesthesia necessary for the collection of blood or tissue samples may profoundly alter the "milieu interieur", thus concealing possible circannual changes.

Reproductive Changes

1. Testis weight is lowest in fall and high in spring and summer (fig. 16). The increase in testis weight seen during winter is like that seen in the Mexican ground squirrel (fig. 17). These data are suggestive of continued activity of the hypothalamic-pituitary axis during hibernation. Thus both species of hibernator, although phylogenetically diverse and widely separated by

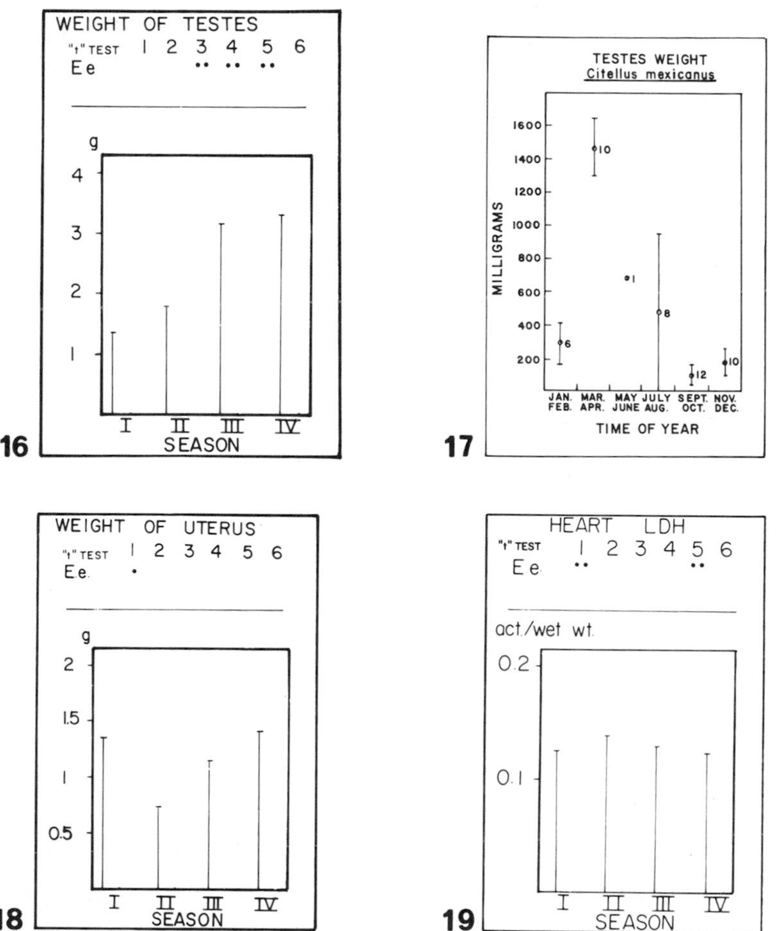

Fig. 16. Testes weight of hedgehogs. See fig. 4 for legend. (From Johansson and Senturia, 1972, reproduced with permission.)

Fig. 17. Weight of testes of *Citellus mexicanus.* Vertical lines represent standard deviation. Number of animals sampled are shown next to each mean point. Conditions the squirrels were shown next to each mean point. Conditions the squirrels were kept under are reported in Senturia, Stewart and Menaker (1970).

Fig. 18. Weight of uterus of th. hedgehog. For legend see fig. 4. (From Johansson and Senturia, 1972, reproduced with permission.)

Fig. 19. Activity of LDH from heart tissue. For legend, see fig. 4. (From Johansson and Senturia, 1972, reproduced with permission.)

latitude (Malmo, Sweden, 55 35'N compared to Austin, Texas, 30 18'N) show similar patterns of testis weight increase. It is assumed that this is an adaptation to provide an effective reproductive response in spring, soon after final arousal from hibernation. It would be exciting to examine levels of gonadotropins and releasing hormones during hibernation.

2. Uterine Weight is lowest in winter (fig. 18).

3. As noted above, uterine sympathetic innervation was highest in spring and summer.

Metabolic Changes

1. There is an increased capability for anaerobic metabolism in winter as evidenced by the increase in LDH activity in heart and liver (figs. 19 & 20).

2. In fall low blood urea (fig. 21) and low alpha-glycerol phosphate dehydrogenase in heart (fig. 22) may be reflections of increased anabolic function. Our absolute levels of urea are higher than those reported by Kristoffersson (1965b) and Clausen (1964) but our conclusions are the same.

3. There is increased lipid metabolism during winter as evidenced by high glucose-6-phosphate dehydrogenase in heart and liver (figs. 23 & 24). Serum lipid is high in winter (fig. 25). The high level of malic dehydrogenase in heart and liver (figs. 26 & 27) in winter might indicate a shift to gluconeogenesis; serum cholesterol (fig. 28) was highest in winter and lowest in summer. There was no seasonal change in degree of saturation in either brown or white fat (Table IV).

4. Brown adipose tissue, which is assumed to play a major role in arousal from

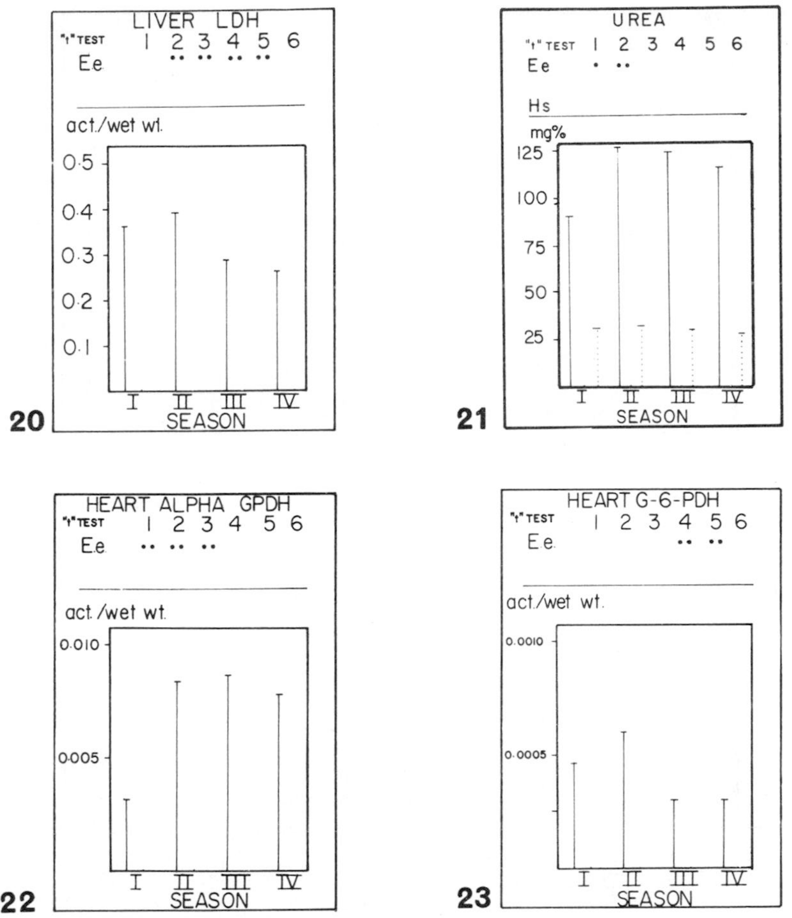

Fig. 20. Activity of LDH from liver tissue. For legend, see fig. 4. (From Johansson and Senturia, 1972, reproduced with permission.)

Fig. 21. Serum urea in milligrams per 100 ml. See fig. 4 for legend. (From Johansson and Senturia, 1972, reproduced with permission.)

Fig. 22. Alpha glycerol phosphate dehydrogenase activity from heart tissue. See fig. 4 for legend. (From Johansson and Senturia, 1972, reproduced with permission.)

Fig. 23. Glucose-6-phosphate dehydrogenase activity from heart tissue. See fig. 4 for legend. (From Johansson and Senturia, 1972, reproduced with permission.)

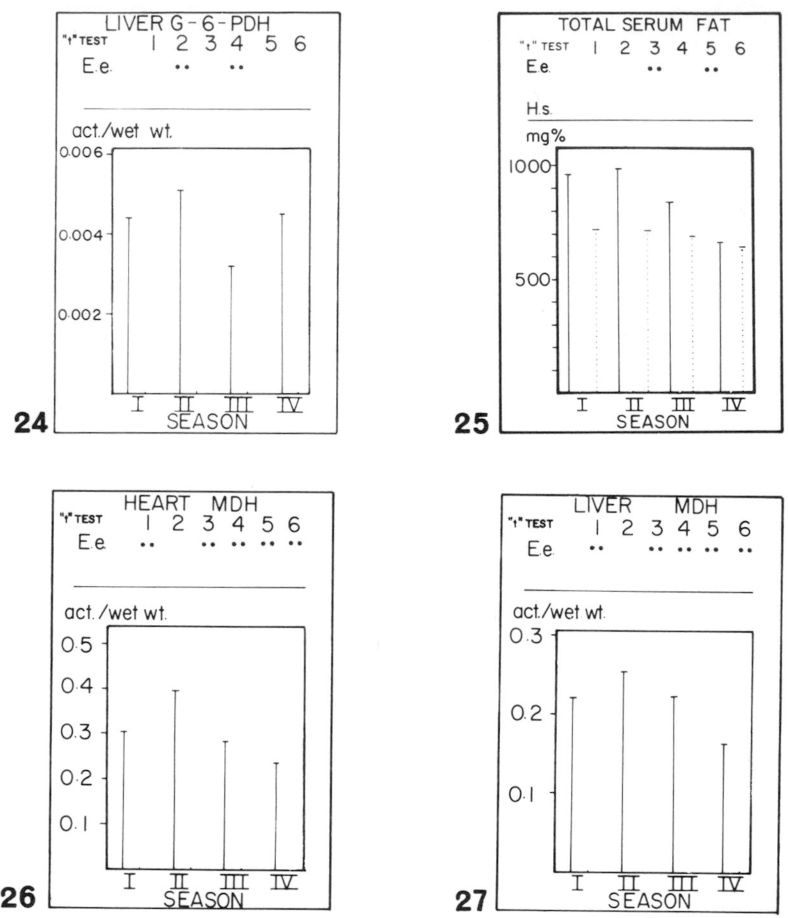

Fig. 24. Glucose-6-phosphate dehydrogenase activity from liver tissue. See fig. 4 for legend. (From Johansson and Senturia, 1972, reproduced with permission.)

Fig. 25. Total serum fat in mg %. See fig. 15 for legend. (From Johansson and Senturia, 1972, reproduced with permission.)

Fig. 26. Malic dehydrogenase activity from heart. See fig. 4 for legend. (From Johansson and Senturia, 1972, reproduced with permission.)

Fig. 27. Malic dehydrogenase activity from liver. See fig. 4 for legend. (From Johansson and Senturia, 1972, reproduced with permission.)

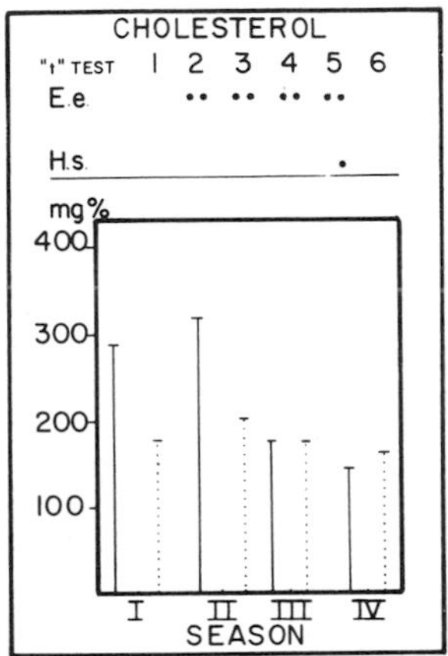

Fig. 28. Serum total cholesterol in milligrams per 100 milliliters. See fig. 15 for legend. (From Johansson and Senturia, 1972, reproduced with permission.)

Table 4

Mean double bond content per mole fatty acid during different seasons (Data from Johansson and Senturia, 1972).

Tissue	Season I	II	III	IV
white fat	0.76	0.74	0.76	0.82
brown fat	0.86	0.83	0.82	0.80

hibernation, changes with the seasons (fig. 29). The tissue is largest in fall and smallest in summer. The number of mitochondria in this tissue is greatest in winter, coordinating well with its thermogenic role.

The above data clearly show that there exist annual variations in the physiology and biochemistry of the hedgehog.

CORRELATIONS IN MAN

Our results from the human material have produced evidence of annual variations. We believe our blood glucose results with lower values in June (fig. 15) show how important it is, when reference values or normal values are established, to take into consideration not only age, sex, fasting state, etc., but also to be aware of the importance of the season when the sample is obtained. These results have been checked against two years of glucose tolerance control values from the daily routine of the Malmo General Hospital, confirming the seasonal variations. Reinberg (this symposium) has suggested that circadian peaks of blood glucose may change on an annual basis. This might give rise to a false annual rhythm of blood glucose. Regardless of the source of the annual variation, the practical implications remain.

Cholesterol is another example (fig. 28) with a higher value in man during winter of 201 mg % against a lowest value of 162 mg % during the summer. Cholesterol is an important parameter in the treatment of coronary heart disease and the therapeutical implication of these large differences are apparent to everybody. The extreme seasonal changes in serum cholesterol in the hedgehog are also of interest. Hibernators might be

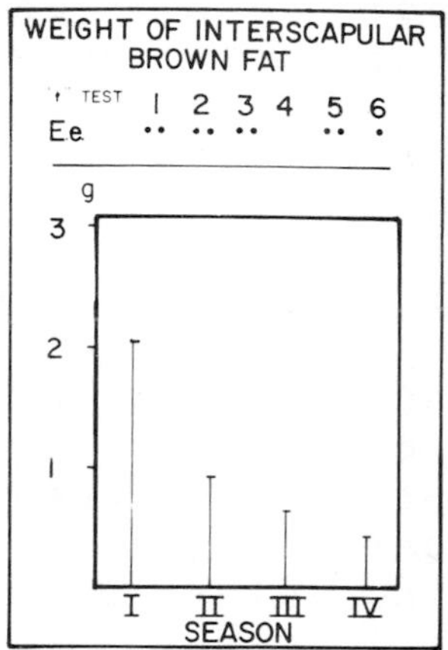

Fig. 29. Weight of superficial lobe of interscapular brown fat from the hedgehog. See fig. 4 for legend. (From Johansson and Senturia, 1972, reproduced with permission.)

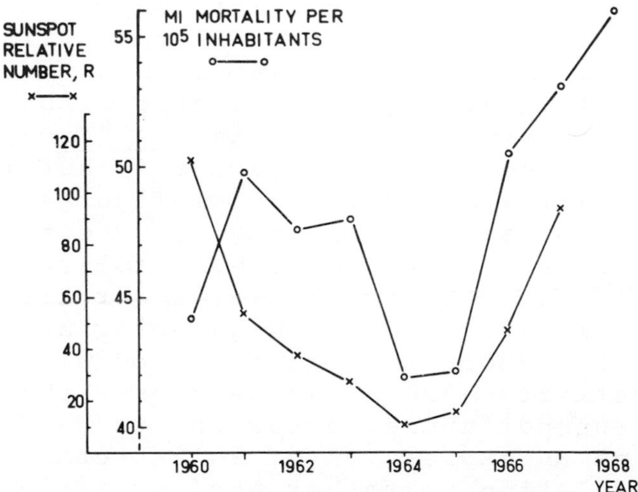

Fig. 30. Mortality figures in acute myocardial infarction in Malmö and the sunspot number according to Waldmeier.

interesting models in the field of experimental coronary heart disease.

Circadian clocks are generally accepted. Circannual clocks have more and more been placed in focus, thanks to the energetic work of Dr. Pengelley. Other longer term rhythms might well play a role in human biology. Verfaillie's experiments with rice seedlings are of interest in this connection (Verfaillie, 1969). He studied the growth rate of rice seedlings and observed a very strong negative correlation between the growth rate and the P indices of the chemical test of Piccardi. This chemical test is a way of measuring solar activity as reflected in, for example, sun flare effects. Are there other fields in which solar activity with its typical eleven year cycle might be of importance, thus superimposing another rhythm on the circannual one? Perhaps there are. Malmo is a town of about a quarter of a million inhabitants which is served by just one hospital. This means that all patients with acute diseases such as acute myocardial infarction are treated in one hospital. This also means that all hospital records are kept in one place and therefore easily available and incidence figures obtained from Malmo are highly representative. Against this background we have followed the mortality in acute myocardial infarction in Malmo from a period with high solar activity --1960 (fig. 30)--through a period with low solar activity--1964. We have used the relative sunspot number according to Waldmeier (1968) as an expression of solar activity. The correlation is rather good.

We are in the lucky situation in Malmo to have incidence figures including mortality for acute myocardial infarction from 1935-1968, i.e. a time period with three

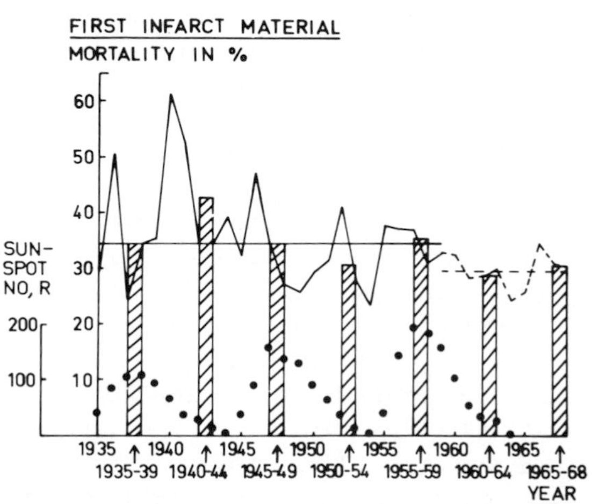

Fig. 31. Mortality figures in acute myocardial infarction in Malmö 1935-1968. Refers to first infarcts only. Hospital mortality per year (curve) and five year period (bars). The horizontal lines denote the mean mortality for the period 1935-1939 (———) and 1960-1968 (- - -). The dotted line indicates the annual sunspot numbers four cycles 17-19 (Dodson *et al,* 1965).

full eleven year periods of solar activity (Johansson, 1972). In fig. 31 these two variables are plotted against each other. It is perhaps premature to assign a causal relation between the two variables.

The reason for including sun flare activity and acute myocardial infarction in Malmo in this paper is to draw your attention to the fact that phenomena such as the sun flare activity with its eleven year period might have biological effects. These long term rhythms might conceal some expressions of the circannual clocks, thus making the results more difficult to interpret.

CONCLUDING COMMENT

We feel that we have raised many more questions than we have answered. The ability to compare changes in variables within one experimental sample makes our approach valuable. We hope that studies such as these can be extended to examine more closely the hypothalamic, endocrine interaction, and we would be excited to see an endogenous circannual clock clearly demonstrated in the European hedgehog. The application of these techniques to the study of human biology should be extended. Further evidence for the influences of the environment on the long term timing of physiological events in man is needed.

SUMMARY

1. A scheme for the interpretation of annual physiological changes in hibernating mammals is presented. It is speculated that seasonal cycles such as those shown in the European hedgehog *Erinaceus europaeus* are the results of synchronization

of an endogenous circannual cycle by environmental stimuli. At the level of the hypothalamus, integration of the immediate environment with the output of the endogenous driving cycle might well take place. Outputs from the hypothalamus are stated to regulate, in large part, the seasonal variations in the physiology and biochemistry of the European hedgehog. Such outputs are apparent as changes in food intake, body temperature, body weight, changes in autonomic nervous control, changes in metabolism as produced by changes in hormone levels, food intake and body temperature, and changes leading to a seasonal reproductive cycle.

2. Some aspects of the natural history of the European hedgehog are discussed.
3. Methods of a study to examine some of the seasonal changes in the physiology and biochemistry of the European hedgehog are presented.
4. Samples were also taken from human subjects for blood glucose, cholesterol and total serum lipid.
5. Results showed seasonal changes in body temperature, body weight, sympathetic nervous system function and adrenal weight.
6. Pancreatic insulin values were highest in summer hedgehogs.
7. Blood glucose was lowest in winter, with another low point in summer.
8. Changes were noted in testes weight which might indicate activity of the hypothalamic-pituitary axis during hibernation.
9. A cycle was also clearly apparent in uterine weight and sympathetic innervation to the uterus.
10. A number of biochemical changes were

noted giving a general effect of increased anabolic activity in fall, a switch to fat metabolism in winter and an increased metabolic tolerance to anoxia in winter.

11. Data for man are presented showing lowered blood glucose and cholesterol values in summer.
12. The possibility of longer term rhythms in man is raised by the presentation of data on the correlation between the mortality in myocardial infarction in Malmo, Sweden (1935-1968) and the cycles of solar activity.

REFERENCES

BOURLIERE, F. (1964). The Natural History of Mammals, third edition, revised. Alfred A. Knopf, New York, 387 pp.

CLAUSEN, G. (1964). Excretion of water, urea and amino acids in the hedgehog *Erinaceus europaeus L.* in fasting, hibernating and normal state. Arb. Univ. Bergen, Mat. Naturv. Serie 1963 No. 24.

CLAUSEN, G. (1965). Plasma glucose in the hibernator *Erinaceus europaeus L.* as related to rectal temperature. Arb. Univ. Bergen, Mat. Naturv. Serie 1964 No. 18.

DODSON, H. W., HEDEMAN, E. R. & STEWART, F. L. (1965). Solar Activity during the First 14 Months of the International Years of the Quiet Sun. Science 148, 1328-1331.

INKOVAARA, P. & SUOMALAINEN, P. (1973). Studies on the physiology of the hibernating hedgehog. 18. On the leukocyte counts in the hedgehog's intestine and lungs. Ann. Acad. Sci. Fenn. A, IV Biologica 200, 1-21.

HEATH, J. E., WILLIAMS, B. A., MILLS, S. H. & KLUGER, M. J. (1972). The responsiveness of the preoptic anterior hypothalamus to temperature in vertebrates. In South, Frank E. *et al*, 1972. Hibernation and Hypothermia, Perspectives and Challenges, Elsevier, New York, pp. 605-627.

JOHANSSON, B. W. (1972). Myocardial infarction in Malmo 1960-1968. Acta Med. Scand. 191, 505-515.

JOHANSSON, B. W. & SENTURIA, J. B., eds. (1972). Seasonal variations in the physiology and biochemistry of the

European hedgehog *Erinaceus europaeus* including comparison with non-hibernators, guinea pig and man. Acta Physiol. Scand. Suppl. 380:1-159.

KAYSER, C. & MALAN, A. (1963). Central Nervous System and Hibernation. Experientia XIX:441-496.

KRISTOFFERSSON, R. (1961). Hibernation of the hedgehog *Erinaceus europaeus*. The ATP and O-Phosphate levels in the blood and various tissues of hibernating and non-hibernating animals. Ann. Acad. Sci. Fenn. A, IV Biologica 50, 1-45.

KRISTOFFERSSON, R. (1965a). Hibernation of the hedgehog *Erinaceus europaeus L.* Blood creatinine and creatinine levels in relation to seasonal and hibernation cycles. Ann. Acad. Sci. Fenn. A, IV Biologica 95, 1-10.

KRISTOFFERSSON, R. (1965b). Hibernation in the hedgehog *Erinaceus europaeus L.* Blood urea levels after known lengths of continuous hypothermia and in certain phases of spontaneous arousals and entries into hypothermia. Ann. Acad. Sci. Fenn. A, IV Biologica 96, 1-8.

KRISTOFFERSSON, R., AHLSTROM, A., & SUOMALAINEN, P. (1966). Studies on the physiology of the hibernating hedgehog. 4. Cerebral free amino acids in hibernating and non-hibernating animals. Ann. Acad. Sci. Fenn. A, IV Biologica 104, 1-9.

KRISTOFFERSSON, R. & SOIVIO, A. (1964a). Hibernation of the Hedgehog *Erinaceus europaeus L.* The periodicity of hibernation of undisturbed animals during the winter in a constant ambient temperature. Ann. Acad. Sci. Fenn.

A, IV Biologica 80, 1-22.

KRISTOFFERSSON, R. & SOIVIO, A. (1964b). Hibernation in the Hedgehog *Erinaceus europeaus*. Changes of respiratory pattern, heart rate and body temperature in response to a gradually decreasing or increasing ambient temperature. Ann. Acad. Sci. Fenn. A, IV Biologica 82, 1-17.

KRISTOFFERSSON, R., SOIVIO, A. & SUOMALAINEN, P. (1965). Studies on the physiology of the hibernating hedgehog. 3. Changes in the water content of various tissues in relation to seasonal and hibernation cycles. Ann. Acad. Sci. Fenn. A, IV Biologica 92, 1-17.

KRISTOFFERSSON, R., SOIVIO A. & SUOMALAINEN, P. (1966). The Distribution of the Hedgehog *Erinaceus europaeus L.* in Finland 1964-1965. Ann. Acad. Sci. Fenn. A, IV Biologica 102, 1-12.

KRISTOFFERSSON, R. & SUOMALAINEN, P. (1964). Studies on the physiology of the hibernating hedgehog. 2. Changes of body weight of hibernating and non-hibernating animals. Ann. Acad. Sci. Fenn. A, IV Biologica 76, 1-11.

KRISTOFFERSSON, R. & SUOMALAINEN, P. (1969). Studies on the physiology of the hibernating hedgehog. 9. Alternations in serum proteins in hibernating and non-hibernating hedgehogs. Ann. Acad. Sci. Fenn. A, IV Biologica 158, 1-9.

LANDELL, N.-E. (1971). Hur gar det med igelkotten? Bonniers, Stockholm, 112 pp.

LAUKOLA, S. & SUOMALAINEN, P. (1971). Studies on the physiology of the hibernating hedgehog. 13. Circum-annual changes of triglyceride fatty acids in white and brown adipose tissue of the hedgehog. Ann. Acad. Sci. Fenn. A, IV Biologica 180, 1-11.

LUECKE, R. H. & SOUTH, F. E. (1972). A possible model for thermoregulation during deep hypothermia. In South, Frank E. *et al*, 1972. Hibernation and Hypothermia, Perspectives and Challenges, Elsevier, New York pp. 577-604.

MIHAILOVIC, L. J. T. (1972). Cortical and subcortical electrical activity in hibernation and hypothermia. A comparative analysis of the two states. In South, Frank E. *et al*, 1972. Hibernation and Hypothermia, Perspectives and Challenges, Elsevier, New York pp. 487 -534.

MROSOVSKY, N. (1971). Hibernation and the Hypothalamus. Appleton-Century-Crofts, New York, 232 pp.

MYERS, R. D. & YAKSH, T. L. (1972). The role of hypothalamic monoamines in hibernation and hypothermia. In South, Frank E. *et al*, 1972. Hibernation and Hypothermia, Perspectives and Challenges, Elsevier, New York, pp. 551-575.

PENGELLEY, E. T. & ASMUNDSON, S. J. (1972). An analysis of the mechanisms by which mammalian hibernators synchronize their behavioral physiology with the environment. In South, Frank E. 1972. Hibernation and Hypothermia, Perspectives and Challenges, Elsevier, New York, pp. 637-661.

ROMER, A. S. (1959). The Vertebrate Story. University of Chicago Press, Chicago.

SAARIKOSKI, P.-L. & SUOMALAINEN, P. (1970). Studies on the physiology of the hibernating hedgehog. 11. Circum-annual changes in the concentration of the blood glucose. Ann. Acad. Sci. Fenn. A, IV Biologica 171, 1-7.

SAARIKOSKI, P.-L. & SUOMALAINEN, P. (1971). Studies on the physiology of the hibernating hedgehog. 12. The glycogen content of the liver, heart, hind-leg muscle and brown fat at different times of the year and in different phases of the hibernating cycle. Ann. Acad. Sci. Fenn. A, IV Biologica 175, 1-7.

SAARIKOSKI, P.-L. & SUOMALAINEN, P. (1972). Studies on the physiology of the hibernating hedgehog. 15. Effects of seasonal and temperature changes on the in vitro glycerol release from brown adipose tissue. Ann. Acad. Sci. Fenn. A, IV Biologica 187, 1-4.

SARAJAS, H. S. S. (1967). Blood glucose studies in permanently cannulated hedgehogs during a bout of hibernation. Ann. Acad. Sci. Fenn. A, IV Biologica 120, 1-11.

SCHALLY, A. V., ARIMURA, A. & KASTIN, A. J. (1973). Hypothalamic Regulatory Hormones. Science 179:341-350.

SENTURIA, J. B., STEWART, S., & MENAKER, M. (1970). Rate temperature relationships in the isolated hearts of ground squirrels. Comp. Biochem. Physiol. 33, 43-50.

SOIVIO, A. (1967). Hibernation in the Hedgehog *Erinaceus europaeus L.* The distribution of blood, the size of the spleen and the hematocrit and hemoglobin values during the annual and hibernating cycles. Ann. Acad. Sci. Fenn. A, IV Biologica 110, 1-71.

SUOMALAINEN, P., & AHLSTROM, A. (1970). Studies on the physiology of the hibernating hedgehog. 8. The concentration of riboflavin and thiamine in liver and heart. Ann. Acad. Sci. Fenn.

A, IV Biologica 157, 1-3.

SUOMALAINEN, P., LAUKOLA, S., & SEPPA, E. (1969). Studies on the physiology of the hibernating hedgehog. 6. The serum and blood magnesium in relation to seasonal and hibernation cycles. Ann. Acad. Sci. Fenn. A, IV Biologica 139, 1-9.

SUOMALAINEN, P. & ROSOKIVI, V. (1973). Studies on the physiology of the hibernating hedgehog. 17. The blood cell count of the hedgehog at different times of the year and in different phases of the hibernating cycle. Ann. Acad. Sci. Fenn. A, IV Biologica 198, 1-8.

SUOMALAINEN, P. & SUVANTO, I. (1953). Studies on the physiology of the hibernating hedgehog. 1. The body temperature. Ann. Acad. Sci. Fenn. A, IV Biologica 20, 1-20.

SUOMALAINEN, P., & SAARKOSKI, P.-L. (1969). Studies on the physiology of the hibernating hedgehog. 7. The concentrations of flavins in the brown fat in relation to seasonal and hibernation cycles. Ann. Acad. Sci. Fenn. A, IV Biologica 140, 1-6.

SUOMALAINEN, P. & SAARIKOSKI, P.-L. (1970). Studies on the physiology of the hibernating hedgehog. 10. Persistence of a circadian rhythm during the hibernation of the hedgehog. Comentat. Biol. 30, 1-5.

SUOMALAINEN, P. & SAARIKOSKI, P.-L. (1971). Studies on the physiology of the hibernating hedgehog. 14. Serum free fatty acid, glycerol and total lipid concentrations at different times of the year and of the hibernating cycle. Ann. Acad. Sci. Fenn. A, IV Biologica

184, 1-6.

SUOMALAINEN, P. & WALIN, T. (1972). Studies on the physiology of the hibernating hedgehog. 16. Variation in the nuclear sizes of cells of the supraoptic nucleus of hedgehog during circadian, hibernation, and annual cycles. Ann. Acad. Sci. Fenn. A, IV Biologica 192, 1-5.

UUSPAA, V. J. & SUOMALAINEN, P. (1954). The Adrenaline and noradrenaline content of the adrenal glands of the hedgehog. Ann. Acad. Sci. Fenn. A, IV Biologica, 1-11.

VERFAILLIE, G. R. M. (1969). Correlation between the rate of growth of rice seedlings and the P-indices of the chemical test of Piccardi. A Solar Hypothesis. Int. J. Biometeor. 1969:13, number 2.

WALDMEIER, M. (1968). Die Beziehung zwischen der Sonnenfleckenrelativzahl und der Gruppenzahl. Astronomische Mitteilungen der Eidgenossischen Sternwarte Zurich, nr. 285, Zurich.

YALOW, R. S. & BERSON, S. A. (1960). Immunoassay of endogenous plasma insulin in man. J. Clin, Invest. 39. 1157.

YAPP, W. B. (1965). Vertebrates, their structure and life. Oxford University Press, New York, 485 pp.

CREDITS

1. Figures 3, 4, 5, 6, 7, 8, 9, 10, 11, 12, 13, 15, 16, 18, 19, 20, 21, 22, 23, 24, 25, 26, 27, 28, 29 and Tables 2, 3, and 4 are reproduced by permission of Acta Physiologica Scandinavica, Karolinska Institutet, Stockholm, Sweden.
2. Figure 2 is reproduced by permission of

Dr. Nils-Erik Landell, Stockholm, Sweden, and originally published in "Will The Hedgehog Survive?": Bonniers Forlag AB, Stockholm, Sweden.

CIRCANNUAL RHYTHM OF REPRODUCTION IN MALE EUROPEAN STARLINGS (*STURNUS VULGARIS*)

JAMES T. RUTLEDGE

(Staff Research Associate)

Division of Wildlife and Fisheries Biology
and the Institute of Ecology,
University of California,
Davis, California, 95616

INTRODUCTION

Many investigations of the photoperiodic control of avian reproductive cycles have been stimulated by the demonstration that seasonal changes in daily photoperiod duration exercise temporal control of reproduction in the slate-colored junco, *Junco hyemalis* (Rowan, 1925). Such studies have documented photosexual response systems in over fifty species of birds (for review, see Lofts & Murton, 1968; Farner & Lewis, 1971). Photoperiodic avian species include the European starling, *Sturnus vulgaris* (Bissonnette, 1931; Burger, 1947, 1949, 1953;

Schwab, 1971).

Most investigations of avian photosexual responses have been conducted on species that breed exclusively in the temperate zone. Selective pressure is strong for annual breeding cycles at these higher latitudes, with their alteration of environments favorable and unfavorable for the survival of offspring. In low latitudes, however, the annual photoperiodic change and the seasonally stable food supplies could make accurate timing of annual reproductive cycles nearly impossible and perhaps unnecessary (Farner, 1970). Under these conditions, birds might then maintain reproductive condition continuously. Such a strategy would have the obvious advantage of maximizing reproductive potential, but the energy required to maintain year-round reproduction, including the behavioral, morphological, and physiological characteristics of this state, might severely restrict other energy requiring stages of the annual life cycle (see King, 1973, for review of the energy requirements for reproduction in birds). Because flight requires morphological and physiological adjustments, there is adaptive value in the temporal separation of such energy-demanding processes as pre and postnuptial molts, fat deposition, reproduction, and (in some cases) migration. Any strategy that birds use to cope with an environment lacking pronounced seasonal fluctuations must include an ability to maintain at least a partial separation of these events. Continuous reproduction would be in violation of this principle.

Since reproductive cyclicity is of distinct survival value for most avian species, a means of generating these cycles under stable equatorial conditions must be operant. This requirement, and the reproductive cycles found in such stable environments (see below), led to a hypothesis that there is an endogenous circannual oscillator, in some cases entrained or overriden by temperate-zone photoperiods but free-running under stable equatorial conditions (see Aschoff, 1955; Farner & Lewis, 1971). Evidence supporting an "annual clock" hypothesis has been obtained in both field and laboratory studies.

Several sea-birds, including the sooty or "wideawake" tern, *Sterna fuscata*, (Chapin, 1954; Chapin & Wing, 1959), the brown booby, *Sula leucogaster*, (Dorward, 1962), and the black noddy, *Anous tenuirostris*, (Ashmole, 1962), breed at eight-to-ten-month intervals under equatorial conditions. The deviation in rhythms from a strictly annual periodicity indicates the absence of a single "most favorable' season for reproduction, and undoubtedly the absence of appropriate seasonal cues necessary for accurate timing of 12-month rhythms. Under such environmental conditions, it can be assumed that reproduction will occur as frequently as is consistent with other physiologically demanding processes such as molt (Lofts & Murton, 1968).

Laboratory studies designed to detect possible circannual rhythmicity are not abundant, presumably because lengthy experiments are required. Circannual cycles of fat deposition, molt, nocturnal restlessness, and reproduction have been reported for several avian species held under chronic daily light/dark schedules (see Lofts, 1964; King, 1968; Gwinner, 1968, 1971; Schwab, 1971). The evaluation of presumed endogenous circannual rhythms under extended periods of constant light (LL) and/or constant darkness (DD) (conditions under which the rhythm should theoretically be free-running; Farner & Follett, 1966) has been controversial. Although Benoit and his associates (1956, 1959) documented repeated testicular cycles in the domestic mallard, *Anas platyrhynchos*, held under these two conditions, the periodicity of the rhythms was somewhat irregular and consistently less than 12 months. Rutledge & Schwab (1974) reported that male starlings held in the absence of daily photostimulation exhibited testicular recrudescence and an extended period of spermatogenesis followed by a gradual, partial testicular regression. They concluded that a circannual reproductive rhythm is not being expressed in these birds, however, since the single testicular "cycle" was still incomplete after 448 days of DD.

The lack of conclusive results from LL and DD experiments is a serious weakness of the endogenous circannual clock hypothesis (Hamner, 1971). Although Farner & Follett (1966; see also Jegla & Poulson, 1970) concluded that an endogenous circannual rhythm can be ultimately documented only from experiments under constant light or constant darkness, most positive evidence for this response comes from experiments under stable daily light cycles even though such conditions permit conflicting interpretations: specific light/dark schedules may actually *cause*, or may only *permit*, the observed circannual rhythm (Schwab, 1971; see below).

Schwab's (1971) data documented the occurrence of two consecutive reproductive cycles in starlings held under 12-hour photoperiods for 27 months. Since the periodicity of the rhythm was close to a year (about 10 months) the response was termed "circannual". Since no overt seasonal or annual cues were available to the birds, the term "endogenous" was also added. During discussion following the report of these data, an important difference of opinion emerged:

PAVLIDIS: 'It seemed that you [Schwab] emphasized the point that the rhythmicity you see is due primarily to the 12-hr photoperiod, while Dr. Gwinner interpreted his results as demonstrating some endogenous circannual oscillator. Am I right?'

ASCHOFF: 'Schwab has a very clear-cut, beautiful endogenous circannual rhythm.'

PAVLIDIS: 'Not necessarily.'

SCHWAB: 'It is a question of whether the LD 12:12 cycle causes, or simply permits, the annual cycle.'

ASCHOFF: 'An LD 12:12 cycle is just as constant an environment with respect to an annual rhythm as the heartbeat is for a circadian rhythm. The free-running cycles are from 9 to 11 months. It has nothing to do with season and nothing to do with multiples of LD 12:12.'

PAVLIDIS: 'It is disturbing to me that you are assuming more and more clocks for the organism.'

ASCHOFF: 'Oh, I like it!'

(from Biochronometry, 1971, pg. 446)

The following results do not conclusively establish the nature of circannual rhythms, but they do shed some light on this rather confusing phenomenon. The experiments were designed to examine circadian rhythms of perch-hopping activity and deep-core body temperature at various times during the reproductive rhythm in male starlings under LD 12:12 and in starlings under constant darkness. If circadian oscillations are involved in the so-called endogenous circannual rhythm of reproduction in the male starling, an annual oscillator might not be needed to explain the reproductive response under LD 12:12 photoperiods.

METHODS AND RESULTS

Juvenile male European starlings, captured in Sonoma County, California, in summer 1970, were maintained in an outdoor aviary at Davis, California (38° N latitude) under natural photoperiods. Food (turkey pellets) and fresh water were provided *ad libitum* during the holding period and subsequent experimental treatments.

On January 30, 1971, these birds were transferred to several 2 X 2 X 4-ft wire cages enclosed in light-tight photochambers and held under chronic LD 12:12 photoperiods, with lights-on at 0800 hours, lights-off at 2000 hours. The initial testicular growth-involution cycle had begun in most birds, as judged by slight yellowing at the base of the beak (see Witschi, 1961) and measurements of testicular widths (discussed below) of a sample of the population. Light intensity from incandescent bulbs was about 640 lux at the midpoint of each cage, and ambient air temperature fluctuated no more than $3.0^{o}C$ about a mean of $20^{o}C$.

During the next four months (February to May, 1971), all birds passed through a reproductive cycle, verified by measurements of the left testis widths following unilateral laparotomy (after the technique of Risser, 1971). Previous histological studies enable a correlation of testicular widths with various stages of spermatogenesis (see Schwab, 1971). These starlings composed the parent population from which birds were selected for study at various times during the next 12 months (June, 1971 to May, 1972).

Experiment 1: Circadian Rhythms under LD 12:12

Characteristics of the daily perch-hopping activity and deep-core body temperature rhythms were determined in male starlings under chronic 12-hour daily photophases, with light intensities between 535 and 640 lux at cage midpoints. A total of 18 birds were removed from the parent population at various times following their initial testicular growth-involution cycle under LD 12:12. Temperature-sensing (± 0.1°C) biotelemetry devices were implanted surgically in the posterior end of the body cavity of each bird. Birds were then cages individually in light-tight photochambers each equipped with an activity-sensitive perch connected electrically to a separate channel of a 20-channel event recorder. Ambient air temperature within the chambers fluctuated between 19 and 22°C. Disturbance from external sounds was reduced by fiberglass insulation attached to the walls of the room containing the photochambers, and masked by continuous noise produced by ventilation fans.

Perch-hopping activity was monitored continuously for over 11 months. The duration of activity (α) and rest (ρ) were determined, as was the period (τ) of the activity rhythms (see Aschoff *et al.*, 1965). Each bird's deep-core body temperature was recorded on magnetic tape at hourly intervals for three consecutive days approximately every two weeks. The transmitter batteries were replaced about every three months by implant surgery. Testicular widths were measured during surgery, permitting comparison of reproductive condition with daily body temperature and activity patterns. More frequent testis measurement was inadvisable since the procedure temporarily decreases activity and alters body temperature in this species (Rutledge, unpublished data).

Results:

Figure 1 presents typical patterns of perch-hopping activity recorded in male starlings under chronic LD 12:12 photoperiods. Thirteen of 15 starlings during the period of reproductive quiescence characteristically confined their activity to the daily 12-hour photophase. The two birds which did not consistently fit this pattern displayed periodic bursts of activity outside the hours of illumination, and also did not exhibit a second testicular cycle after prolonged exposure to LD 12:12. Four of five starlings monitored during testicular recrudescence (Left testis widths between 2.0 mm and 5.5 mm) and spermatogenesis (testis widths above 5.5 mm) extended their daily activity into the scotophase, resulting in an average daily activity period of 15 hours and 59 minutes (S.E. ± 19.9 min). Two of those starlings consistently extended perch-hopping beyond lights-off; one starling generally began activity before lights-on, with occasional perch-hopping extensions beyond lights-off; another extended activity

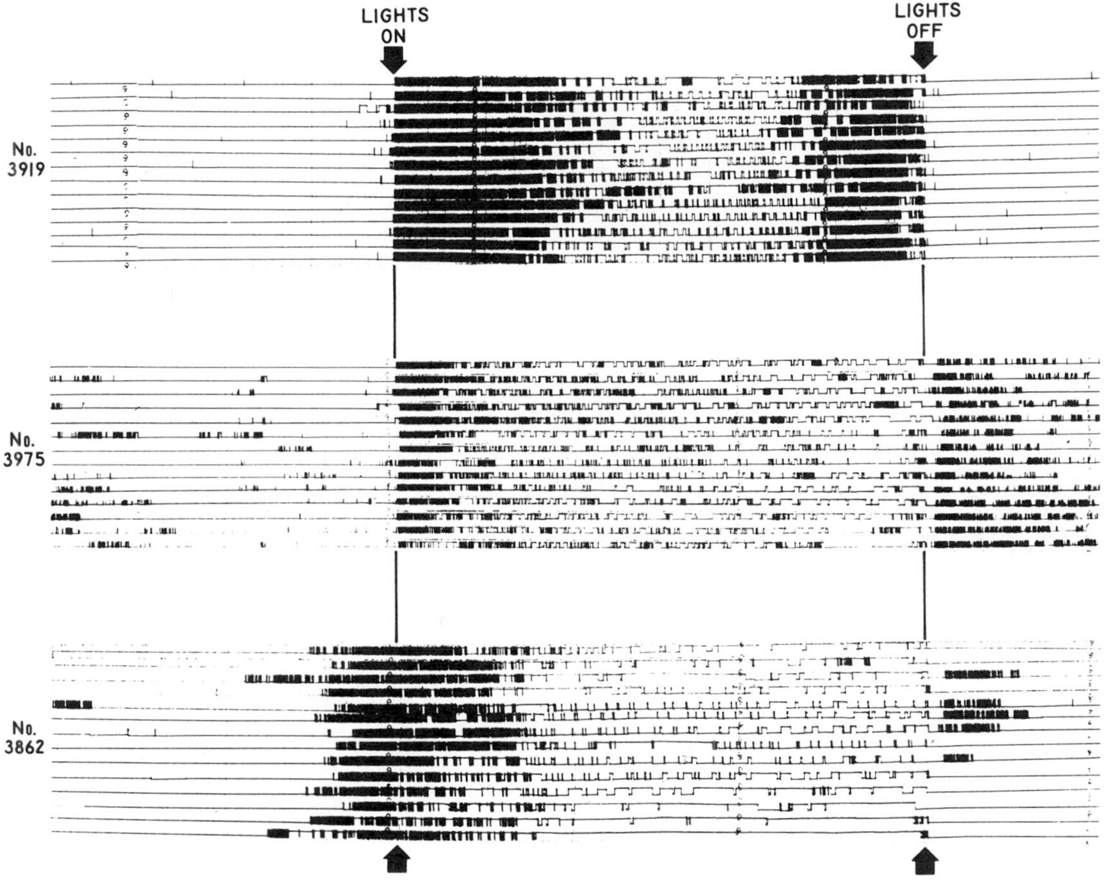

LIGHTS ON

LIGHTS OFF

No. 3919

No. 3975

No. 3862

Fig. 1: Selected two-week perch-hopping records from three birds indicating the typical activity patterns associated with stages of the reproductive rhythm in male starlings under LD 12:12. Solid vertical arrows at top and buttom indicate lights-on (0800 hrs) and lights-off (2000 hrs). In each of the three examples, successive daily 24-hr records are positioned one below the other. Bird No. 3919: Recorded during testicular quiescence; activity is confined to the daily 12-hr photophase. This pattern was characteristic of 13 of 15 starlings monitored during the quiescent stage of the reproductive rhythm. Bird No. 3975 and No. 3862: Recorded during the period of spermatogenesis; activity no longer confined to the daily 12-hr photophase but extended into the scotophase following lights-off (3975), preceding lights-on (3862), or occasionally during both periods (also 3862). These patterns (3975 and 3862) were characteristic of four of five starlings monitored during the period of spermatogenesis and during the preceding period of testicular recrudescence.

after lights-off, with occasional extensions before lights-on. A fifth bird did not extend his activity outside the photophase despite testicular enlargement. Figure 2 indicates a 100-day record of one bird during the period of spermatogenesis and subsequent testicular involution. The top two-thirds of the figure indicates the extensions of perch-hopping activity characteristic of starlings during the reproductive stage. The solid horizontal arrow designates the approximate time that activity patterns reverted to conformity with the daily 12-hour photophase as testicular involution commenced.

Table 1 presents a statistical analysis of the 3,064 deep-core body temperature measurements recorded during four stages of the reproductive rhythm in 15 male starlings under LD 12:12. Early testicular quiescence (the two-month period following completion of the previous testicular cycle under daily 12-hour photophases) is characterized by core temperatures significantly elevated above levels recorded during other stages of the reproductive cycle at all hours of the day and night (one-way ANOVA, Dixon, 1971; Scheffe's method of testing for significant differences between groups, $P<0.05$). As the period of testicular quiescence proceeds, temperatures decline, reaching lowest values just prior to testicular recrudescence (noted as "late quiescence"; the six-week period before testis growth). As testicular recrudescence begins, core temperatures are somewhat higher during the photophase (statistically significant elevations, $P<0.05$, during nine of the 12 hours of illumination) but remain relatively low during the scotophase. Spermatogenesis is characterized by a significant increase ($P<0.05$) in all scotophase core temperatures above levels estab-

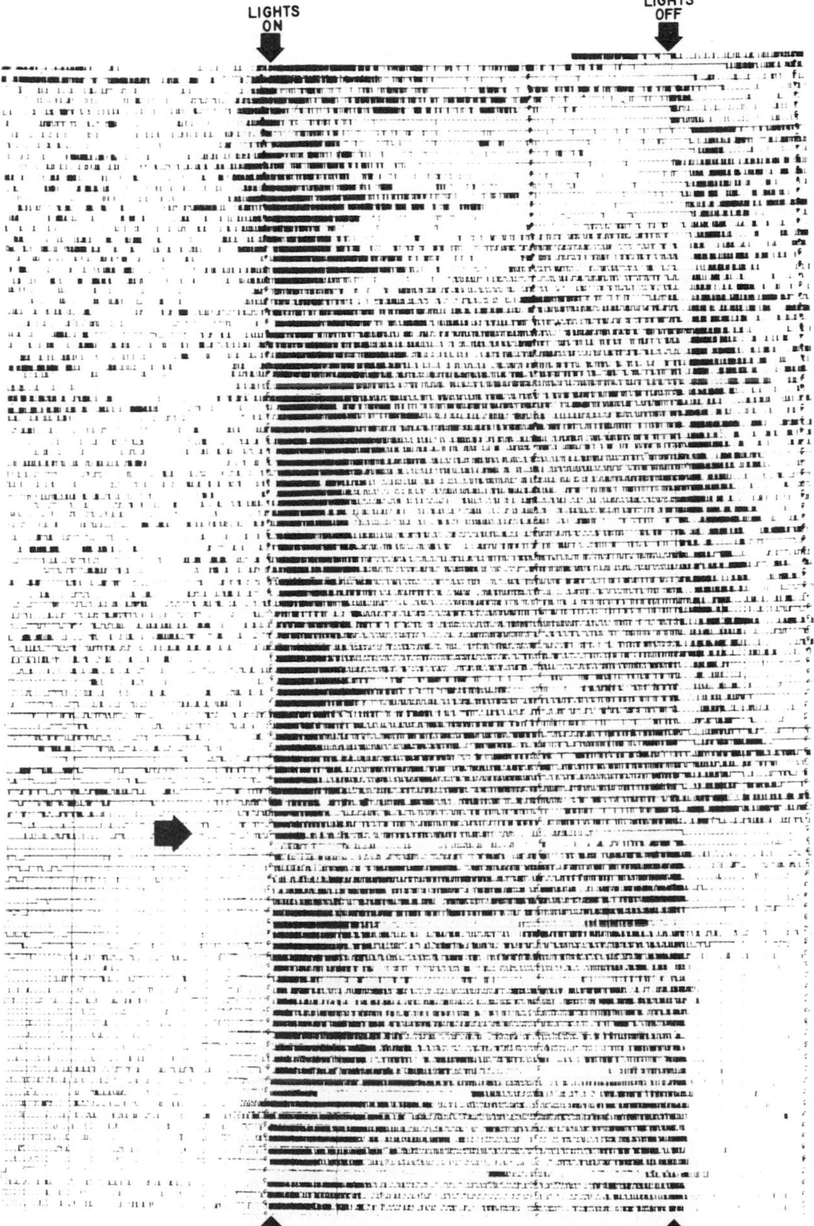

Fig. 2: A 100-day record of one starling indicating extended perch-hopping activity during the spermatogenic phase of the reproductive rhythm under LD 12:12 (top two-thirds of records above solid horizontal arrow), and reversion of activity patterns to conformity with the daily 12-hr photophase during testicular involution (bottom records below horizontal arrow). Solid vertical arrows at top and bottom indicate lights-on (0800 hrs) and lights-off (2000 hrs).

lished during the preceding two stages, although still significantly lower ($P<0.05$) than is characteristic of early quiescence. Photophase temperatures during eight of 12 hours are significantly higher ($P<0.05$) than in the previous two stages, although significantly lower ($P<0.05$) than during early quiescence. Figure 3 presents the hourly deviations from the mean core temperature (calculated by the unweighted averaging of hourly means from each stage of the reproductive rhythm) for the 15 birds during each stage of the testicular cycle under LD 12:12. Again, temperatures were highest during early quiescence, and were also above average for most hours during the spermatogenic stage. The period of testicular recrudescence is characteristically below average, with photophase temperatures somewhat higher than during the preceding stage (late quiescence). Scotophase temperatures are low during late quiescence and recrudescence. Pronounced deviations from the mean temperature near the dark-light interface (0800 hours) appear in birds during all stages except late quiescence, with most notable deviation occurring during early quiescence. The period of spermatogenesis is characterized by an anticipatory rise in core temperature one hour prior to lights-on.

The method used for recording perch-hopping activity required the birds to be caged individually for the 12-month period following the initial testis cycle under LD 12:12. Such isolation either abolished or delayed the second testicular cycle in 11 of the 18 birds studied. A deterioration of reproductive synchrony in group-caged starlings during prolonged exposure to daily 12-hour photophases, reported by Schwab (1971), was less than observed here. A control

TABLE 1 Statistical description of mean deep-core body temperatures (°C) recorded from male starlings during four stages of the reproductive rhythm under LD 12:12 (see text). Heading labeled "N" indicates the number of separate daily temperature recordings used in each analysis. Underlined temperature averages within any row are not significantly different from one another in any one hour (one-way ANOVA & Scheffé's test, $P < 0.05$). The last column indicates the grand mean of the hourly means calculated for each reproductive stage (see "0" line, Fig. 3).

TIME	EARLY QUIESCENCE			LATE QUIESCENCE			RECRUDESCENCE			SPERMATOGENESIS			$\bar{X}$ of $\bar{x}$
	N	$\bar{x}$	SE	N	$\bar{x}$	SE	N	$\bar{x}$	SE	N	$\bar{x}$	SE	
1200	48	42.9	0.05	18	42.1	0.05	15	42.2	0.10	47	42.5	0.04	42.4
1300	48	43.0	0.06	18	42.2	0.07	15	42.3	0.06	47	42.5	0.04	42.5
1400	48	42.9	0.06	18	42.1	0.07	15	42.5	0.10	47	42.5	0.05	42.5
1500	48	43.0	0.06	18	42.3	0.13	14	42.4	0.09	47	42.6	0.06	42.6
1600	48	42.9	0.06	18	42.2	0.06	14	42.6	0.07	47	42.7	0.05	42.6
1700	48	43.0	0.06	18	42.4	0.10	15	42.7	0.13	47	42.6	0.05	42.7
1800	47	43.1	0.05	18	42.3	0.07	15	42.5	0.10	48	42.4	0.06	42.6
1900	48	43.1	0.05	18	42.3	0.08	15	42.3	0.10	48	42.3	0.06	42.5
2000[a]	47	42.6	0.07	18	41.7	0.13	15	41.8	0.12	48	41.9	0.06	42.0
2100	46	41.1	0.11	18	40.2	0.08	15	40.5	0.20	48	40.4	0.06	40.6
2200	47	40.6	0.10	18	39.6	0.09	15	39.8	0.09	48	40.0	0.04	40.0
2300	48	40.3	0.10	18	39.4	0.10	15	39.5	0.11	48	39.8	0.04	39.8
2400	48	40.3	0.10	17	39.3	0.10	15	39.2	0.06	48	39.7	0.04	39.6
0100	48	40.2	0.10	18	39.2	0.13	15	39.2	0.08	48	39.6	0.04	39.6
0200	48	40.1	0.09	17	39.2	0.11	15	39.2	0.05	48	39.6	0.03	39.5
0300	47	40.1	0.10	16	39.0	0.12	15	39.1	0.04	48	39.5	0.04	39.4
0400	48	40.0	0.10	17	39.0	0.11	15	39.1	0.07	48	39.5	0.04	39.4
0500	47	40.1	0.09	16	39.1	0.11	15	39.1	0.06	47	39.5	0.05	39.5
0600	45	40.2	0.09	15	39.1	0.12	15	39.1	0.06	48	39.7	0.06	39.5
0700	47	40.8	0.09	18	39.8	0.15	15	39.6	0.10	48	40.6	0.10	40.2
0800[b]	48	42.5	0.09	18	39.9	0.16	15	40.5	0.11	48	41.6	0.09	41.4
0900	48	43.1	0.08	18	42.4	0.13	15	42.3	0.15	48	42.7	0.03	42.6
1000	48	43.2	0.07	18	42.3	0.11	15	42.5	0.13	48	42.7	0.04	42.7
1100	46	43.1	0.05	18	42.2	0.11	15	42.3	0.07	48	42.4	0.05	42.5

[a] lights-off
[b] lights-on

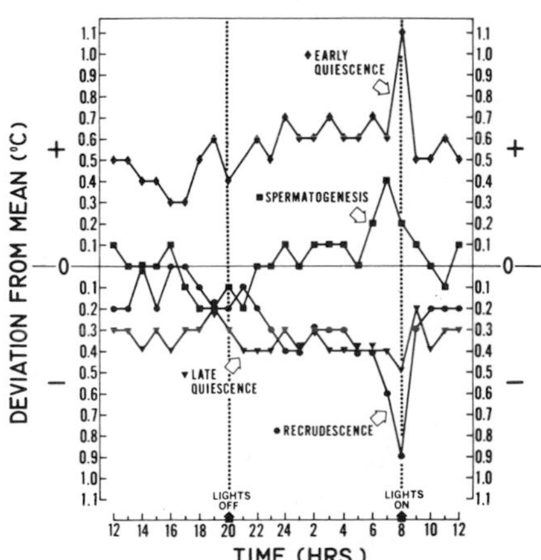

Fig. 3: Hourly deep-core body temperature deviations from mean body temperatures in each of four stages of the reproductive rhythm in male starlings under LD 12:12 (see text). The 0 line was determined by averaging the four mean temperatures for each reproductive stage for each hour. See table 1 for statistical description of the data.

group of starlings, caged together under LD 12:12, were implanted with mock telemetry devices and subjected to biweekly testis width measurements for one year. Despite the transmitter-sized object in the body cavity, testicular recrudescence to spermatogenesis occurred in all birds, indicating that the lack of a second testis cycle in a majority of birds in the experimental group was not due to the transmitters.

The following points summarize the relationship of perch-hopping activity and deep-core body temperatures to the reproductive rhythm under 12-hour photoperiods:

1) During testicular quiescence, perch-hopping activity was almost exclusively confined to the daily 12-hour photophase, producing an α/ρ value (the ratio of activity to rest) of 1.0.

2) Both the duration and extent of daily perch-hopping increased during testicular recrudescence and spermatogenesis, with an α/ρ value of approximately 2.0.

3) Deep-core body temperature was significantly higher in both the photophase and scotophase during the first portion of testicular quiescence following involution.

4) As the period of spontaneous testicular recrudescence nears (late quiescence), core temperatures decline significantly during both the photophase and scotophase, reaching lowest levels just prior to spontaneous recrudescence.

5) As testicular recrudescence begins, core temperatures are higher during the photophase, while remaining low during the

scotophase until testicular enlargement is nearly complete.

6) The spermatogenic phase is characterized by relatively high core temperatures during both the photophase and the scotophase with an anticipatory rise in temperatures one hour prior to lights-on.

Experiment 2: <u>Circadian Rhythms under Constant Darkness</u>

The circadian rhythms of perch-hopping activity and deep-core body temperature were determined in starlings under DD. Six birds, preconditioned to DD for six to nine months, were surgically implanted with temperature-sensing biotelemetry devices and placed in individual light-tight chambers equipped with activity sensing perches. Perch-hopping activity was monitored continuously for 27 days. Body temperature of four of the six birds (two transmitters malfunctioned) was recorded at hourly intervals for six consecutive days by the procedure described in the previous section.

<u>Results</u>:

Figures 4 & 5 respectively indicate perch-hopping activity and core-temperature patterns of one of the six starlings under DD. These results are representative of the patterns seen in the other five birds. Average periodicity of the free-running activity rhythms was 24 hours and 29 minutes (S.E. ± 9.3 min) measured from the midpoints of successive activity periods. The activity portion of each cycle averaged 11 hours and 3 minutes, with inactivity (rest) averaging 13 hours and 26 minutes. The resultant α/ρ value was 0.83. Core temperature rhythms displayed the same free-running character-

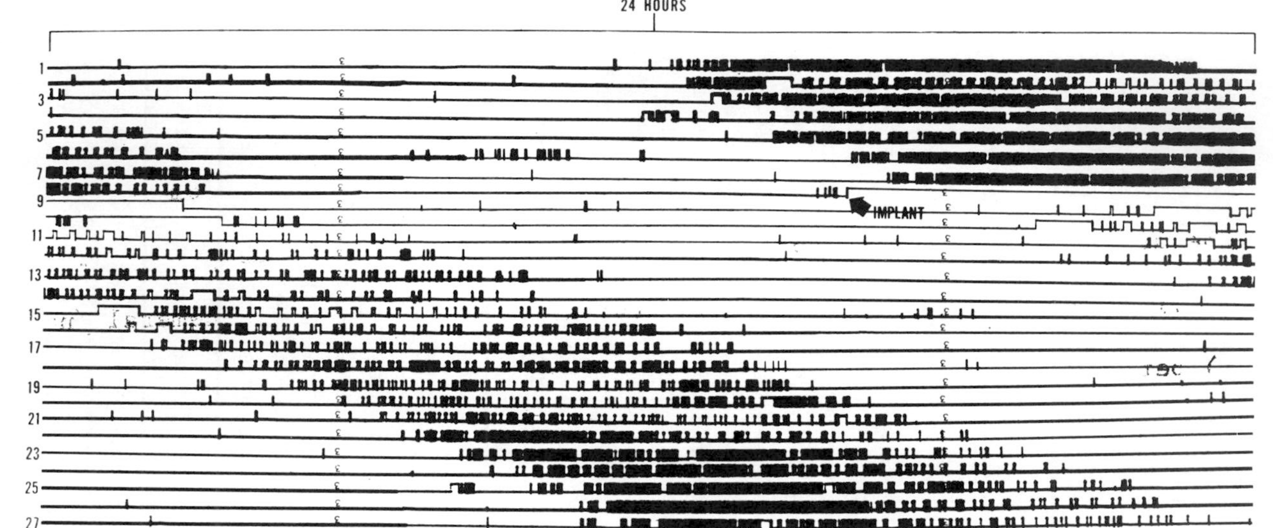

Fig. 4: Perch-hopping activity records of one starling during 27 days of constant darkness. Surgical implantation of a temperature-sensing biotelemetry device on day 8 is indicated by the solid arrow. This procedure decreases the extent of activity for several days but does not alter the free-running periodicity of the rhythm.

istics found in the activity cycles, suggesting that control was by the same circadian oscillator or by two such oscillators phase-locked under constant darkness. During the 27-day experiment the activity (and also temperature) rhythms of individual starlings shifted out of phase with one another, verifying the absence of an external "zeitgeber."

Experiment 3: Testis Responses to an Ahemeral Photoperiod Determined by Activity Rhythms under Constant Darkness

The ratio of perch-hopping activity to rest (α/ρ) determined in experiment 2 under constant darkness was used to construct the following photoperiod schedule. The average duration of activity per free-running "day" of the DD rhythm (each "day" being 24 hours and 29 minutes long) determined the length of the photophase (11 hours and 3 minutes) and the average duration of rest per "day" the scotophase (13 hours and 26 minutes). This resulted in ahemeral (non-24-hour) light cycles with a photoperiod schedule of LD 11^{03}:13^{26}, simulating the α/ρ value determined under DD. Sixteen adult male starlings were removed from the outdoor aviary on 13 September 1972 (during the naturally occurring photorefractory period, see Schwab, 1971), placed in a cage enclosed in a light-tight photochamber, and held under the ahemeral photoperiod for 127 days. Widths of the left testes were periodically measured.

Results:

Photorefractory starlings held under an ahemeral photoperiod schedule of LD 11^{03}: 13^{26} exhibited testicular development to, and above, testis widths associated with spermatogenesis (5.5 mm in this species;

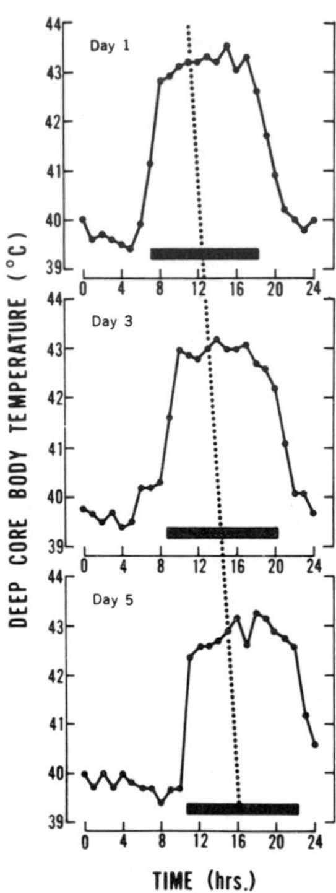

Fig. 5: Deep-core body temperature patterns of one starling recorded during constant darkness (solid lines), with solid horizontal bars at bottom representing the durations of perch-hopping activity recorded simultaneously (see Fig. 4). Dotted line connects midpoints of activity bars to emphasize the ahemeral (non-24-hr) periodicity of the free-running rhythms (days 2 and 4 omitted for clarity).

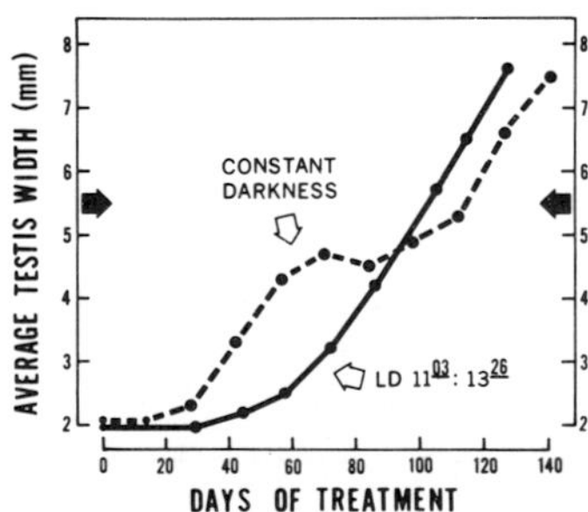

Fig. 6: Testicular growth in European starlings held under constant darkness (dashed line; modified from Rutledge & Schwab, 1974) and LD $11^{03}:13^{26}$. This latter ahemerel photoperiod simulates the average activity/rest ratio recorded from starlings held under constant darkness (see Fig. 4 and text). Left testis widths were measured *in situ* following unilateral laparotomy. Solid horizontal arrows at 5.5 mm on the vertical axes indicate left testis widths at or above which histologically-determined spermatogenesis is complete. See table 2 for statistical analyses of these data.

TABLE 2 Statistical description of measurements of the left testis widths of 16 starlings maintained under the ahemeral photoperiod LD $11^{03}:13^{26}$. Measurements were taken following unilateral laparotomy (see text). Widths of less than 2.0 mm cannot be accurately determined by this technique, and are reported below as "<2.0" mm.

Assay date	Days of treatment	Average testis width (mm)	Standard error (mm)	Range (mm)
13 Sept. '72	0	<2.0	-	-
13 Oct.	30	<2.0	-	-
27 Oct.	44	2.2	0.08	<2.0-3.2
10 Nov.	58	2.5	0.22	<2.0-4.9
24 Nov.	72	3.2	0.38	<2.0-7.0
8 Dec.	86	4.2	0.46	2.7-7.7
27 Dec.	105	5.7	0.46	3.2-9.5
6 Jan. '73	114	6.5	0.30	4.8-8.2
19 Jan.	127	7.6	0.31	6.2-9.9

fig. 6, table 2). Average time to spermatogenesis was 101 days (S.E. ± 4.7 days). This was the same as the average time to spermatogenesis of 101 days (S.E. ± 10.6 days) documented in starlings held under DD (Rutledge & Schwab, 1974). Thus, an ahemeral photoperiod schedule, presumably entraining the free-running perch-hopping activity/rest rhythm of starlings under DD, resulted in an average time to spermatogenesis identical to that found under DD, though with a sizable reduction in variability (decreased standard error).

DISCUSSION

There are a variety of different photoperiodic response strategies whereby ambient seasonal photoperiodic fluctuations facilitate ecologically proper timing of avian reproductive cycles. For example, some species begin annual gonadal metamorphosis under the relatively short daily light durations temporally associated with the winter solstice, whereas others may require much longer "summertime" photoperiods before the gonadal growth cycle is induced. Similarly, some species terminate the annual sexual cycle as a function of relatively long early-summer photoperiods, whereas others do not cease reproduction until late summer or early fall, when photoperiods are much shorter. Various species-specific photoperiodic response strategies are given an in-depth discussion in the comprehensive review by Lofts & Murton (1968).

Photoperiodic control of the annual reproductive cycle in European starlings inhabiting the temperate zone involves seasonal oscillations in daily photoperiod duration above and below LD 12:12, i.e., the vernal and autumnal equinoxes (Schwab, 1971). Schwab's experiments indicate that the timing of starling reproductive cycles is photoperiodically facilitated in the following manner:

- Daily photoperiods of less than 12 hours, such as in the temperate-zone latitudes from the autumnal to the vernal equinox, initially dispel the photorefractory state, followed ultimately by testicular recrudescence. The period of spermatogenesis is considerably prolonged, and testicular involution inhibited, when starlings are maintained under chronic daily photoperiods of less than 12 hours.

- Daily photoperiods longer than 12 hours, occurring naturally from the vernal to the autumnal equinox, will induce testicular recrudescence in photosensitive starlings. However, they ultimately lead to testicular involution and maintenance of the photorefractory state as long as starlings remain under these long-day light regimens.

This interpretation of the role of photoperiod in the timing of starling sexual cycles is generally consistent with earlier results by Bissonnette (1931) and Burger (1947, 1949, 1953). Although it is likely that the role of daily light/dark cycles proposed by Schwab (1971) accurately reflects the photoperiodic control of annual testicular cycles in starlings breeding in temperate-zone environments, it is unlikely that the same role can be ascribed to the slight seasonal photoperiodic fluctuations associated with equatorial latitudes.

Schwab (1971) documented repeated sexual cycles in European starlings under photoperiods simulating those at the equator (chronic LD 12:12). He found that starlings undergo at least two complete testicular growth-involution cycles in the absence of any seasonal or annual change in the length of the daily photoperiod. The periodicity of this reproductive rhythm was about 10 months, however, a deviation from the normal 12-month periodicity characteristic of this species under temperate-zone photoperiods. Subsequent studies by Schwab & Rutledge (1973) suggest a plausible explanation for the reduced periodicity of the reproductive rhythm in starlings under chronic LD 12:12. During their first testicular cycle under this light regimen, starlings show no overt testicular development after transfer to constant light at various times during the initial three-month period following testicular involution. Thus, a period of photorefractoriness of the same duration as that occurring in starlings under temperate-zone photoperiods following testicular involution (Schwab, 1971) was demonstrated between testis cycles under LD 12:12. Under LD 12:12, this three-month photorefractory period is immediately followed by spontaneous testicular recrudescence. There is no intervening three-month period (as with starlings under temperate-zone photoperiods) during which birds gradually regain "photosensitivity" prior to spontaneous testicular growth. This phase of the starling reproductive cycle is comparable to the "preparatory phase" described by Wolfson (1959) and to the "relative refractory period" noted by Hamner (1968) in other avian species. Starlings under LD 12:12 appear not to require a separate "relative refractory period," although this phase does eventually occur in conjunction with spontaneous testicular recrudescence. The partial overlap of two phases (photosensitivity acquisition and spontaneous testis growth), which are characteristically separate and distinct under temperate-zone photoperiods, explains reduction of a normally 12-month reproductive rhythm to one of only 10 months under chronic LD 12:12.

Although these studies of starling sexual cycles under a simulated equatorial photoperiod present considerable information on the occurrence and chronology of a "circannual" reproductive rhythm, they were not designed to investigate the phenomenon at the operational photoperiodic control level.

Circadian Control of Avian Reproductive Responses

Investigations of avian reproductive responses strongly implicate circadian control. Circadian control of photoperiodically induced gonadal recrudescence has been demonstrated in the house finch, *Carpodacus mexicanus*, (Hamner, 1963, 1964, 1966); white-crowned sparrow, *Zonotrichia leucophrys gambelii*, and *Z. l. pugetensis*, (Farner, 1965; Turek, 1972); house sparrow, *Passer domesticus*, (Menaker, 1965); slate-colored junco and white-throated sparrow, *Zonotrichia albicolis*, (Wolfson, 1965); Japanese quail, *Coturnix coturnix japonica*, (Follett & Sharp, 1969); greenfinch, *Chloris chloris*, (Murton *et al*, 1969, 1970); and golden-crowned sparrow, *Zonotrichia atricapilla*, (Turek, 1972). It should be emphasized, however, that these studies were concerned primarily with factors responsible for either the dispelling of refractoriness and/or gonadal recrudescence. Mechanisms controlling termination of the reproductive phase, i.e., gonadal involution and photorefractoriness, are not completely understood.

Drawing from their experimental work with circadian rhythms in insects, Pittendrigh & Minis (1964) developed a control scheme consistent with many of the responses obtained in experiments on the birds cited above. They propose a dual role for daily photoperiod in the control of photoperiodic responses: 1) entrainment of a circadian oscillation in photosensitivity; and 2) photostimulation of the response system. Hamner (1968) expanded these concepts, and those of Bunning (1936, 1964) and other botanists, to clarify his previous results on house finches. The circadian control hypothesis proposed by Hamner is currently accepted as the best explanation for the photosexual responses of temperate-zone avian species.

Essentially Hamner suggests that a photosensitive or photoinducible phase of a circadian clock is temporally situated "midway" between the light-dark and dark-light interfaces of the scotophase during early autumn. This photosensitive phase moves progressively closer to the daily photoperiodic interfaces as the winter solstice is approached as a function of decreasing daily photophases naturally occurring during this season. When daily light duration increases following the winter solstice, the photosensitive phase is illuminated, thereby inducing gonadal growth. Several investigators, including Murton *et al*, (1969, 1970); Meier *et al*, (1971); and Follett & Nicholls (1973), have demonstrated a circadian basis for specific gonadotropin release in relation to daily photoperiod duration in birds inhabiting temperate latitudes, thus providing physiological support for Hamner's hypothesis. More complete reviews of the rather extensive literature on this subject are given by Farner & Lewis (1971) and Van Tienhoven & Planck (1971).

Recent findings on the testicular responses of European starlings under constant darkness (Rutledge & Schwab, 1974) add support that circadian oscillations are involved in the control of avian reproductive responses, albeit in a form slightly modified from that of Hamner (1968). Photorefractory male starlings were placed in light-tight chambers and held without daily photostimulation (DD) for 448 days. Brief exposure to dim light twice weekly was necessary for maintenance procedures. The authors also caged a small number of refractory starlings in complete and uninterrupted darkness for 127 days. Testicular metamorphosis and spermatogenesis in starlings under both experimental protocols conclusively demonstrates that light is not necessary for sexual development in this species (see Hamner, 1971, for the initial suggestion that light, *per se*, is not a prerequisite for testicular growth in starlings). Prolongation of the spermatogenic phase in starlings in the 448-day experiment, similar to results reported by Schwab (1970) for starlings under LD 11:13 and 10 1/2:13 1/2, suggested that daily photophases of sufficient duration are required for normal testicular involution in this species. Rutledge & Schwab (1974) concluded from their experiments that spermatogenesis in starlings under constant darkness may result as a function of free-running circadian stimulation of hypophyseal gonadotropin secretion. The response cannot be explained adequately by proposing involvement of a hypothetical circannual oscillator.

The perch-hopping activity patterns (fig. 4) and the daily core temperature rhythms (fig. 5) of starlings held under constant darkness are consistent with the hypothesis assigning circadian oscillations a role in the reproductive response. That the free-running activity-rest ratio (0.83) occurring in DD represents a stimulatory condition for reproductive development was demonstrated by photoperiodically entraining such a rhythm in post-involution starlings (fig. 6, table 2). Birds maintained on an LD $11^{03}:13^{26}$ regimen attained spermatogenesis in 101 days, precisely the same period as in constant darkness. The reduction of standard error in time to spermatogenesis in the ahemeral photoperiod group (from 10.6 days under DD, to 4.7 days under LD $11^{03}:13^{26}$) suggests that the photoperiodic regimen more precisely entrained the stimulatory circadian oscillations, resulting in a less variable reproductive response. While it is likely that slight deviations from this ahemeral schedule may result in similar rates of gonadal growth (experiments which should be conducted), not all inductive ahemeral light cycles appear to facilitate the same rate of testicular metamorphosis in the European starling (Schwab, personal communication).

Interpretation of the results concerning the ahemeral photoperiodic induction of testicular growth in the European starling involved the assumption that locomotor activity rhythms are somehow indicative of, or perhaps determined by, the circadian oscillator controlling hypophyseal gonadotropin secretion rates. The next section reviews the historical basis for the existence of such a relationship in a variety of photoperiodic avian species.

Locomotor Activity and Avian Photosexual Responses

Rowan (1925) experimentally documented the relationship between increasing spring photoperiods and seasonal gonadal recrudescence in birds. He was aware, however, that demonstration of this relationship did not prove conclusively that increased daily photoperiods had a direct stimulatory effect on the reproductive system. Rowan suggested the possibility that extended periods of light merely allowed birds to be more active, thus expending additional energy each day, which factor was the direct cause of gonadal metamorphosis. This thought stimulated many investigations designed to elucidate the respective roles of light and activity in the control of sexual cycles in birds. Two major types of studies have been conducted: those in which locomotor activity is induced; and those involving analysis of spontaneous activity with respect to the reproductive cycle.

Induced-Activity Experiments:

Investigations by Rowan (1929) on the slate-colored junco, by Bissonnette (1931, 1933) on the European starling, and by Riley (1940) and Kendeigh (1941) on the house sparrow were designed to determine whether light or activity was the primary factor inducing annual gonadal recrudescence in these "photoperiodic" avian species. Each of these studies, however, involved enforcing activity in constant or near-constant darkness by enclosing the experimental animals in cages which could be revolved, or with perches which could be rotated a predetermined number of hours each day. Typically, the birds were placed in such cages and exposed to non-stimulatory short-day photoperiods supplemented with periods of cage-rolling following lights-off. Control groups were held under a stimulatory photoperiod equal to the combined daily period of illumination and induced activity in the experimental group. Except in Rowan's study, gonadal development either was not induced or was significantly lower with forced activity than with the stimulatory daily photoperiods. Rowan's induced activity ("wakefulness") took place under dim light, however, which may have been responsible for the gonadal recrudescence observed.

Two major objections to these experiments should be noted: 1) It is possible, perhaps even likely, that enforced activity is not physiologically comparable to activity occurring naturally in response to appropriate environmental conditions, so that they might not be expected to induce identical gonadal responses. Unless at least one physiological parameter is measured (e.g., body temperature, heart rate, or respiratory rate), it seems inappropriate to derive a firm conclusion regarding the respective roles of light and activity on the sexual cycle. 2) Since these experiments employed daily illumination besides periods of forced activity, it is likely that circadian functions were being entrained by the light. Numerous studies indicate that the effects of light (the primary "zeitgeber") override the effects of other environmental variables (see Bunning, 1964). Thus, if the short-day photoperiods entrained circadian oscillations which were not immediately stimulatory for gonadal enlargement, the enforced periods of activity could not be expected to modify these oscillations. The relative lack of gonadal development found in the starlings and house sparrows can thus be explained, although the findings on the slate-colored junco are less easily interpreted. It is possible that the "dim" background lighting Rowan employed during cage-rolling periods may have caused the observed gonadal growth.

Induced activity studies, by the nature of the methods employed, do not clearly delineate the respective roles of light and activity in spring resurgence of sexual development in birds.

Spontaneous-Activity Experiments:

Few studies have been conducted specifically to determine the relationship between freely occurring activity patterns and gonadal cycles in birds. From experiments conducted under various daily durations of flashing light (e.g., 5 seconds of light, 15 seconds of darkness, 5 seconds of light, etc., for 14 hours each day) Burger *et al* (1942) suggest that activity was as great as in starlings under uninterrupted daily photophases. Testicular recrudescence, however, was not induced consistently under these flashing light regimens. Since activity was "measured" only by observation of the birds and since no other physiological parameters were recorded, it is difficult to be certain that the degree of activity was the same in each case. Burger and his associates did not believe this objection to be serious.

More recent experiments, particularly those of Menaker (1965); Wolfson (1966); Hamner & Enright (1967); Menaker & Eskin (1967); Farner (1970); and Farner *et al* (1973), demonstrate a relationship between daily periods of activity and gonadal recrudescence. Hamner & Enright (1967) noted some exceptions to this conclusion, the significance of which is difficult to determine. Although these studies differ markedly from one another in design, and their conclusions are often unrelated, sexual activation was generally marked in each case by extended or altered daily perch-hopping activity. Results of Wolfson (1966) are particularly striking. Activity patterns were entrained in slate-colored juncos and white-throated sparrows by brief exposure to various "stimulatory" light cycles. After verifying that these activity patterns persisted following transfer of the birds to constant darkness, Wolfson noted that: "...the activity in darkness appeared to be just as effective 'gonadotropically' as long photoperiods." These results, coupled with those of the investigations cited above, indicate a clear relationship between activity patterns and gonadal recrudescence. To be sure, assumption that this relationship is causal is inappropriate without additional experimentation. Nevertheless, it is possible to formulate a hypothesis based, not on activity *per se*, but upon the underlying control of activity patterns--presumably the same circadian oscillator determining gonadotropin secretion rates of the hypophysis (Farner & Lewis, 1971). Wolfson (1966) states this hypothesis:

> The role of motor activity...is not known, but the duration of activity is well correlated with the gonadotropic response, and one cannot help but speculate on the significance of this correlation. The duration of activity could be a reflection of a time-measuring portion of the brain which uses a circadian system and the duration of the light period to measure the length of the day, and then controls both the hypothalamohypophy-

seas system and the duration of motor activity in accordance with the decision reached.

An excellent test of this hypothesis is offered by starling reproductive rhythms under LD 12:12. Since the daily photophase remains constant, the relationship between activity patterns and testicular development can be examined without the modifying effects of changing photoperiod lengths. Subsequent sections of this paper examine this relationship.

Control of the "Circannual" Reproductive Rhythm in Starlings

A 12-hour daily photophase is the longest daily duration of light which will both dispel refractoriness and lead to spontaneous testicular recrudescence in European starlings (Schwab, 1971). However, unlike with daily photophases of less than 12L (as well as constant darkness), LD 12:12 does not facilitate a significant prolongation of spermatogenesis. Testicular involution is complete, followed by a period of photorefractoriness and a subsequent testicular growth-involution cycle. This paradox may be explained by the extensions in locomotor activity and the elevations in body temperature occurring in starlings during testicular recrudescence and spermatogenesis.

Immediately following the initial gonadal cycle under LD 12:12, perch-hopping activity temporally conforms to the daily 12-hour photophase (see fig. 1), producing an α/ρ value (ratio of activity to rest) of 1.0. This pattern persists through the period of testicular quiescence, reflecting precise entrainment by the light/dark schedule. However, once testicular recrudescence begins (detected by yellowing of the beak and measurement of left testis widths), locomotor activity is altered pronouncedly. It is no longer confined to the daily photophase but extends into the period following lights-off, the period before lights-on, or into both periods. The result is an α/ρ value consistently greater than 1.0, as might be found in birds held under photoperiods greater than 12 hours. Significantly, the resulting testis response is also one which would occur under longer photoperiods, i.e., testicular involution. This result is similar to results of Wolfson (1966), who demonstrated that gonadal recrudescence can be induced by creating a situation in which locomotor activity is of a "stimulatory" duration even in the absence of light. The results reported here for starlings under LD 12:12 add to Wolfson's concepts the fact that testicular involution ultimately follows daily locomotor activity in excess of 12 hours. That one starling failed to show pronounced activity extenstions despite passing through a complete testis cycle does not necessarily invalidate this conclusion. It is entirely possible that: 1) the "threshold" for testicular involution in this bird was nearer to 12 hours, and thus extended activity periods were not a necessary event leading to involution; or 2) the method used to determine the extent of activity was inadequate in this case, i.e., the bird was "physiologically active" either before or after the daily photophase, but "chose" not to depress his perch during these hours of darkness.

From current photoperiodic theory (see Pittendrigh & Minis, 1964; Hamner, 1968) it might have been predicted that the locomotor activity of starlings would phase-lead the onset of the daily photophase as testicular recrudescence began under LD 12:12. Such an occurrence would provide an explanation for recrudescence if locomotor activity patterns reflect the phase of the circadian oscillator responsible for the temporal position of a daily photosensitive phase (the apparent validity of this assumption is reviewed by Farner & Lewis, 1971). For example assume that when activity begins at lights-on, the photosensitive phase in the starling is outside the daily 12-hour photophase, and therefore not subject to photostimulation. The testes would remain quiescent under this condition. If activity began prior to lights-on (phase lead), however, this might indicate that circadian oscillations controlling the position of the photosensitive phase have shifted, bringing the phase into the daily period of illumination and thus promoting the photostimulation of testicular recrudescence. Although several studies under a variety of photoperiodic schedules have demonstrated the validity of such an explanation for other avian photosexual responses (see Farner *et al*, 1973, for example) the concept is inadequate for results reported herein with European starlings under LD 12:12. Only one of five birds showed phase-lead activity consistent enough to explain gonadal growth on this basis. However, since this bird, and three of the other four starlings monitored during testicular recrudescence and spermatogenesis did exhibit extended activity, it is suggested that prime importance resides in extension in activity *per se*, not in the time when activity occurs in relation to the daily period of illumination.

Interpretation of the deep-core body temperature measurements of male starlings is slightly more difficult. Essentially no comparable data have been reported on other feral avian species concerning possible correlations between body temperature patterns and annual gonadal cycles. This lack of comparative information is due largely to the difficulty of obtaining accurate temperature measurements without disturbing the experimental animals excessively. Recent advances in miniature biotelemetry devices provide a solution to this problem.

It was initially expected that if differences in body temperature patterns were detectable in relation to the LD 12:12 reproductive rhythm in male starlings, the highest temperatures would occur during the period of testicular activity. Such a prediction was based on the "excitatory" effects of gonadal steroids and hypophyseal gonadotropins on behavior and physiology in a variety of organisms including the European starling (see Van Tienhoven, 1968). This prediction was not confirmed, however. Elevations in temperature are greatest following testicular involution (see fig. 3, table 1). Since that is the period when the post-nuptial molt takes place, elevated core temperatures may reflect aspects of this process. Not only is a great deal of energy expended in feather regeneration (Rawles, 1970), but thermoregulatory difficulties due to decreased insulation may necessitate increased heat production. Core temperatures should, and apparently do, reflect the physiological demands imposed during this period. Wieselthier (1969) indicates that thyroxin levels are high in starlings following testicular involution, a factor likely contributing to increased body temperature (see Turner & Bagnara, 1971).

The statistically significant reduction

in body temperature in starlings by the end of the period of testicular quiescence would appear to reflect a return to a "normal" thermal state without influence from the process of molt or reproduction. The post-nuptial molt is complete, thereby reducing demands for protein synthesis and providing a fully developed plumage to prevent excessive heat dissipation to the external environment. It is an interesting possibility that such a reduction in body temperature may be necessary before testicular recrudescence can occur. Riley (1937) indicates such a belief, substantiated by studies of gonadal cycles in the house sparrow. Although his temperature measurements were crudely taken by today's standards, results suggest that low body temperatures recorded prior to sexual development reflect a decreased metabolic rate. Such a reduction in metabolism might allow "recharging" of some aspect of the reproductive system, much as proposed by Dol'nik (1964). It should be added that refractoriness in the starling and many other avian species is terminated under photoperiods below a species-specific threshold level. These reduced photoperiods may facilitate a generally low level of metabolism and an overall reduction in body temperature, factors perhaps associated with the reacquisition of photosensitivity.

As testicular recrudescence begins in starlings under LD 12:12, body temperature rises as anticipated, presumably as a function of increased physiological-endocrinological activity. Photophase temperatures are slightly elevated during this stage of the reproductive rhythm, followed by further elevations during the photophase and scotophase as spermatogenesis is attained. It is likely that this response reflects an overall increase in metabolic rate, perhaps due to the hormonal characteristics of breeding condition (Van Tienhoven, 1968). In addition, the possibility should not be overlooked that the altered physiological state detected by temperature elevations is related to subsequent testicular involution.

Hypothesis: Circadian Control of the Circannual Reproductive Rhythm

The perch-hopping activity patterns and deep-core body temperatures of male starlings during the reproductive rhythm under LD 12:12 make possible a hypothesis explaining the testicular cycles characteristic of this photoperiodic treatment: Initially, LD 12:12 photoperiods (and perhaps all daily photophases less than 12L) entrain circadian oscillations which result in a physiological state conducive to the termination of photorefractoriness and spontaneous testicular recrudescence. Entrainment of the controlling circadian oscillator is reflected by the conformity of locomotor activity patterns to the daily 12-hour photophase. During this period, body temperatures reflect the physiological state of the birds as they complete the post-nuptial molt, and may later indicate a reduced metabolic rate that allows the termination of photorefractoriness. As testicular recrudescence begins and spermatogenesis is attained, the circadian oscillator controlling activity, and presumably the gonadotropic output of the hypophysis, is altered. It seems likely that this alteration occurs as a consequence of reproductive condition via the feedback action of gonadal steroids and/or hypophyseal gonadotropins. Extensions of daily activity patterns reflect this alteration, and elevations in core temperatures are indicative of an "excited" metabolic state. Phase alteration of the circadian oscillator during this period ultimately results in testicular involution, such as occurs in starlings maintained under photoperiods greater than 12L. Elevated body temperatures (increased metabolic rate) may in some way contribute to the eventual reduction of gonadotropin synthesis and secretion, leading to testicular regression. Again, this effect may be similar to that occurring in starlings maintained under daily light/dark schedules greater than LD 12:12. Apparently the spermatogenic phase is prolonged in starlings under daily photophases shorter than 12L because the gonadotropin-controlling circadian oscillator under the entraining influence of "short" photoperiods is not sufficiently altered by the reproductive state to result in testicular involution. Once testicular involution is complete under LD 12:12, the entire sequence can be, and apparently is repeated, thus generating the so-called endogenous circannual reproductive rhythm in the male starling. There is absolutely no need to involve a hypothetical endogenous circannual oscillator in this response!

According to the hypothesis just proposed, it is only coincidental that the circannual reproductive rhythm in starlings under LD 12:12 has a periodicity of "about" one year. In fact, this periodicity probably represents evolutionary selection of the optimum rate for reproductive cycles to occur under an essentially nonoscillating equatorial photoperiod, as suggested for other avian species by Lofts & Murton (1968). The period of testicular quiescence between testis growth cycles allows time for the energetically demanding, biologically necessary, post-nuptial molt to take place without temporally overlapping the reproductive phase of the "circannual" rhythm.

As indicated in the introduction, circannual rhythms of reproduction with periodicities of eight to ten months have been documented in several resident equatorial species. It would appear that the circannual reproductive rhythm in European starlings represents retention of a basic equatorial strategy, perhaps reflecting relatively recent extension of their geographic distribution to the temperate zone. Whether the documentation of endogenous circannual periodicities in other birds can also be explained by the hypothesis developed herein for the starling has not been determined. It seems likely, however, particularly with regard to experiments conducted under LD 12:12, that such an explanation should be seriously considered and experimentally tested. While I do not accept the virtually impossible task of refuting the existence of an endogenous circannual oscillator in the European starling, I do believe that the reproductive cycles observed under LD 12:12 can best be explained by the interaction of photoperiodically entrained daily rhythms and physiological conditions associated with the reproductive state. This hypothesis is based on established biological concepts and physiological realities, and thus is amenable to further testing.

SUMMARY

1. Circannual rhythms have been documented in several avian species with regard to physiological processes including molt, fat deposition, nocturnal restlessness, and reproduction. Most of this evidence was

obtained from experiments conducted under stable light/dark schedules rather than continuous light or dark.

2. Reproductive cycles in European starlings under natural north-temperate photoperiods are controlled by seasonal fluctuations in daily photophase durations above and below a 12-hour "threshold". These fluctuations assure the relatively accurate 12-month periodicity of the reproductive rhythm in temperate-zone environments.

3. Starlings held under stable LD 12:12 photoperiods exhibit a circannual reproductive rhythm, with testicular growth-involution cycles repeated about every ten months. Previous interpretation of this result has involved the assumption that an endogenous circannual oscillator is controlling the response.

4. Free-running circadian oscillations, as reflected by rhythms in activity and body temperature in starlings held in constant darkness, are implicated as the controlling stimulus for testicular recrudescence and spermatogenesis in starlings maintained in constant darkness. It is unnecessary to invoke a hypothetical circannual oscillator to explain this response.

5. During periods of testicular quiescence under LD 12:12, perch-hopping activity in starlings is confined almost entirely to the daily 12-hour photophase. Daily periods of perch-hopping are extended into the scotophase in starlings undergoing testicular recrudescence and spermatogenesis. These modified daily rhythms presumably reflect alteration of a controlling circadian

oscillator.

6. Elevated core body temperatures recorded during the portion of the testicular rhythm just following complete gonadal involution under LD 12:12 are presumably due to the simultaneous occurrence of the post-nuptial molt. Prior to testicular recrudescence, starling core temperatures drop to significantly lower levels. The post-nuptial molt has been completed by this time, and temperatures apparently reflect a return to normal nonreproductive levels. As testicular recrudescence begins, core temperatures during the daily 12-hour photophase rise slightly. Scotophase temperatures remain at the low levels established prior to recrudescence. Spermatogenesis is characterized by relatively elevated core temperatures during the photophase and the scotophase. These elevations are presumably due to the effects of gonadal steroids, hypophyseal gonadotropins, or both.

7. The circannual rhythm of reproduction in male starlings under LD 12:12 photoperiods can be best explained by the interaction of circadian oscillations and the reproductive state induced by these oscillations. A hypothesis is proposed which suggests that LD 12:12 initially entrains the circadian clock responsible for the gonadotropic output of the hypophysis and locomotor activity rhythms. This entrainment is reflected by conformity of activity to the daily 12-hour photophase during the period of testicular quiescence, and subsequent termination of photorefractoriness leading to spontaneous testicular recrudescence. Once the reproductive state is achieved, extensions in daily activity periods and elevated core

body temperatures occur. These modifications likely reflect alteration of the underlying circadian control system, an alteration ultimately causing testicular involution. A return to the initial quiescent state allows the reproductive cycle to repeat, producing the circannuan rhythm of reproduction in the male European starling.

8. The circannual periodicity of the reproductive rhythm in starlings under LD 12:12 most likely represents evolutionary selection of the optimum reproductive periodicity consistent with other mutually exclusive biological events such as molt under this simulated equatorial photoperiod.

ACKNOWLEDGEMENTS

This paper has been submitted to the Graduate Division, University of California, Davis, in partial fulfillment for the degree of Doctor of Philosophy. I am especially indebted to Dr. Robert G. Schwab for guidance in the design of experiments and assistance in the preparation of the manuscript. I also thank Dr. William M. Hamner and Dr. Wilbor O. Wilson for critical appraisal of the manuscript, and Dr. Hiram W. Li for help in the statistical analysis of the data. Financial support was provided by the State of California Starling Control Program.

REFERENCES

ASCHOFF, J. (1955). Jahresperiodik der Fortpflanzung bei Warmblutern. Stad. Gen. 8, 742-776.

ASCHOFF, J., KLOTTER, K. & WEVER, R. (1965). Circadian vocabulary. Circadian Clocks. North-Holland Publ. Co., Amsterdam. pp. x-xix.

ASHMOLE, N. P. (1962). The black noddy *Anous tenuirostris* on Ascension Island. Part 1. General biology. Ibis 103b, 235-273.

BENOIT, J., ASSENMACHER, I. & BRARD, E. (1956). Apparition et maintien de cycles sexuels non saisonniers chez le Canard domestique place pendant plus de trois ans a l'obscurite totale. J. Physiol., Paris 48, 388-391.

BENOIT, J., ASSENMACHER, I. & BRARD, E. (1959). Action d'un eclairement permanent prolonge sur l'evolution testiculaire du canard Pekin. Arch. Anat. Microscop. Morphol. Exptl. 48, 5-11.

BISSONNETTE, T. H. (1931). Studies on the sexual cycle in birds. IV. Experimental modification of the sexual cycle in males of the European starling (*Sturnus vulgaris*) by changes in daily period of illumination and of muscular work. J. Exp. Zool. 58, 2810320.

BISSONNETTE, T. H. (1933). Light and sexual cycles in starlings and ferrets. Quart. Rev. Biol. 8, 201-208.

BUNNING, E. (1936). Die endonome Tagesperiodik als Grundlage der photoperiodischen Reaktion. Ber. Deut. Bot. Ges. 54, 590-607.

BUNNING, E. (1964). The Physiological Clock. Academic Press, New York.

BURGER, J. W. (1947). On the relation of day length to the phases of testicular involution and inactivity of the spermatogenic cycle of the starling. J. Exp. Zool. 105, 259-268.

BURGER, J. W. (1949). A review of experimental investigations on seasonal reproduction in birds. Wilson Bull. 61, 211-230.

BURGER, J. W. (1953). The effect of photic and psychic stimuli on the reproductive cycle of the male starling, *Sturnus vulgaris*. J. Exp. Zool. 124, 227-239.

BURGER, J. W., BISSONNETTE, T. H. & DOOLITTLE, H. D. (1942). Some effects of flashing light on testicular activation in the male starling (*Sturnus vulgaris*). J. Exp. Zool. 90, 73-82.

CHAPIN, J. P. (1954). The calendar of wideawake fair. Auk 71, 1-15.

CHAPIN, J. P. & WING, L. W. (1959). The wideawake calendar 1953-1958. Auk 76, 153-158.

DIXON, W. J. (1971). BMD Biomedical Computer Programs. University of California Press, Berkeley.

DOL'NIK, V. R. (1964). Photoperiodic control of endogenous rhythm of avian sex cycle. Zool. Zh. 43, 720-734.

DORWARD, D. F. (1963). Comparative biology of the white booby and the brown booby *Sula spp*. at Ascension. Ibis 103b, 174-220.

FARNER, D. S. (1965). Circadian systems in the photoperiodic responses of vertebrates. Circadian Clocks. North-Holland Publ. Co., Amsterdam. pp. 357-369.

FARNER, D. S. (1970). Predictive functions in the control of annual cycles. Environ.

Res. 3, 119-131.
FARNER, D. S. & FOLLETT, B. K. (1966). Light and other environmental factors affecting avian reproduction. J. Anim. Sci. 25, 90-115.
FARNER, D. S. & LEWIS, R. A. (1971). Photoperiodism and reproductive cycles in birds. Photophysiology, vol. VI. Academic Press, New York. pp. 325-370.
FARNER, D. S., LEWIS, R. A. & DARDEN T. (1973). Photostimulation of gonads in birds: The Bunning hypothesis and responses to very short daily photoperiods. International Congr.: Le soleil au service de l'homme, Paris.
FOLLETT, B. K. & SHARP, P. J. (1969). Circadian rhythmicity in photoperiodically induced gonadotropin release and gonadal growth in the quail. Nature 223, 968-971.
FOLLETT, B. K. & NICHOLLS, T. J. (1973). Daily rhythms of gonadotropin secretion in quail when gonadal development is initiated by long daylengths. International Congr.: Le soleil au service de l'homme, Paris.
GWINNER, E. (1968). Circannuale Periodik als Grundlage des jahreszeitlichen Funktionswandels bei Zugvogeln. Untersuchungen am Fitis (*Phylloscopus trochilus*) und am Waldlaubsanger (*P. sibilatrix*). J. Ornithol. 109, 70-95.
GWINNER, E. (1971). A comparative study of circannual rhythms in warblers. Biochronometry. National Academy of Sciences, Wash., D.C. pp. 405-427.
HAMMOND, J., JR. (1954). Light regulation of hormone secretion. Vitamins Hormones 12, 157-204.
HAMNER, W. M. (1963). Diurnal rhythm and photoperiodism in testicular recrudes-

cence of the house finch. Science 142, 1294-1295.

HAMNER, W. M. (1964). Circadian control of photoperiodism in the house finch demonstrated by interrupted-night experiments. Nature, Lond. 203, 1400-1401.

HAMNER, W. M. (1966). Photoperiodic control of the annual testicular cycle in the house finch, *Carpodacus mexicanus*. Gen. Comp. Endocrinol. 7, 224-233.

HAMNER, W. M. (1968). The photorefractory period of the house finch. Ecology 49, 211-227.

HAMNER, W. M. (1971). On seeking an alternative to the endogenous reproductive rhythm hypothesis in birds. Biochronometry. National Academy of Sciences, Wash., D.C. pp. 448-462.

HAMNER, W. M. & ENRIGHT, J. T. (1967). Relationships between photoperiodism and circadian rhythms of activity in the house finch. J. Exp. Biol. 46, 43-61.

JEGLA, T. C. & POULSON, T. L. (1970). Circannian rhythms. I. Reproduction in the cave crayfish, *Orconectes pellucidus inermis*. Comp. Biochem. Physiol. 33, 347-355.

KENDEIGH, S. C. (1941). Length of day and energy requirements for gonad development and egg-laying in birds. Ecology 22, 237-248.

KING, J. R. (1968). Cycles of fat deposition and molt in white-crowned sparrows in constant environmental conditions. Comp. Biochem. Physiol. 24, 827-837.

KING, J. R. (1973). Energetics of reproduction in birds. Breeding Behavior of Birds. National Academy of Sciences, Wash., D.C. pp. 78-107.

LOFTS, B. (1964). Evidence of an autonomous

reproductive rhythm in an equatorial bird (*Quelea quelea*). Nature, Lond. 201, 523-524.

LOFTS, B. & MURTON, R. K. (1968). Photoperiodic and physiological adaptations regulating avian breeding cycles and their ecological significance. J. Zool., Lond. 155, 327-394.

MEIER, A. H., MARTIN, D. D. & MACGREGOR, R. (1971). Temporal synergism of corticosterone and prolactin controlling gonadal growth in sparrows. Science 173, 1240-1242.

MENAKER, M. (1965). Circadian rhythms and photoperiodism in *Passer domesticus*. Circadian Clocks. North-Holland Publ. Co., Amsterdam. pp. 385-395.

MENAKER, M. & ESKIN, A. (1967). Circadian clock in photoperiodic time measurement: A test of the Bunning hypothesis. Science 157, 1182-1185.

MURTON, R. K., BAGSHAWE, K. D. & LOFTS, B. (1969). The circadian basis of specific gonadotropin release in relation to avian spermatogenesis. J. Endocrinol. 45, 311-312.

MURTON, R. K., LOFTS, B. & WESTWOOD, N. J. (1970). The circadian basis of photoperiodically controlled spermatogenesis in the greenfinch *Chloris chloris*. J. Zool., Lond. 161, 125-136.

PITTENDRIGH, C. S. & MINIS, D. H. (1964). The entrainment of circadian oscillations by light and their role as photoperiodic clocks. Amer. Nat. 98, 261-294.

RAWLES, M. E. (1960). The integumentary system. Biology and Comparative Physiology of Birds. Academic Press, New York, pp. 189-240.

RILEY, G. M. (1937). Experimental studies on

spermatogenesis in the house sparrow, *Passer domesticus (Linnaeus)*. Anat. Rec. 67, 327-351.

RILEY, G. M. (1940). Light versus activity as a regulator of the sexual cycle in the house sparrow. Wilson Bull. 52, 73-86.

RISSER, A. C. (1971). A technique for performing laparotomy on small birds. Condor. 73, 376-379.

ROWAN, W. (1925). Relation of light to bird migration and development. Nature 115, 494.

ROWAN, W. (1929). Experiments on bird migration. I. Manipulation of the reproductive cycle: Seasonal histological changes in the gonads. Proc. Boston Soc. Nat. Hist. 39, 151-208.

RUTLEDGE, J. T. & SCHWAB, R. G. (1974). Testicular metamorphosis and prolongation of spermatogenesis in starlings (*Sturnus vulgaris*) in the absence of daily photostimulation. J. Exp. Zool. 187, 71-76.

SCHWAB, R. G. (1970). Light-induced prolongation of spermatogenesis in the European starling, *Sturnus vulgaris*. Condor 72, 466-470.

SCHWAB, R. G. (1971). Circannian testicular periodicity in the European starling in the absence of photoperiodic change. Biochronometry. National Academy of Sciences, Wash., D.C. pp. 428-447.

SCHWAB, R. G. & RUTLEDGE, J. T. (1973). Effects of natural and artificial illumination on testicular metamorphosis in the European starling (*Sturnus vulgaris*): Maturation, involution, and photorefractory phases. International Congr.: Le soleil au service de l'homme, Paris.

TUREK, F. W. (1972). Circadian involvement

in termination of the refractory period in two sparrows. Science 178, 1112-1113.

TURNER, C. D. & BAGNARA, J. T. (1971). General Endocrinology. W.B. Saunders Co., Philadelphia.

VAN TIENHOVEN, A. (1968). Reproductive Physiology of Vertebrates. W.B. Saunders Co., Philadelphia.

VAN TIENHOVEN, A. & PLANCK, R. J. (1971). The effect of light on avian reproductive activity. Handbook of Physiology. American Physiological Society, Wash., D.C. pp. 79-107.

WIESELTHIER, A. S. (1969). The stimulation and prolongation of testicular activity in the European starling, *Sturnus vulgaris*, by thyroidectomy and reduced photoperiod. (Ph.D. thesis). Ithaca, N.Y.: Cornell University.

WITSCHI, E. (1961). Sex and secondary sexual characters. Biology and Comparative Physiology of Birds. Academic Press, New York. pp. 115-168.

WOLFSON, A. (1959). Role of light and darkness in the regulation of spring migration and reproductive cycles in birds. Photoperiodism and Related Phenomena in Plants and Animals. American Association for the Advancement of Science, Wash., D.C. pp. 679-716.

WOLFSON, A. (1965). Circadian rhythm and the photoperiodic regulation of the annual reproductive cycle in birds. Circadian Clocks. North-Holland Publ. Co., Amsterdam. pp. 370-387.

WOLFSON, A. (1966). Environmental and neuroendocrine regulation of annual gonadal cycles and migratory behavior in birds. Recent Progr. Hormone Res. 22, 177-239.

WURTMAN, R. J. (1967). Effects of light and visual stimuli on endocrine function. Neuroendocrinology, vol. II. Academic Press, New York. pp. 19-59.

THE ADAPTATIONAL VALUE OF INTERNAL ANNUAL CLOCKS IN BIRDS

HELMUT KLEIN

Max-Planck-Institut für
Verhaltensphysiologie
Abteilung Aschoff, Erling-Andechs/Obb.
West Germany

I. INTRODUCTION

Precise timing of annual events like breeding or molting is of great adaptational value (Moreau, 1950; Lack, 1950 and 1954; Aschoff, 1955; Nalbandov, 1958), and in recent years it has been well established that in several bird species an endogenous annual rhythm contributes to the timing of seasonal events. For detailed information see Berthold (this volume). Such a "circannual" clock enables birds and other animals to anticipate ensuing changes of conditions in the environment, i.e. to be prepared in advance to react properly to these conditions. (Brehm, 1881; Moreau, 1950; Aschoff, 1955 and 1967; A.J. Marshall 1959,

1960a and 1960b; Immelmann, 1963 and 1967; Pengelley and Kelley, 1966; Gwinner, 1972a and 1972b). To realize the advantage of this adaptive mechanism, the clock has to be in synchrony with, and in a certain phase-relationship to, the periodically changing environment. Synchronization and phase-control are provided by periodic cues from the environment. These cues, in addition, may have a direct influence on some of the biological events within the whole cycle which are coupled to the basic clock or triggered by it the way an event is set off by the switch slides on the dial of a time-switch. For more information see Immelmann (1963, 1967 and 1971) and Farner and Lewis (1971 and 1973). As will be shown in the following pages, the adaptive value of the system depends on the rigidity or "inertia" of the clock, as well as on the flexibility of some of the processes coupled to it. The feature of temporal organization, discussed here mainly for a circannual clock and its synchronization mechanisms, can also be applied to parts of the full cycle or to temporal program respectively which have to be set off each year anew, i.e. for birds which do not have a "true" circannual clock.

II. THE SIGNIFICANCE OF A CIRCANNUAL CLOCK FOR PROPER TIMING

2.1 Irregularities of the photoperiod.

For most birds living not too close to the equator, the main annual timing factor is the seasonally changing photoperiod. This has been demonstrated in a variety of experiments in which the normal temporal course of seasonal events was altered by artificially manipulating the photoperiod (Rowan, 1929; Burger, 1947; Damste, 1947; Wolfson, 1954; Emlen, 1969; Gwinner, 1971b and 1973; Farner and Lewis, 1971 and 1973). Such effects could be explained as a result of direct responses of the organism to photoperiodic "release" stimuli. Pursuing this kind of reasoning it could be argued that proper timing is achieved by such responses to the photoperiod with sufficient accuracy, since photoperiod is the most reliable source of information about the season. However,the regular changes of photoperiod in the course of the year can be obscured by irregularities caused by overcast skies. Depending on latitude and season, clouds can reduce the effective day length by several hours. For instance, at 50 degrees latitude in min June (fig. 1), effective day length can be shortened by about one hour, simulating photoperiodic conditions of more than a month earlier or 1.5 months later. In addition, weather factors may even temporarily invert the sign of changes of the photoperiod which provides the information as to whether it is the first or second half of the annual cycle. Analogous considerations can be made for temperature, precipitation or other environmental factors as potential time cues.

It is therefore obvious that precision of timing of annual events could be increased if an organism had an internal time-measuring device which would run for at least several months and which would be inert enough to ensure that irregular fluctuations in environmental conditions had only small immediate effects on the timing. The extreme case of such a system would be a continuously running, self-sustaining oscil-

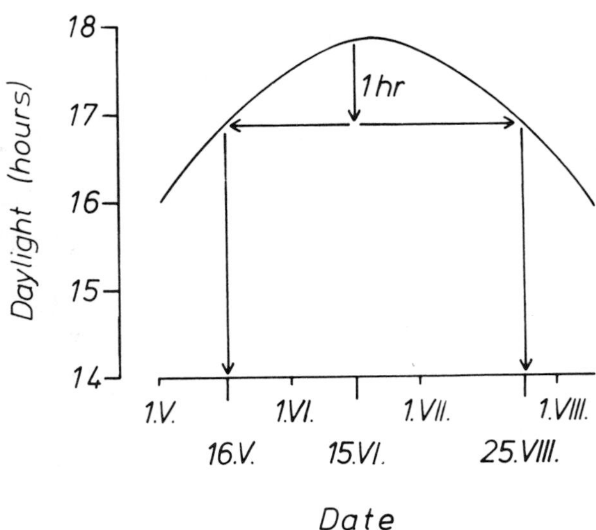

Fig. 1: Temporal course of the civil daylight at 50° N from May to August. The downwards arrow at June 15th demonstrates the effect that a heavy cloud cover would have on actual daylength. The following arrows indicate what dates would be indicated by that actual daylength.

lation, i.e. a circannual clock. While release-mechanisms are more likely to be influenced by the "noise" of the environmental time cues, a circannual clock which runs by itself and is only synchronized by zeitgebers, such as the photoperiod, is less susceptible to irregular variations of the zeitgeber.

2.2 Timing in long distance migrants and in equatorial breeders.

Transequatorial migrants. Besides the problems discussed above which are caused by characteristics of the environment, there are cases where birds themselves produce "irregularities" in the environment by actively changing the photoperiodic conditions to which they are exposed in the course of the year. Many migratory species have their wintering grounds on the hemisphere opposite to that of their breeding grounds. These birds are exposed to two cycles of relatively long and short days per year. Since it has been shown that in birds it is possible to induce several breeding cycles per year by exposing them to more than one artificial photoperiodic cycle within 12 months (Miyazaki, 1934; Burger, 1947; Damste, 1947; Wolfson, 1954; Emlen, 1969), the question arises why these transequatorial migrants do not breed during the "wrong" season and why the gonads of these birds do not even grow during wintering time (Berthold, *et al*. 1971 and 1972b for *Sylvia borin*; Marshall and Serventy, 1959; Marshall, 1960b for *Puffinus tenuirostris*; Engels, 1964; Hamner and Stocking, 1970 for *Bobolinks*; Curry-Lindahl, 1963 for *Motacilla Flava*). The necessary non-responsiveness to the increasing photoperiods experienced in the winter quarters can be due to an extended refractory period (Immelmann, 1971; Farner and Lewis, 1971 and 1973) even in a species without a circannual clock. But it is certainly less surprising in a species where the whole cycle can be controlled by an endogenous program as is the case for the Garden warbler (Berthold *et al.* 1972b).

Birds wintering close to the equator. Some populations of migratory birds spend part of the year in areas close to the equator. It has been pointed out that in this zone there are very poor, if any, environmental factors indicating the phase of the annual cycle (Marshall and Williams, 1959; Miller, 1959a). This means that birds wintering there cannot depend on environmental information for the release and timing of their spring migration. But many of them do return with only very little variation in the dates of arrival from year to year (Cooke, 1913; Phillips, 1913; Bretscher, 1920; Marshall and Williams, 1959, and Klein, *et al.* 1973). It was assumed that this requires an inert clock mechanism which runs with good accuracy at least for the time these animals live in the areas without reliable annual time-cues (Rowan, 1926; Chapin, 1932 and 1954; Baker and Baker, 1936; Baker and Ranson, 1938; Moreau, *et al.* 1941; Moreau, 1950 and 1972; Marshall, 1960b and Gwinner, 1972a). Again, a circannual clock serves this purpose and guarantees the adaptational value of returning to the breeding ground and of breeding at an optimal time as for instance shown in the Willow warbler (Gwinner, 1968a).

Equatorial breeders. The synchronization of the individuals of a population in a certain phase relation to an environmental cycle is also the main mechanism of synchronizing the individuals with one another. In areas close to the equator environmental annual cycles, especially that of the photoperiod, are weak and we have information on many breeding populations of birds, especially from tropical islands, that these birds can breed during all seasons, but that breeding nevertheless occurs in synchrony within one population, and that it often occurs every 9 to 10 months (Chapin, 1954; Chapin and Wing, 1959; Stonehouse, 1962; Snow, 1965; Ashmole, 1965 and 1968; Snow and Snow, 1967). These observations can be interpreted by assuming that these birds have:

a) a mechanism providing social synchronization of internal clocks of the whole population, or
b) an internal clock, like in case a), which is synchronized by non-annual cues such as the lunar cycle, providing a mechanism called "synchronization by demultiplication", or
c) a mechanism using other than annually cycling environmental factors to build a time measuring device based on counting relatively short environmental cycles like that of the moon.

We do not know whether a mechanism classified under a) is realized. The only indications that it could be, are the cases where we know that social influences can affect the timing of the sexual cycle (Lewis and Orcutt, 1971).

With respect to b) and c), there are strong indications, discussed by Chapin and Wing (1959) and Ashmole (1963) for *Sterna fuscata* from Ascension Island that these birds always return after 10 lunar cycles. But so far there is no way to decide whether we are dealing with a real counting mechanism (case c) or a system where an internal circannual clock is running with a period length of about 10 lunar cycles, which is triggered by the moon (case b).

In these cases the circannual clock would not serve the function of adapting the biological rhythm to an environmental cycle but rather would provide the basis for a mechanism of temporal coordination within the population, improving reproductive success.

In addition it could be an adaptational mechanism to increase the number of offspring per year in a situation where the maximum possible number of offspring per breeding season is already reached, as was discussed by Ashmole (1965) and Immelmann (1967).

III. ADAPTIVE FEATURES FOR MIGRATION

3.1 Temporal programs for distance and direction.

Control of distance. To reach its species- and population-specific winter quarters a first year bird not migrating together with experienced con-specifics needs information about direction and distance of travel. According to a hypothesis proposed by Gwinner (1968a and b), the distance traveled on the first migration is the result of an endogenous time program that induces as many hours of migratory flight as are necessary to bring the bird from its breeding grounds to its wintering area. This hypothesis is supported by the results of several experiments which I will summarize here.

a) In Figure 2, where data from Gwinner (1968a and b), Berthold *et al.* (1972c), and Berthold (1973) have been used, it is shown that for *Sylvia* and *Phylloscopus* species there is a good correlation between the number of hours with migratory restlessness as measured in caged birds and the distance normally traveled during migration. But this method does not allow conclusions to be drawn about the absolute distance traveled.

b) Absolute distance may be estimated by the method of Gwinner (1968a), where he used data of *Phylloscopus* warblers to find out the distance covered during a particular part of the migration. After determining the distance equivalent of an hour of restlessness from banded and recovered birds, he could calculate the distance the birds would have traveled, had they been flying in the right direction. The result is presented in Figure 3. The average points of termination of migration are well within the natural wintering areas of the two species.

c) The next step was to test whether the number of hours with restlessness would represent the number of hours the birds would spend in actual migratory flight. Therefore, the speed of migratory flight (speed of the maximum range) was calculated for different species after Pennycuick (1969). Now as the total time of restlessness and the flying speed of our warblers is known, Gwinner (unpublished) could calculate the distance these birds would have covered, had they spent all the time of restlessness in actual

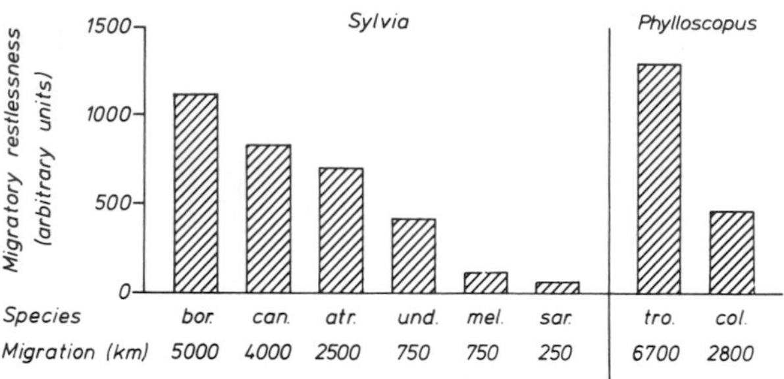

Fig. 2: The total amount of nocturnal restlessness exhibited by several species of *Sylvia* and *Phylloscopus* warblers during their first fall migratory season in an LD 12:12 photoperiod. Migratory restlessness is expressed as the total number of half-hour intervals during which a bird showed nocturnal activity; mean values for the 6 to 18 individuals of each experimental group are given. The numbers underneath each column indicate the approximate distance between breeding grounds and wintering area of the various species. The species are *Sylvia borin, S. cantillans, S. atricapilla, S. undata, S. melanocephala, S. sarda, Phylosocopus trochilus, P. collybita).* Data from Berthold (1973) and Gwinner (1972b).

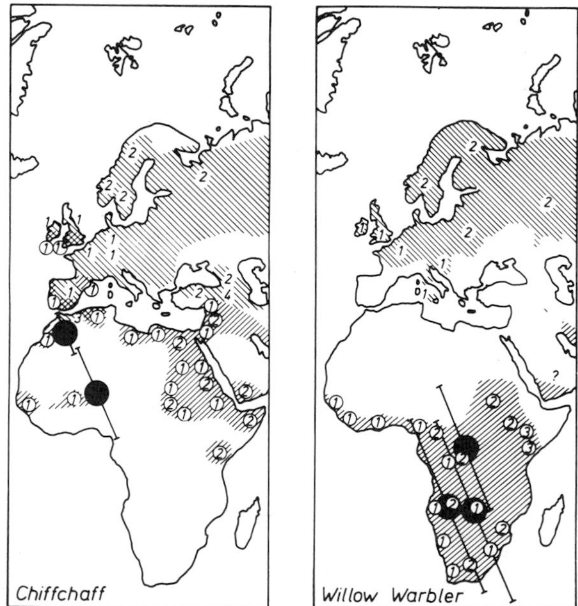

Fig. 3. Breeding areas and winter quarters of Chiffchaff *(Phylloscopus collybita)* and Willow warbler *(P. trochilus).* Numbers refer to various races of either species (numbers encircled represent winter quarters). Large solid circles are calculated endpoints of migration (with standard errors) for birds kept under different experimental conditions. (From Gwinner 1972)!

flight. This approach also yields satisfactory results.

d) After these experiments it was not yet clear whether the underlying mechanism is a direct time measuring system or whether it measures time indirectly, via the energy consumption during restlessness or flight respectively. Gwinner (1974) made use of the fact that "total darkness" drastically reduces nocturnal activity in Garden warblers. He recorded migratory restlessness of two groups of *Sylvia borin* throughout the fall migratory season, exposing the birds to a constant 12:12 hour light-dark cycle with bright light of 200 lux during L and about moon-light (0.01 lux) during D.

For one group, the night lights were switched off for 8 weeks, so that the restlessness was reduced to almost zero. This together with the fact that the body weight of this group was higher during the time with night lights off, makes it very likely that these birds during the 8 weeks spent much less energy than the control birds which had night lights and were normally restless. Nevertheless, both groups terminated restlessness at the same time, strongly suggesting the dominant significance of an internal temporal program.

It is obvious that a mechanism like the time program for distance orientation described above must be relatively inaccurate, because there will always be unpredictable morphological and physiological irregularities within the birds and irregularities in the environment that lead to variations. So one would expect that a system like this is linked up with some mechanisms to correct for possible errors.

Gwinner (1971b) transferred one group of Willow warblers before the end of their autumnal period of restlessness from photoperiods of about LD 13.5:10.5 to relatively short days (LD 12:12), simulating conditions of areas further north than the proper winter quarter. A second group was transferred from identical conditions as the others into a relatively long day (LD 18:6), simulating conditions of areas further south than the proper winter quarter. Then the beginning of the postnuptial molt and the termination of restlessness were recorded. Relative to the total mean value this termination was about 30 days early in the long-day-group and about 30 days late in the short-day-group. This suggests that there is a photoperiodic influence on the termination of migration, assuming that birds are kept migrating while still too far north and brought down when they overshoot the proper latitude.

Control of directional changes. Many migratory birds, especially during autumn migration in the European - African system, do not migrate over the shortest possible route from the breeding to the wintering grounds (for examples, see Klein, *et al.*, 1973; Schüz, 1971; Zink, 1973). Many of them fly first SW and later turn towards the SE or vice versa. It seems possible that this change of direction might also be a result of interaction with the internal annual clock (Gwinner, 1971a).

Earlier it was expected that the fact that migrants of the northern hemisphere fly southwards in autumn and northwards in spring was due to changes in the environment (Sauer, 1957; Sauer and Sauer, 1960). But

Emlen (1969) showed that the differences in migration directions of the *Passerina cyanea* are also dependent on the phase of an internal time measuring system. He trapped birds during autumn migration, kept one group under a simulated natural photoperiod, and manipulated the photoperiod for group 2 in such a way that the "clock" was phase shifted, i.e. that the birds came into autumnal migration condition at the time of natural spring migration. When tested for orientational preference, the control group showed a clear orientational tendency towards north, while the phase shifted birds were oriented southwards.

3.2 Adaptation to biotopes crossed during migration.

Characteristic of migration intensity. If one accepts the idea that migratory restlessness represents the bird's intention to migrate, then one has to pose the more quantitative question, whether changes in intensity of migratory restlessness represent an adaptational pattern. The problem has already been discussed by Berthold, *et al.* (1972c) and Gwinner (1972a).

To test this, I used data from:

a) *Sylvia borin* of S-Germany (Berthold, *et al.*, 1972c), normally wintering close to the equator (Klein, *et al.*, 1973).
b) *Sylvia atricapilla* from S-Finland, normally wintering in Africa about 5° to 10° N (Klein, *et al.*, 1973).
c) *Sylvia atricapilla* from S-Germany (Berthold, *et al.*, 1972c), most of which normally winter in N-Africa between the Mediterranean Sea and the Sahara Desert (Klein, *et al.*, 1973).

d) *Phylloscopus trochilus* from S-Germany (Gwinner, 1972b), wintering normally about 10° S.

e) *Phylloscopus collybita* from S-Germany (Gwinner, 1972b), most of which normally winter in Africa, north of the Sahara Desert.

The curves of intensity of restlessness, originally drawn as a function of season, were transformed into curves reflecting intensity of migratory drive at different locations along the migratory route (fig.4).

On the route from Europe to Africa the Mediterranean Sea and the Sahara Desert are especially difficult to cross, because there is practically no resting biotope and no food. As can be seen in figure 4, all curves reach high values during the part that corresponds to the migration over the Mediterranean Sea and the Sahara Desert, regardless whether they start at 60° N or at 49°N. Most of the S-German Blackcaps and Chiffchaffs, wintering in the narrow band of favourable land between the Mediterranean Sea and the Sahara Desert, show distributions which reach their maxima at about the same latitude as the long distance migrants. Then they fall back to zero very fast, thus preventing the animals from "being pushed" into the desert. This suggests that the temporal course of migration-intensity in these species may be controlled by an internal program that adapts speed of migration to ecological conditions of the countries crossed.

Amount of fat deposited. Most migratory birds build up fat deposits before or during migration. This deposition is at least partly under the control of the inter-

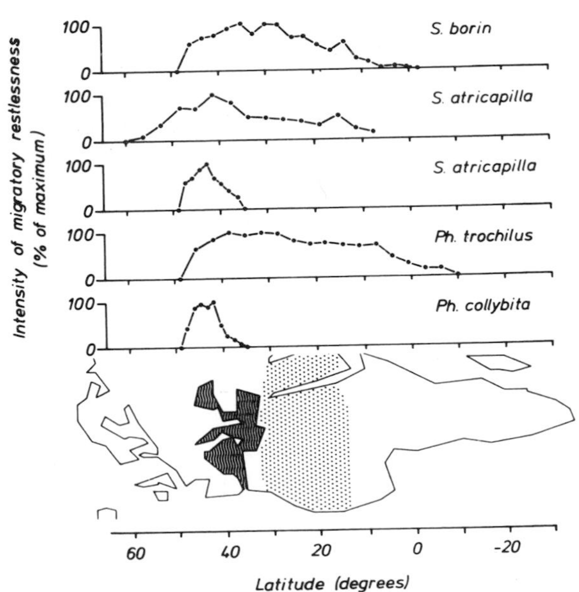

Fig. 4. Curves of intensity of migratory restlessness of five groups of European warblers of the genera *Sylvia* and *Phyllosocopus.* Data were obtained from groups of 6 to 18 individuals kept under photoperiodic conditions similar to those experienced by these birds during actual migration; From *S. atricapilla* individuals of a northern and of a southern population were studied. Originally the curves gave intensity of migratory restlessness as a function of season. For this graph they were redrawn as a function of latitude instead of time by stretching the curves along the abscissa so that they span the latitudes from the breeding grounds of the experimental birds to those of their winter quarters. The bottom figure gives a map of Europe and Africa to illustrate latitudes and the position of the Mediterranean Sea and the Sahara Desert, (Data from Berthold *et al.* 1973, Gwinner, 1972b and original).

nal annual clock, because it may also be observed in experimental groups under constant laboratory conditions (King, 1968; Gwinner, *et al.*, 1971 and 1972; Berthold, *et al.*, 1972a, and Berthold, this volume). In addition, it is known that long distance migrants build up larger deposits than short distance migrants (Berthold, 1971 and in press). This difference is especially obvious between species that do or do not cross large ecological barriers like the Sahara Desert or the Gulf of Mexico.

For an intraspecific comparison, data from three populations of *S. atricapilla* with different migratory habits were used (Klein *et al.*, 1973).

a) Birds from Finland, which cross the Mediterranean Sea and the Sahara Desert on their migration;
b) birds of S-Germany, which cross only the Mediterranean Sea, and
c) birds from the Cape Verde Islands, which do not migrate at all.

The Finnish birds have 3.5 g fat by the time when they would reach the Mediterranean Sea. The German birds only gain 1.5 g during the time of autumn migration, and those from the Cape Verde Islands do not build up any fat depots at all. This suggests also that the amount of fat deposited under the control of the internal annual clock, in interaction with the photoperiod, is adaptive.

Time of fat deposition. If the amount of fat deposited is adaptive, then it seems reasonable that the timing of this deposition could also be adaptive. King and Farner (1965) have pointed out that this is the case. In addition, we have an example from European warblers shown in figure 5. It gives the numbers of birds trapped at a banding station in S-Germany and average body weight of the birds caught. We know that at the beginning of the time of passage the Scandinavian birds come through, while the S-German birds have not yet begun migration. The German birds are the last to leave, when all their northern conspecifics have passed. The farther north the origin of the birds, the earlier is their passage (Klein, *et al.*, 1973).

The average body weight of Scandinavian Blackcaps during breeding time is 18 g, and after the first 1000 km of migration, when they reach S-Germany, they average about 18.2 g (figure 5, left side). This means that as long as they travel over favorable land with food and resting biotops they do not carry extra fat. S-German Blackcaps average about 16 g during summer (Berthold, *et al.*, 1970). When they leave their breeding grounds, around the beginning of October, their weight is about 19.5 g (figure 5, right side); this means that they already have 3.5 g deposited when they leave. Figure 6 shows only data from the resident birds of the area (Klein, *et al.*, 1973). They build up the fat depot during the last 10 weeks before they leave.

If we investigate the situation further south, at the northern shore of the Mediterranean Sea (Station Biologique de Tour du Valat, Dr. L. Hoffmann), then we see (figure 7) that, during the first two thirds of passage there, the average body weight is constant 19.4 g; these are all northern breeding birds which are going to cross the two big barriers during the following nights. Later on come successively those

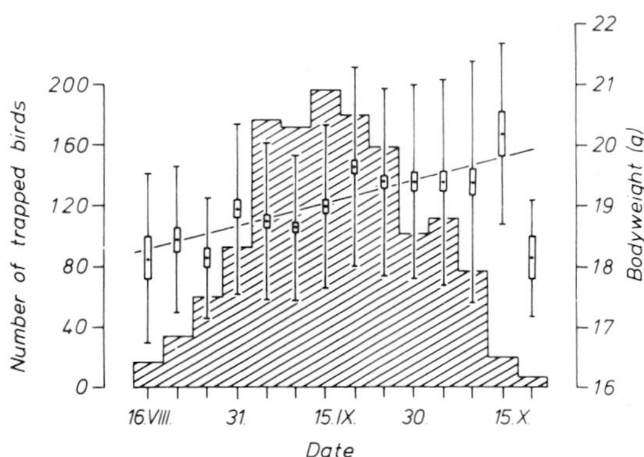

Fig. 5: Number of Blackcaps *(Sylvia atricapilla)* trapped during autumn passage, during 5-day-intervals in S-Germany (histogram) and mean bodyweight of these birds together with the regression-line for these weights as a function of time. (N=1402, b=0.0246 (g/d), r=0.218, the 1 % confidence limits for b are b ± 0.0068 (g/d).

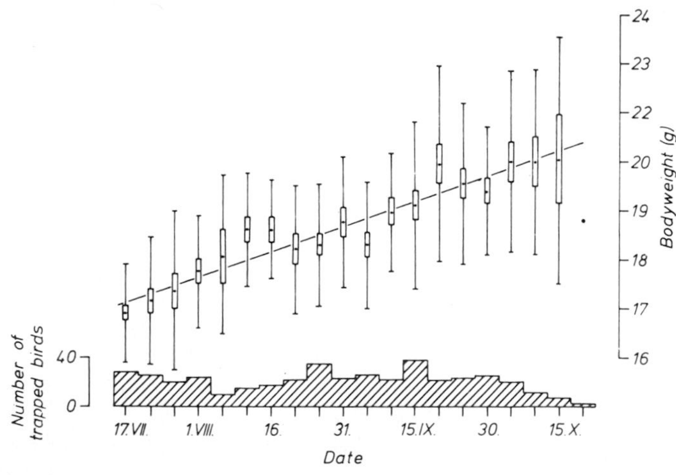

Fig. 6: As figure 5, but for breeding birds of S-Germany before they leave for autumn migration, around the middle of October. (N=414, b=0.0342 (g/d), r=0.517, the 1% confidence limits for b are b ± 0.0063 (g/d).

birds from S-Germany, Switzerland, etc. which cross only the sea, and finally, there is a population which winters there (Klein, *et al.*, 1973). So, not only the amount of fat built up, but also the phase during which it is deposited is different for different populations. A fat deposit such as that necessary to cross ecological barriers, is disadvantageous as long as it is not needed. So it is of adaptive value that these deposits are built up just before they are required.

Time of day of migration. A special mechanism providing adaptation to climatic conditions via behavioural patterns, was discussed by Dorka (1966). Based on field data on passerines, he considers it likely that birds migrating in the early autumn, while it is still very warm, tend to fly during the cooler nights. At that time it is easier for the heat, produced by the exertion of flying, to dissipate. Later in the fall, migration takes place more during daytime. The later in the year, and hence the cooler the time of migration, the greater the tendency to diurnal flight. Here the most adaptive circadian phase of migration may differ between different species and populations.

3.3 Adaptation to differences in climatic conditions.

Differences between eastern and western Europe. If all birds of a species were reactive in the same way to photoperiodic conditions, and as photoperiodic conditions are independent of longitude, one might expect that all birds of a species breed and migrate at the same time in one latitude.

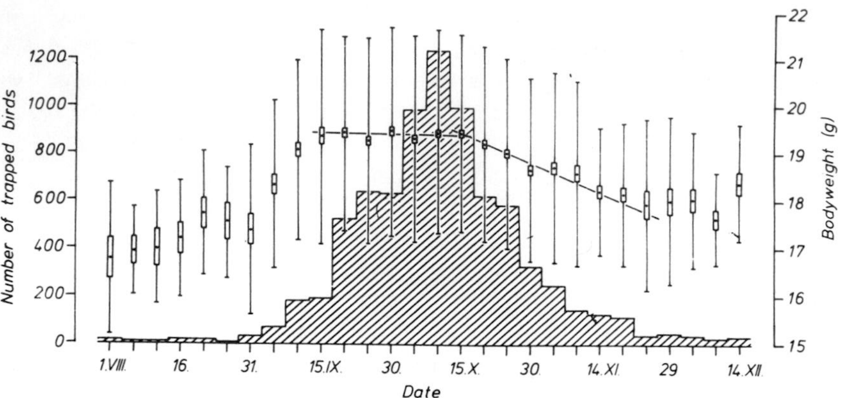

Fig. 7: As figure 6, but for a banding station in S-France (Dr. L. Hoffmann, Station Biologique, Tour du Valat B.d.R.). First regression from September 13th through October 17th: N=5177, b=0.0022 g/d, r=0.0083, the 1 % confidence limits for b are b ± 0.0096 (g/d) second regression from October 13th through November 26th: N=3179, b=–0.0418 g/d, r=–0.2061, the 1 % confidence limits for b are ± 0.0090.

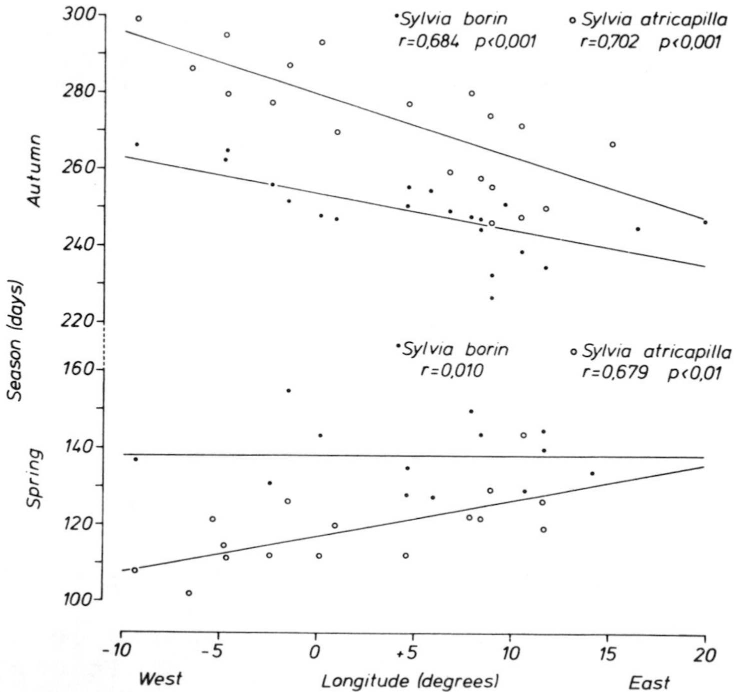

Fig. 8: Sylvia borin (•) and *S. atricapilla* (○): Dependence of the date of the median of passage on the longitude of the locality of banding stations (from Klein, *et al.* 1973 in Vogelwarte 27, 73-134). Ordinate gives numbers of days within the year.

As a result of the influence of the Atlantic Ocean, the climate in western Europe is more of a sea climate, whereas in eastern Europe it is more continental. Consequently, the west part of Europe cools off much later in autumn than the east, and it warms up again earlier in spring. Consequently returning and breeding should be earlier in spring in the west, and migration should take place earlier in the east if populations were adapted to make use of the full period with permissive weather. Indeed we (Klein, *et al.*, 1973) could show (figure 8) that there are clear differences in the time of passage of *Sylvia atricapilla* in autumn and spring, and of *S. borin* at least in autumn. These differences are such that the birds leave later and return earlier in the warmer west than in the east. It is also known that breeding occurs earlier in western Europe (Britain) than on the European continent (Lack, 1950). In coastal California, with its milder climate, breeding is earlier than in more central parts of the American continent (Lindale, 1933; Johnston, 1954; Immelmann, 1971). This suggests that in this case also there are adaptive differences in the interaction between photoperiod and the internal annual clock.

Differences between higher and lower latitudes. Another, more universal, pattern in medium and higher latitudes is that the lower temperatures occur earlier in autumn in the higher latitudes than in lower ones. Thus, the question also arises whether this might have some impact on migratory timing.

Sylvia borin leaves its breeding grounds with only minor differences between northern and southern Europe (about 15° to

20° E), at the beginning of August (Klein, *et al.*, 1973). At that time the temperatures are lower in the north than in the south, but at about their annual maximum in all latitudes. Hence, ambient temperature is not likely to be a releasing factor of autumn migration in this species. The photoperiod (civil daylight) at that date is 16.6:7.4 at 50° N (Frankfurt), and 18.9:5.1 at 60° N (Oslo). It is longer in the northern part of the breeding area and clearly decreasing on the whole northern hemisphere. Since autumn migration starts at the same time in the different parts of the population, we have to conclude that either the reactions to temperature or photoperiod as releasing factors are different, or that the timing of fall migration depends more on internal timing than on proximate factors of the environment.

Sylvia atricapilla leaves later in autumn and returns earlier in spring, so that it "comes closer" to unfavorably low temperatures. The time of beginning autumn migration of *S. atricapilla* in Europe correlates well with the time when the 10° C-isotherm reaches the birds (Klein, *et al.*, 1973). When the local breeding population leaves around September 1st, the photoperiod at 60° N is 15.75:8.25, while it is about 12:12 at 50° N when the population of that latitude leaves on about October 15th.

With these data two interpretations are possible:

a) In *S. atricapilla* ambient temperature, compared to photoperiod, is to a higher degree a proximate factor for the timing of autumn migration.
b) If the efficacy of photoperiod compared to ambient temperature as a proximate

factor (or zeitgeber) is the same, than the reaction to photoperiod is even more different in the different populations of *S. atricapilla* than in *S. borin*. The findings on both species, regardless of whether possibility a) or b) applies for *S. atricapilla*, show that the interactions between the environmental factors and the internal temporal organizations are, in an adaptive way, different in northern and southern breeding-populations.

Another result of differences between climates in different latitudes is that in lower latitudes as compared to higher latitudes, the period with permissive weather for breeding begins earlier and ends later. An interspecific, as well as an intraspecific, comparison shows that breeding starts earlier in the south than in the north (Moreau, 1931; Baker, 1938a, b, and c; Lack, 1950 and 1954; Johnston, 1954; Klopfer, 1962; Haartman, 1963; McArthur, 1964). This means breeding starts as early as possible in the different latitudes. Consequently the young birds are as old as possible at the beginning of winter. In addition, the early termination of the vernal reproductive time allows the adult birds in medium latitudes to go through a second reproductive cycle in early autumn (Serventy and Marshall, 1957; Keast, 1959; Miller, 1959a and b, 1962; Orians, 1960; Selander and Micholson, 1962; Winterbottom, 1963; Serventy and Whittell, 1967; Immelmann, 1971).

In southern Australia, autumnal breeding can amount to up to 5 or 10% of the nesting in spring (Immelmann, 1971). This has the effect that the reproductive rate is increased. So far, this phenomenon and its less complete manifestation, the autumnal singing and display of many birds such as the Blackcaps, were usually attributed to the termination of photorefractoriness, at a time when the photoperiods are still long. Our results from *S. atricapilla* (Berthold, *et al.*, 1972) however, show that the gonadal size of S-German Blackcaps can have a second maximum during autumn, even with constant photoperiods, which is another demonstration of the significance of endogenous temporal programming. For the breeding population of *S. atricapilla* on the Cape Verde Islands (15° N) it is reported by Bannerman (1968, p. 423) that there are two distinct breeding seasons of the species, in January and September respectively, and I have field data from the islands supporting the idea that individual birds can breed twice a year during both seasons (Klein, unpublished).

Climatic differences between summer and winter quarters. Thermal insulation is highly adaptive in endothermic animals (Scholander, *et al.*, 1950a, b, c; Calder and King, 1974). It is known from several species that there are differences in adaptation to winter and summer climate (for literature, see Calder and King, 1974). In birds most of the insulation is provided by plumage. at least in some species, molting and therefore plumage insulation is under the control of the internal annual clock.

A group of Blackcaps was kept in S-Germany, under the naturally changing photoperiod of that latitude, and in a constant temperature of 22° ± 2°C. *Sylvia atricapilla* would breed at about 22°C and winter in N-Africa at about 12°C daily mean temperature. Oxygen consumption was measured as a function of ambient temperature during summer and winter. The results are presented in figure 9. It shows that at temperatures above zero, *S. atricapilla* consumes less oxygen per hour in winter than in summer. This means that if there is no change in body temperature, *S. atricapilla* has better mechanisms to conserve warmth during winter than during summer. In addition, it seems that during summer the birds are not able to increase their metabolic rate above that of +5°C. This is an indication that the lower lethal temperature for these birds may be higher in summer than in winter.

In a parallel experiment with *Sylvia borin* we did not find such differences. This species normally spends the summer in areas with similar macro-climatic conditions as *S. atricapilla*, but during winter it is found in the even warmer central part of Africa. Very recently Biebach (in preparation) demonstrated that in European Blackbirds (*Turdus merula*), thermal insulation is higher during winter than during summer, when they are kept like the warblers mentioned before.

In these cases we find differences in the animals' adaptations to seasonal environmental changes that are not a result of thermal acclimatisation. Rather are they a result of adaptive differences in the reaction to different photoperiods or a result of adaptive changes induced by a circannual clock, possibly together with reactions to photoperiodic changes. These ideas are supported by reports of other authors who found similar seasonal differences in non-migrating birds, from medium or high latitudes, kept under seminatural conditions. Authors investigating birds from lower latitudes, with less pronounced seasonal changes of climate, or birds that do migrate to warm

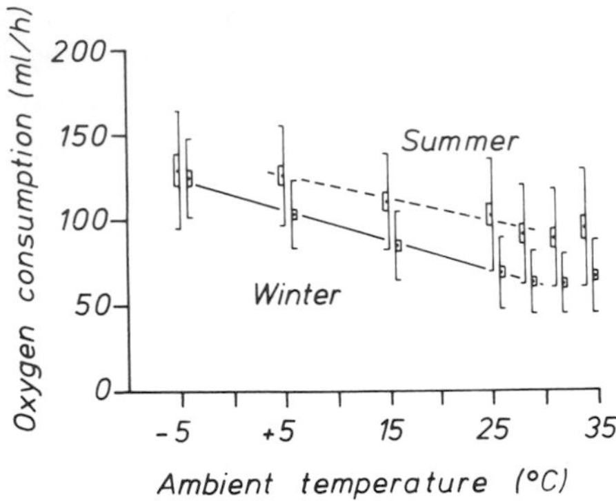

Fig. 9. Oxygen consumption as a function of ambient temperature during summer and winter of Blackcaps *(Sylvia atricapilla)* kept under the natural photoperiod of 48° N. Mean values with standard deviations and standard errors are given (drawn to the left for summer and to the right for winter). Regression line for summer values dashed, for winter values continuous.

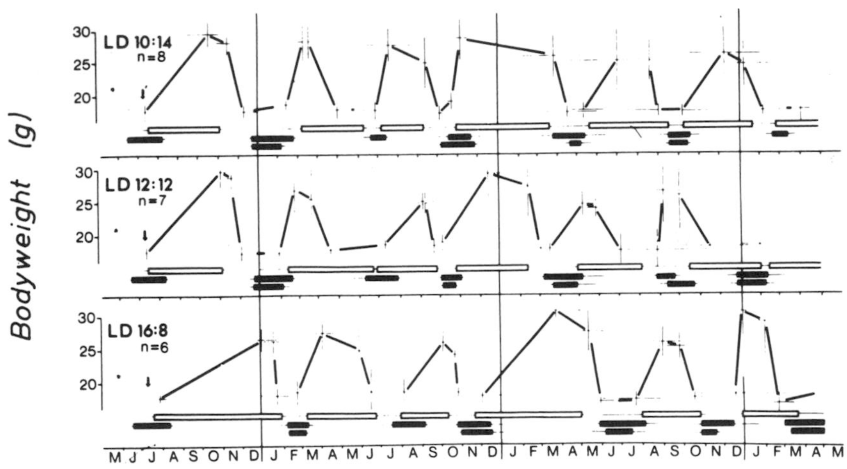

Fig. 10: Circannual rhythms of body weight, migratory restlessness and molt of 3 groups of Garden warblers kept for 2 1/2 years under 3 different constant photoperiods. The curves representing changes in body weight connect the means of the last minima before migration obesity, the means of the first maxima during migration obesity, the means for the last maxima during migration obesity, and the means for the first minima after migration obesity (with the standard errors of the weights and the time). Open bars: time of nocturnal restlessness; black bars: times of molts (top rows body feather molts, lower rows flight feather molts). (From Berthold, *et al.* 1972).

areas did, as a rule, not find seasonal differences in thermal insulation (for citations, see Calder and King, 1974).

IV. CONTROL OF SEQUENCE OF EVENTS

4.1 Juvenile development.

Breeding of birds can take place during very different phases of the environmental annual cycle, between early spring and late summer. In addition, birds can leave their breeding grounds in late summer (mostly late breeders) or late autumn (mostly early breeders). In these two groups, the young birds do not have the same time to finish their juvenile development and to be prepared for the autumn migration. Therefore, differences in the speed of juvenile development are to be expected.

To test this we hand raised warblers of different species from about the 4th day on, under identical conditions (Gwinner, 1969; Berthold, *et al.*, 1970; and Gwinner, 1971b). There was always within the systematic groups a positive correlation between the time spent in the breeding area and the time needed for juvenile development. In addition, the total reproductive season of some species can span several months (compare fig. 11 and 12). Consequently, birds of one species hatched late in summer have considerably less time until they have to leave in autumn than those hatched earlier.

The positive correlation between the date of hatching and speed of development which again can be expected, has been demonstrated for several species (Berthold, *et al.*, 1970). This could be the result of either effects of different photoperiods--early hatched birds being exposed to longer

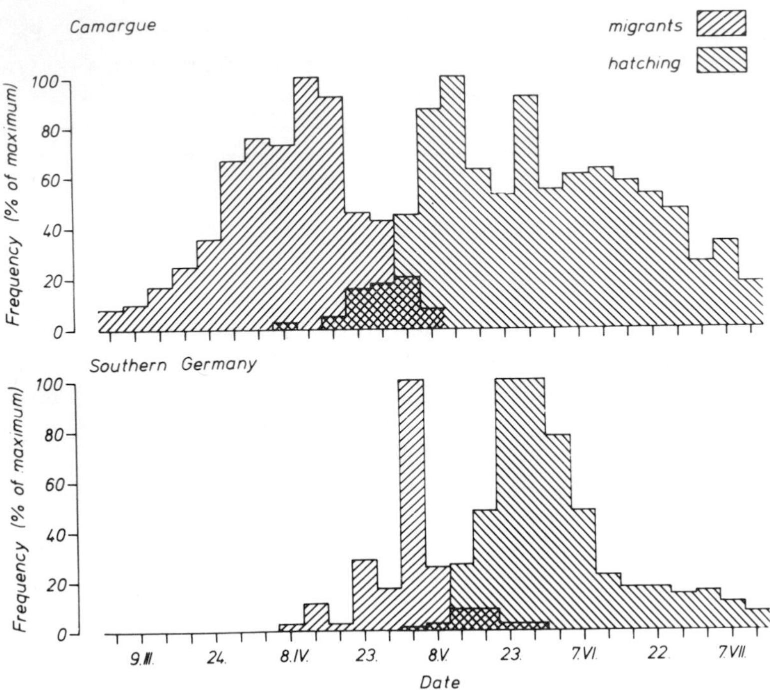

Fig. 11: Relative frequency of birds trapped during spring passage on a banding station in the Camargue, S-France (N=3506) and in S-Germany (N=74). The relative frequency of hatching for S-France is calculated from the number of trapped young birds, assuming that they can be trapped from their 25th day on. (N=343). For S-Germany it is calculated from nest-records of 518 clutches. (The data from S-France are from Dr. L. Hoffmann, Station Biologique, Tour du Valat, B.d.R.. The nest-records are previously unpublished data from the Max-Planck-Institut für Verhaltensphysiologie. The curves of passage are after Klein, *et al.* 1973).

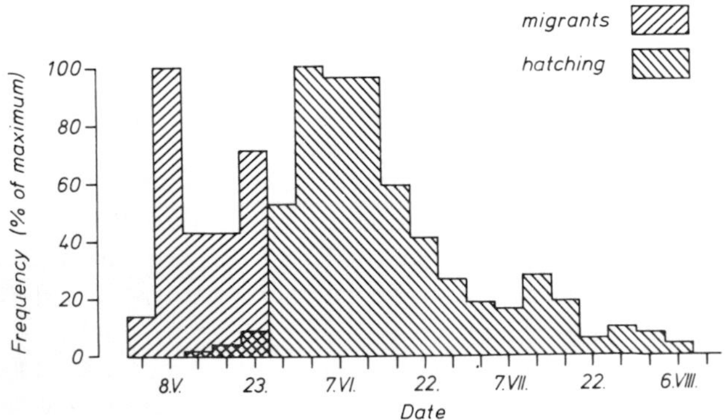

Fig. 12: As figure 11, for *Sylvia borin* in S-Germany, N=347.

photoperiods than late hatched birds--or to differences of an internal temporal program with or without interaction with the photoperiod. Only the first hypothesis has been tested rigorously so far. Results from Berthold, *et al.* (1970) and Gwinner *et al.* (1971) show that in the genera *Sylvia* and *Phylloscopus* a relativ short daylength--simulating late summer--can accelerate development.

Besides these mechanisms working via the interaction with the photoperiod, Berthold, *et al.* (in press) could show that in Garden warblers there are also inborn intraspecific differences independent of environmental stimuli that make the speed of juvenile development faster in northern populations that naturally have less time between egglaying and migration. These differences can be based on endogenous timing mechanisms or they can be a result of interaction between the endogenous time measuring system and environmental cues.

4.2 Molt.

During the course of the annual cycle of birds there are several particularly demanding phases. They are: the two migration times, the time of reproduction, and the one or two molts, especially those of flight feathers. It was often mentioned that intensive molt and intensive migration must be, and are mutually exclusive (see Stresemann and Stresemann, 1966 and Payne, 1972), even though they can change their phase relations (Berthold, *et al.*, 1971 and figure 10). Migratory birds molt flight feathers either during summer or during winter; there are presumably ultimate factors which have determined the selection of one of these.

During summer the birds have to reproduce; this is also a relatively demanding task (see King, 1973). As a result, molt can only occur simultaneously with breeding, if the environment provides plenty of food and favorable weather, and if there is enough time to finish both reproduction and molt early enough to avoid interference with autumn migration. In addition, some migrants that leave late in autumn and return early in spring, do not have extended periods when they remain at one wintering place; like in *S. atricapilla*, where autumn and spring migrations can merge (Klein, *et al.*, 1973). Hence, as a rule, flight feathers are molted during summer only in species that have relatively long stays at the breeding grounds, such as *Sylvia atricapilla* (figure 10). Species such as *S. borin*, that remain for only a relatively short time at the breeding grounds, molt flight feathers in the winter quarters (see Stresemann and Stresemann, 1966).

For an intraspecific comparison, we can use species that have a wide distribution including populations that spend long periods of time in the breeding area and other populations that remain for only a short time. It is known in *Sylvia communis* (Stresemann and Stresemann, 1966), *Lanius isabellinus* (Stresemann and Stresemann, 1972) and *Hirundo rustica* (Stresemann and Stresemann, 1968) that those populations staying for prolonged periods molt their flight feathers in summer, while their opposite extreme tend to do so in the winter.

Of special interest is the case of *Sylvia communis*, because the British population stays longer and molts during summer, while the continental birds of the same latitude molt in the African winter quarters.

As photoperiodic conditions are the same along a latitude, this indicates that there are either genetical differences, or that there is a complex system of interactions between internal and environmental factors.

We see therefore that an internal temporal program that can be responsible for the phase relations of molt and other annual processes can permit or induce adaptive differences of these processes.

4.3 Reproduction.

As was mentioned earlier, there can be considerable interspecific and intraspecific differences in the time of breeding, which is defined as the time of development of gonads and copulation. It has been stated that there should be, and are, adaptive principles in the timing of the breeding of birds of different latitudes (Lack, 1950; Moreau, 1950; Aschoff, 1955; Nalbandov, 1958; Immelmann, 1963 and 1971). For the intraspecific considerations we have to remember it has been expected that breeding times would stagger out, the more farther south the birds breed, because the season with permissive weather and available food is longer (Klopfer, 1962 and MacArthur, 1964). This was however objected to, especially by Ricklefs (1966).

In most of these discussions, an important factor was neglected. Not only weather and absolute food availability are important, but the amount of food available per bird. If one considers that during the spring migratory passage of *S. atricapilla* in S-France the density of Blackcaps in that area is up to 50 times greater than during breeding time (Klein, *et al.*, 1973), and that about 5,000,000,000 (Moreau, 1972) migrants spend the winter in Africa, it is obvious that in view of the limited food supply, competition for food becomes an important factor. Consequently, birds with similar food requirements might run into difficulties when raising offspring, when either wintering or migrating competitors share their breeding grounds.

If one considers this for an interspecific comparison, then it becomes understandable that even in tropical areas, in which one would expect favorable breeding conditions all year round, many birds have distinct breeding times. Betts (1952) gave data on the breeding of different groups of birds in S-India, where many migrants of northern Asia have their wintering areas, and where weather and vegetation would seem to allow breeding throughout the year (Betts, 1952). His data show that those local groups breeding while the migrants are present, are those that have hardly any migrant food competitors. Among them are the Woodpeckers and the fruit and nectar eating species (breeding November - March). Some raptors may even take advantage of an increased food supply during the period when the migrants are present. Some species with particularly adaptable feeding habits breed all year round (*Molpastes cafer*, *Otocompsa emeria*, *Streptopelia sinensis* and *Uroloncha striata*). On the other hand, Thrushes and the group of aerial insect hunters breed when the migrants have left.

Skutch (1950) gives data from middle America; and he also considers migrants as a possible ultimate factor for the timing of reproduction of local birds. Furthermore this hypothesis is supported by the fact that we find breeding all year round especially on islands and/or far enough out in the sea, where only very few or no migrants visit them, like the Ascension Island or Christmas Islands.

For the intraspecific comparison we use data presented in figures 11 and 12, where we see that in S-France and in S-Germany the breeding time seems to be adapted to the vernal passage of conspecifics, because hatching just sets in when passage is practically over.

All this shows that there are also adaptive principles in the temporal pattern of reproduction which is controlled mainly by reactions to the photoperiod and/or by an internal annual clock.

V. SUMMARY

Internal circannual clocks or an internal timing mechanism in birds can be or is of adaptational value for:

1) proper timing of annual events,
2) distance orientation during the first autumn migration of young *Sylvia* and *Phylloscopus* warblers,
3) for directional orientation in *Passerina cyanea*,
4) adaptation to ecological conditions in areas crossed during autumn migration,
5) adaptation to summer and winter conditions,
6) adaptive speed of juvenile development,
7) adaptive phase relations between molt and migration,
8) adaptive timing of reproduction.

VI. ACKNOWLEDGEMENTS

Previously unpublished data included here are from investigations supported by grants from the Deutsche Forschungsgemeinschaft, Bonn-Bad Godesberg, and from the Station Biologique de Tour du Valat (Dr. L. Hoffmann). I thank Professor J. Aschoff and Dr. E. Gwinner, Max-Planck-Institut für Verhaltensphysiologie, Abteilung Aschoff, Erling-Andechs, for numerous suggestions and critical remarks with respect to this manuscript.

REFERENCES

ASCHOFF, J. (1955). Jahresperiodik der Fortpflanzung bei Warmblütern. Studium Generale 8, 742-776.

ASCHOFF, J. (1967). Adaptive Cycles: Their Significance for defining environmental hazards. Int. J. Biometeor. 11, 255-278.

ASHMOLE, N.P. (1963). The biology of the Wideawake or Sooty tern *Sterna fuscata* on Ascension Island. Ibis 103b, 297-364.

ASHMOLE, N.P. (1965). Adaptive Variation in the breeding regime of a tropical sea bird. Proc. of the National Academy of Sciences 53, 311-318.

ASHMOLE, N.P. (1968). Breeding and molt in the white tern (*Gygis alba*) on Christmas Island, Pacific Ocean. Condor 70, 35-55.

BAKER, J.R. (1938a). The evolution of breeding seasons. In: Beer, G.R. Evolution pp 161-177. University Press, Oxford.

BAKER, J.R. (1938b). Latitude and egg season in old world birds. Tabulae Biol. 15, 333-370.

BAKER, J.R. (1938c). The relation between latitude and breeding seasons in birds. Proc. Zool. Soc. London A 108 557-582.

BAKER, J.R. & BAKER, I. (1936). The seasons in a tropical rain-forest. Part II. Botany. J. Linn. Soc. Zoology. London 39, 507-519.

BAKER, J.R. & RANSOM R.M. (1938). The

breeding seasons of southern hemisphere birds in the northern hemisphere. Proc. Zool. Soc. London A 108, 101-141.

BANNERMAN, D.A. (1968). Birds of the Atlantic Islands. Vol. 4. Oliver & Boyd, London.

BERTHOLD, P. (1971). Physiologie des Vogelzugs. In: Schüz, E. Grundriss der Vogelzugskunde. Pp. 257-299. Parey: Berlin and Hamburg.

BERTHOLD, P. (1973). Relationship between migratory restlessness and migration distance in 6 *Sylvia* species. Ibid 115, 594-599.

BERTHOLD, P. (in press). Migration--Control and Metabolic Physiology. In: Farner D.S. & King J.R. Avian Biology. Academic Press: London & New York.

BERTHOLD, P. (this volume). Circannual rhythms in birds with different migratory habits.

BERTHOLD, P., GWINNER, E., & KLEIN, H. (1970). Vergleichende Untersuchung der Jugendentwicklung eines ausgeprägten Zugvogels, *Sylvia borin*, und eines weniger ausgeprägten Zugvogels, *S. atricapilla*. Vogelwarte 25, 297-331.

BERTHOLD, P., GWINNER, E., & KLEIN, H. (1971). Circannuale Periodik bei Grasmücken (*Sylvia*). Experimenta 27, 399.

BERTHOLD, P., GWINNER, E. & KLEIN, H. (1972a). Circannuale Periodik bei Grasmücken. I. Periodik des Körpergewichts, der Mauser und der Nachtunruhe unter verschiedenen konstanten Bedingungen. J. Orn. 113, 170-190.

BERTHOLD, P., GWINNER, E. & KLEIN, H. (1972b). Circannuale Periodik bei

Grasmücken. II. Periodik der Gonadengrösse bei *Sylvia atricapilla* und *S. borin* unter verschiedenen konstanten Bedingungen. J. Orn. 113, 407-417.

BERTHOLD, P., GWINNER, E., KLEIN, H. & WESTRICH P. (1972c). Beziehungen zwischen Zugunruhe und Zugablauf bei Garten- und Mönchsgrasmücke (*Sylvia borin* und *S. atricapilla*). Z. Tierpsychol. 30, 26-35.

BERTHOLD, P., GWINNER, E. & QUERNER U. (in press). Vergleichende Untersuchung der Jugendentwicklung südfinnischer und südwestdeutscher Gartengrasmücken, *Sylvia borin*. Oikos.

BETTS, F.N. (1952). The breeding seasons of birds in the hills of south India. Ibis 94, 621-628.

BIEBACH, H. (in preparation). Physiologische Anpassungen der Amsel (*Turdus merula*) an den Winter.

BREHM, C.L. (1881). Die Wanderungen der Vögel. Th. Griebens Verlag, Leipzig.

BRETSCHER, K. (1920). Der Vogelzug in Mitteleuropa. Innsbruck.

BURGER, J.W. (1947). On the relation of day-length to the phases of testicular involution and inactivity of the spermatogenetic cycle of the starling. J. Exp. Zool. 105, 259-267.

CALDER, W.A. & KING, J.R. (1974). Thermal and caloric relations of birds. In: Farner, D.S. & King, J.R.: Avian Biology, Vol. 4, 259-413. Academic Press: New York & London.

CHAPIN, J.P. (1932). The birds of the Belgian Congo. Part I. Amer. Mus. Nat. Hist. Vol. 65.

CHAPIN, J.P. (1954). The calendar of Wideawake Fair. Auk 71, 1-15.

CHAPIN, J.P. & WING, L.W. (1959). The Wide-awake calendar 1953 to 1958. Auk 76, 153-158.

COOKE, W.W. (1913). The relation of bird migration to the weather. Auk 30, 215-221.

CURRY-LINDAHL, K. (1963). Molt, Body Weights, Gonadal Development, and Migration in *Motacilla flava*. Proc. 13th Int. Orn. Congr. 1963, pp. 960-973.

DAMSTE, P.H. (1947). Experimental modification of the sexual cycle of the greenfinch. J. Exp. Biol. 24, 20-35.

DARLING, F. (1938). Bird flocks and the breeding cycle. Cambridge University Press.

DORKA, V. (1966). Das jahres- und tageszeitliche Zugmuster von Kurz- und Langstrechenziehern nach Beobachtungen auf den Alpenpässen Cou/Bretolet (Wallis). Orn. Beobachter 63, 165-223.

EMLEN, S.T. (1969). Bird Migration: Influence of Physiological State upon Celestial Orientation. Science 165, 716-718.

ENGELS, W.L. (1962). Day-length and termination of photorefractoriness in the annual testicular cycle of the transequatorial migrant *Dolichonyx* (The Bobolink). Biol. Bul. 123, 94-104.

ENGELS, W.L. (1964). Further observations on the regulation of the annual testicular cycle in Bobolinks (*Dolichonyx oryzivorus*). Auk 81, 95-96.

FARNER, D.S. & LEWIS, R.A. (1971). Photoperiodism and reproductive cycles in birds. In: Giese, A.C.: Photophysiology 6, 325, Academic Press, New York & London.

FARNER, D.S. & LEWIS, R.A. (1973). Field and experimental studies of the annual cycles of white-crowned Sparrows. J. Reprod. Fertility Supplement No. 19, 35-50.

GWINNER, E. (1968a). Circannuale Periodik als Grundlage des jahreszeitlichen Funktionswandels bei Zugvögeln. Untersuchungen am Fitis (*Phylloscopus trochilus*) und am Waldlaubsänger (*P. sibilatrix*). J. Orn. 109, 70-95.

GWINNER, E. (1968b). Artspezifische Muster der Zugunruhe und ihre mögliche Bedeutung für die Beendigung des Zuges im Winterquartier. Z. Tierpsychol. 25, 843-853.

GWINNER, E. (1969). Untersuchungen zur Jahresperiodik von Laubsängern. J. Orn. 110, 1-21.

GWINNER, E. (1971a). Orientierung. In: Schüz, E.: Grundriss der Vogelzugskunde. pp. 299-348, Parey, Berlin & Hamburg.

GWINNER, E. (1971b). A comparative study of circannual rhythms in Warblers. In: Menaker, M. Biochronometry. Nat. Acad. Sci. Wash. pp. 405-427.

GWINNER, E. (1972a). Adaptive functions of circannual rhythms in warblers. Proc. 15th Int. Orn. Congr. 218-235.

GWINNER, E. (1972b). Endogenous timing factors in bird migration. In: Galler, S.R.; Schmidt-Koenig, K.; Jacobs G.J. & Belleville, R.E.: Animal Orientation and Navigation. NASA: Washington D.C., pp. 321-338.

GWINNER, E. (1973). Circannual rhythms in birds: Their interaction with circadian rhythms and environmental Photoperiod. J. Reprod. Fert., Suppl. 19, 51-65.

GWINNER, E. (1974). Adaptive significance of circannual rhythms in birds. In: Vernberg, T.J. Physiological Adaptation to the Environment. Intext Educational Publ.: New York.

GWINNER, E. (in press). Endogenous temporal control of migratory restlessness in warblers. Naturwissenschaften 61.

GWINNER, E., BERTHOLD, P. & KLEIN, H. (1971). Untersuchungen zur Jahresperiodik von Laubsängern II. J. Orn. 112, 253-265.

GWINNER, E., BERTHOLD, P. & KLEIN, H. (1972). Untersuchungen zur Jahresperiodik von Laubsängern III. J. Orn. 113, 1-8.

HAARTMAN, L. von (1963). The nesting times of Finnish birds. Proc. 13th Int. Orn. Congr., pp. 611-619.

HAILMAN, J.P. (1964). Breeding Synchrony in the Equatorial Swallow-tailed Gull. American Naturalist 98, 79-83.

HAMNER, W.M. & STOCKING, J. (1970). Why don't Bobolinks breed in Brazil? Ecology 51, 743-751.

IMMELMANN, K. (1963). Tierische Jahresperiodik in ökologischer Sicht. Zool. Jahrb. Syst. 91, 91-200.

IMMELMANN, K. (1967). Periodische Vorgänge in der Fortpflanzung tierischer Organismen. Studium Generale 20, 15-33.

IMMELMANN, K. (1971). Ecological aspects of periodic reproduction. In: Farner, D.S. & King, J.R.: Avian biology Vol. I, Academic Press, New York & London.

JOHNSTON, R.F. (1954). Variation in breeding season and clutch size in Song Sparrows of the Pacific coast. Condor 56, 268-273.

KEAST, A. (1959). Australian Birds: Their zoogeography and adaptations to an arid environment. In: Keast, A., Crocher, R.L. & Christian, C.S. Biogeography and Ecology in Australia. Junk Publ. The Hague.

KING, J.R. (1968). Cycles of fat deposition and molt in White-crowned Sparrows in constant environmental conditions. Comp. Biochem. Physiol. 24, 827-837.

KING, J.R. (1973). Energetics of Reproduction in Birds. In: Breeding biology of birds. National Academy of Sciences, Washington D.C.

KING, J.R. & FARNER, D.S. (1965). Studies of fat deposition in migratory birds. Ann. N.Y. Acad. Sci. 131, 422-440.

KLEIN, H., BERTHOLD, P. & GWINNER, E. (1973).. Der Zug europäischer Garten- und Mönchsgrasmücken. Vogelwarte 27, 73-134.

KLOPFER, P.H. (1962). Behavioural aspects of ecology. Prentice-Hall Inc. Englewood Cliffs, New Jersey.

LACK. D. (1950). The breeding seasons of euripean birds. Ibis 92, 288-316.

LACK, D. (1954). The natural regulation of animal numbers. Oxford. The Clarendon Press.

LEWIS, R.A. & ORCUTT, F.S. (1971). Social behaviour and avian sexual cycles. Scientia 65, 447-472.

LINDALE, J.M. (1933). The nesting season of birds in Doniphan County, Kansas. Condor 35, 155-160.

MacARTHUR, R.H. (1964). Environmental factors affecting bird species diversity. American Naturalist 98, 387-397.

MARSHALL, A.J. (1960a). Annual periodicity in the migration and reproduction of birds. Cold Spring Harbour Symp. Quant. Biol. 25, 499-505.

MARSHALL, A.J. (1960b). The role of the

internal rhythm of reproduction in the timing of avian breeding season including migration. Proc. 12th Int. Orn. Congr., pp. 475-482.

MARSHALL, A.J. & SERVENTY, D.L. (1959). Experimental demonstration of an internal Rhythm of reproduction in a transequatorial migrant (the Short-tailed Shearwater *Puffinus tenuirostris*). Nature 184, 1704-1705.

MARSHALL, A.J. & WILLIAMS, M.C. (1959). The pre-nuptial migration of the yellow wagtail (*Motacilla flava*) from latitude 0.04 N. Proc. Zool. Soc. London 132, 313-320.

MILLER, A.H. (1959a). Reproductive cycles in an equatorial sparrow. Proc. Nat. Acad. Sci. 45, 1095-1100.

MILLER, A.H. (1959b). Response to experimental light increments by andean sparrows from an equatorial area. Condor 61, 344-347.

MILLER, A.H. (1962). Bimodal occurence of breeding in an equatorial sparrow. Proc. Nat. Acad. Sci. 48, 396-400.

MIYAZAKI, H. (1934). On the relation of the daily period to the sexual maturity and to the molting of *Zosterops palperosa japonica*. Sci. Rep. Tohoku Imp. Univ. 4th Series Biol. 9, 183-203.

MOREAU, R.E. (1950). The breeding seasons of African birds. 1. Land birds. Ibis 92, 223-267.

MOREAU, R.E. (1972). The Palaearctic-African bird migration system. Academic Press: London and New York.

NALBANDOV, A.V. (1958). Reproductive Physiology. Freeman & Co., San Francisco.

ORIANS, G. (1960). Autumnal breeding in the

tricolored blackbird. Auk 77, 379-398.

PAYNE, R.B. (1972). Mechanisms and control of molt. In: Farner, D.S. & King, J. R.: Avian Biology Vol. II, 101-155. Academic Press, New York & London.

PENGELLEY, E.T. & KELLEY, K.H. (1966). A "circannian" rhythm in hibernating species of the genus *Citellus* with observations on their physiological evolution. Comp. Biochem. Physiol. 19, 603-617.

PENNYCUICK, C.J. (1969). The Mechanics of Bird Migration. Ibis 111, 525-556.

PHILLIPS, J.C. (1913). Bird migration from the standpoint of its periodic accuracy. Auk 30, 191-204.

RICKLEFS, R.E. (1966). The temporal component of diversity among species of birds. Evolution 20, 235-242.

ROWAN, W. (1926). On photoperiodism, reproductive periodicity and the annual migration of birds and certain fishes. Proc. Boston Soc. Nat. Hist. 38, 147-189.

ROWAN, W. (1929). Experiments in bird migration. I. Manipulation of the reproductive cycle: Seasonal histological changes in the gonads. Proc. Boston Soc. Nat. Hist. 39, 151-208.

SAUER, E.G.F. (1957). Die Sternorientierung nächtlich ziehender Grasmücken (*Sylvia atricapilla, borin* und *curruca*). Z. Tierpsychol. 14, 29-70.

SAUER, E.G.F. (1961). Further studies on the stellar orientation of nocturnally migrating birds. Psychol. Forschg. 26, 224-244.

SAUER, E.G.F. & SAUER, E.M. (1960). Star navigation of nocturnal migrating birds. The 1958 planetarium experiments. Cold Spring Harbour Symp. Quant. Biol. 25, 463-473.

SCHOLANDER, P.F., WALTERS, V., HOCK, R. & IRVING, L. (1950a). Body insulation of some arctic and tropical mammals and birds. Biol. Bull. 99, 225-235.

SCHOLANDER, P.F., HOCK, R., WALTERS, V., JOHNSON, F. & IRVING, L. (1950b). Heat regulation in some arctic and tropical mammals and birds. Biol. Bull. 99, 237-258.

SCHOLANDER, P.F., HOCK, R., WALTERS, V. & IRVING, L. (1950c). Adaptation to cold in arctic and tropical mammals and birds in relation to body temperature, insulation and basal metabolic rate. Biol. Bull. 99, 259-271.

SCHÜZ, E. (1971). Grundriss der Vogelzugskunde. Parey, Berlin & Hamburg.

SELANDER, R.K. & NICHOLSON, D.J. (1962). Autumnal breeding of Boat-tailed Grackles in Florida. Condor 64, 81-91.

SERVENTY, D.L. & MARSHALL, A.J. (1957). Breeding periodicity in western Australian birds: With an account of unseasonal nesting in 1953 and 1955. Emu 57, 99-126.

SERVENTY, D.L. & WHITTELL, H.M. (1967). Birds of Western Australia. Lamb Publ.: Perth, Australia.

SKUTCH, A.F. (1950). The nesting seasons of Central American birds in relation to climate and food supply. Ibis 92, 185-222.

SNOW, D.W. (1965). The breeding of Audubon's Shearwater (*Puffinus lherminieri*) in the Galapagos. Auk 82, 591-597.

SNOW, D.W. & SNOW, B.K. (1967). The breeding cycle of the Swallow-tailed Gull

(*Creagrus furcatus*). Ibis 109, 14-24.

STONEHOUSE, B. (1962). The tropic birds (genus *Phaeton*) of Ascension Island. Ibis 103 b, 123-161.

STRESEMANN, E. & STRESEMANN, V. (1966). Die Mauser der Vögel. J. Orn. 107, Sonderheft.

STRESEMANN, E. & STRESEMANN, V. (1968). Im Sommer mausernde Populationen der Rauchschwalbe, *Hirundo rustica*. J. Orn. 109, 475-484.

STRESEMANN, E. & STRESEMANN, V. (1972). Über die Mauser in der Gruppe *Lanius isabellinus*. J. Orn. 113, 60-75.

WINTERBOTTOM, J.M. (1963). Avian breeding seasons in southern Africa. Proc. 13th. Int. Orn. Congr., pp. 640-648.

WOLFSON, A. (1954). Production of repeated gonadal, fat, and molt cycles within one year in the Junco and White-crowned Sparrow by manipulation of day length. J. Exp. Zool. 125, 353-376.

ZINK, G. (1973). Der Zug europäischer Singvögel. Ein Atlas der Wiederfunde beringter Vögel. Vogelwarte Radolfzell, Schloss Möggingen.

CREDITS

Figure 3. Reproduced by permission of the National Aeronautics and Space Administration from Animal Orientation and Navigation, pp. 321-338, 1972.

Figure 8. Reproduced by permission of Vogelwarte 27, Heft 2, p. 88, 1973.

Figure 10. Reproduced by permission of Deutsche Ornithologen Gesellschaft from Journal für Ornithologie 113, Heft 2, pp. 170-190, 1972.

EXPRESSION AND SUPPRESSION OF THE CIRCANNUAL ANTLER GROWTH CYCLE IN DEER

RICHARD J. GOSS, CHARLES E. DINSMORE, L. NICHOLS GRIMES and JEFFREY K. ROSEN[1]

Division of Biological and Medical Sciences,
Brown University
Providence, Rhode Island, 02912

[1]Present address: University of Dar es Salaam, Tanzania

I. INTRODUCTION

The year of the deer is punctuated by the annual replacement of its antlers. Most species native to temperate zones tend to cast their old antlers in the spring and grow new ones through the summer. When fully grown, the antlers become firmly ossified and the velvety skin in which they are enveloped is shed in time for the autumn mating season.

These extraordinary outgrowths are renewed at yearly intervals by virtually all of the world's species of deer (Whitehead, 1973) from the diminutive pudu, dwelling in forests on the slopes of the Andes where it grows tiny spikes a centimeter or two long, to the North American elk, moose and caribou whose antlers commonly extend more than a meter from the frontal pedicles by which they are attached to the skull. According to the fossil record, the giant Irish elk grew antlers nearly two meters long (Gould, 1973a, b), trophies which must have been irresistible to our Pleistocene ancestors who may have contributed to their extinction. Despite these extremes, all antlers are regenerated each year in approximately 4 months, the larger ones elongating two or more centimeters a day during the steepest inflections of their growth curves (Goss, 1970).

II. THE GROWTH OF ANTLERS AND HORNS

Even before the old antlers drop off, signs of renewed growth can often be seen in the swollen pedicle skin subjacent to the base of the antler. It is this skin which grows over the raw stump of the pedicle once the antlers have been shed following osteoclastic erosion of their osseous attachments to the pedicle bone (Wislocki, 1942; Waldo and Wislocki, 1951). Within a couple of weeks the healing pedicle skin rounds up into an antler bud which, by mechanisms reminiscent of the apical meristem in plants, becomes the growing tip responsible for the rapid elongation of the antler and the sprouting of branches from time to time (fig. 1). It is no coincidence that the French call them *les bois*.

Histologically, the apical growth zone consists of proliferating fibroblasts embedded in a dense network of collagen fibers. The overlying epidermis is richly adorned with velvety hairs which give the antler skin its name and which arise from follicles that differentiate *de novo* on the ends of the growing branches (Billingham, *et al*, 1959). The velvet is also endowed with sensory innervation enabling even the bull moose to trot through deep woods without injuring his sensitive antlers. Arteries entwine the growing antler, climbing distally in the skin to deliver a copious supply of blood to the growing tip (Waldo, *et al*, 1949). Here they ramify into a profusion of capillaries which then course down into the venous channels that honeycomb the core of the antler.

Elongation is achieved by terminal proliferation of undifferentiated connective tissue cells. Descendants of these cells, left behind by the advancing growth zone, then differentiate into chondrocytes in a region just proximal to the tip (Noback, 1929; Wislocki, *et al* 1947). Farther down, this cartilage becomes calcified, and below that it gradually gives way to spongy bone (Molello, *et al*, 1963). All of these tissues are permeated by innumerable blood vessels, but as ossification thickens the bony trabeculae the veins in between become progressively constricted. It is this vascular restriction which is responsible for the demise of the antler as its bone solidifies. Once the velvet itself is rubbed off the antlers are burnished but dead status symbols coordinated with the onset of spermatogenesis, testosterone secretion and the

Fig. 1: Successive photographs of antler growth in a sika deer as he appeared on April 6, May 7, June 5, July 2, July 29, August 23 and September 6. The previous year's antlers were lost a day or two prior to the first photograph and the velvet was shed shortly before the last picture was taken.

belligerent temperament that anticipates the rutting season.

Not to be confused with antlers are the horns of other ungulates. In these headpieces it is the cornification of the outer epidermis, rather than the ossification of the bony core, which gives these weapons the hardness they need when their bearers lock horns. Unlike antlers, these structures grow from the base, are unbranched and do not replace themselves annually. Nevertheless, they are not insensitive to the change in seasons. Species native to the temperate zones add to their horns only during those times of year when they are reproductively inactive. It is the cessation in horn growth during the mating season that is responsible for the demarcation separating one annual increment from the next, a feature best seen in the magnificent horns of bighorn sheep. Less conspicuous yearly delineations can be detected in the horns of other species of sheep and goats, but ungulates of the tropics lack signs of seasonal variations in horn growth.

III. HORMONAL REGULATION OF ANTLER DEVELOPMENT

The proximate control of the antler cycle lies in the annual fluctuations in sex hormone secretion (Tachezy, 1956; Wislocki, 1956). The onset of antler growth, for example, is correlated with a natural decline in testosterone output during the spring, a decline paralleled by the cessation of spermatogenesis at that time of year. Indeed, castration in the fall or winter will precipitate the precocious shedding and regrowth of antlers within a few weeks owing to the abrupt depletion of sex hormones (Wislocki, *et al*, 1947a). In this way it is possible to induce the growth of antlers even in the middle of winter.

The antlers of castrated deer grow quite normally at first, but are unable to complete their development in the sense that they must remain permanently in velvet. Each year new growth occurs, but since the old antlers have not been lost they add to what has already been formed. The results can become quite grotesque. In some species, extra branches may sprout. In others, amorphous growths bulge out onto the head (fig. 2). Winter temperatures may freeze the more distal parts, but in the spring the necrotic regions drop off and may be replaced by new outgrowths.

The European roe deer normally grows antlers in the winter. This is correlated with the fact that they have a summer mating season, followed by delayed implantation until December, with births taking place in the spring. When the males are castrated (Tachezy, 1956) their antlers sometimes grow to enormous proportions as tissue masses cover the head like a wig and may even obstruct vision. Since these antlers normally grow in the winter, they are not susceptible to freezing and therefore attain greater, albeit more bizarre, dimensions than do those of other species.

These permanently viable antlers of castrated deer testify to the important role normally played by testosterone in bringing about the maturation of antlers at the end of the growing season. As autumn approaches male sex hormones normally cause the solidification of the spongy bone leading to constriction of the vascular channels, death of the entire antler and peeling off of the desiccated velvet.

What happens naturally in the late

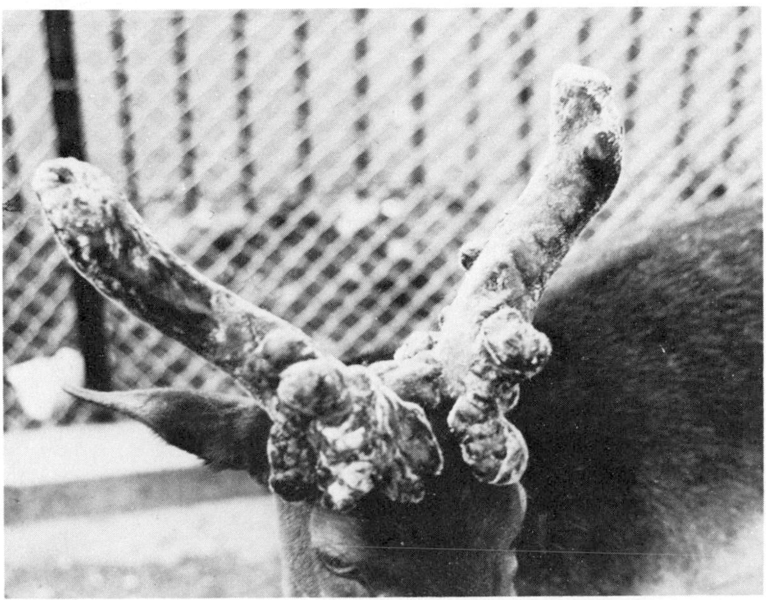

Fig. 2: Abnormal antlers of a fallow deer castrated several years earlier. The distal portions of the antlers have been lost as a result of freezing. The main beams are thickened and new growths have been produced around the base. (Specimen courtesy of Regent's Park Zoo, London.)

summer can be triggered prematurely by untimely injections of testosterone (Jaczewski, and Galka, 1967; Goss, 1968). Given early in the spring before the old antlers are lost, shedding and regrowth can be totally precluded. Administered in the summer while the antlers are in velvet, testosterone aborts further growth by inducing maturation of the partly grown antlers. Testosterone can also permit the antlers of castrated deer to complete their development by turning to bone and shedding their velvet. When the effects of the hormone wear off, however, the old antlers drop off and new ones grow out as abnormally as before.

Curiously, estrogen injections exert the same effects as testosterone (Goss, 1968). Female hormones can prevent the initiation of antler growth or cut short development already in progress. Only male deer grow antlers, but perhaps the potential for estrogen to mimic testosterone may help explain why antlered does are encountered on rare occasions, or why, in reindeer and caribou, even the females normally grow antlers.

IV. RELATION TO THE LIGHT CYCLE

A. PHASE AND FREQUENCY OF THE PHOTOPERIOD

Just as it is easy to manipulate the antler growth cycle hormonally, it is not difficult to do so by changing the annual light cycles to which the hormones themselves are responsive (Jaczewski, 1954, 1973; Goss, 1969a, b). In experiments on sika deer (*Cervus nippon*), reversal of the annual increases and decreases in day lengths shifts the antler growth cycle 6 months out of phase. If the frequency of the photoperiodic rhythm is changed, the frequency of antler replacement is altered accordingly--up to a point (fig. 3). A two-, three- or fourfold acceleration of the annual light cycle results in the production of a new set of antlers every 6, 4 or 3 months, respectively. Since the rate of individual antler growth does not accomodate to the accelerated change in day lengths, such antlers are progressively foreshortened with the increasingly abbreviated growing periods. Perhaps this is the explanation for the failure of the antler growth cycle to keep pace with a sixfold increase in the frequency of the photoperiod. When the annual light cycle is made to go by in only two months, the antlers fail to respond altogether. Instead, they revert to their natural rhythm and replace themselves the following spring irrespective of the rapid fluctuations in day lengths to which they have been exposed since the previous growing season.

At the opposite extreme, it is possible to lengthen the annual light cycle from its normal period of 12 months to an artificially prolonged "year" of 24 months. Under these unearthly conditions, deer respond in one of two ways, depending upon their age. Mature bucks tend to replace their antlers every 12 months regardless of the 24-month light cycle. Yearlings, however, replace their antlers only every other year in accordance with the artificial light cycle to which they have been exposed (fig. 3). Although the length of time spent in velvet is extended about 30 percent beyond normal, the antlers do not grow any larger than usual. These results suggest that young deer adjust to whatever light cycle they are exposed, while adults may have been conditioned to the natural year and tend to express this rhythm whenever the frequency

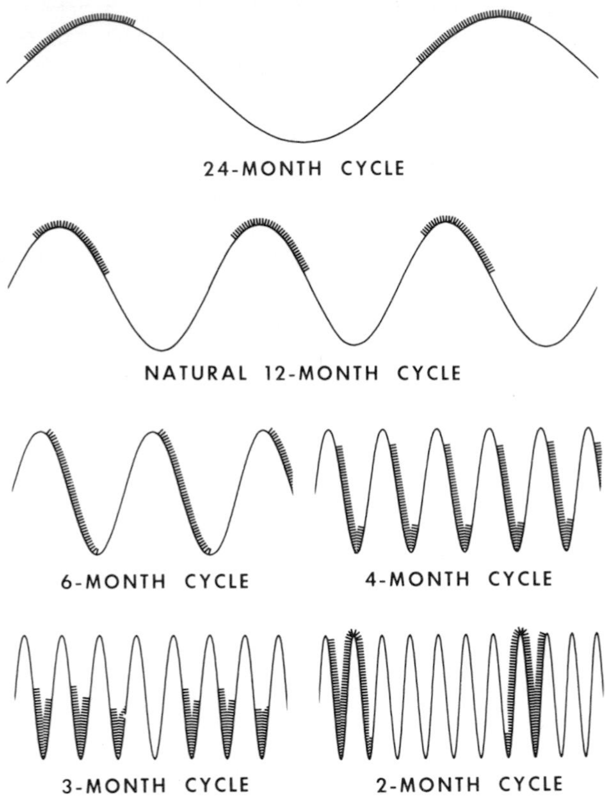

Fig. 3: Responses of the antler growth cycle to "annual" photoperiods of different frequencies. Each curve represents a cycle of day length changes to which sika deer were exposed. Their periods in velvet, when antlers were in the growing phase, are indicated on each of the curves. Deer were able to synchronize to cycles as long as 24 months and as short as 3 months, although their periods in velvet shifted out of phase and were longer or shorter than normal in direct relation to the cycle lengths. Deer were unable to respond to cycles of 2 months' duration, reverting instead to an inherent annual rhythm.

of the artificial photoperiod is too long or too short.

The family of curves illustrated in figure 3 reveals some interesting relationships between the antler growth cycle and the fluctuations in day lengths to which it responds. Normally antlers begin to grow while the days are still lengthening and their velvet is shed as the autumnal equinox approaches. When the cycle is protracted, however, the period in velvet is displaced to the left. In shortened cycles the growth phase shifts to the right. Thus, in a 6-month cycle antler growth occupies nearly half of the light curve between the summer and winter solstices. In the 4-month cycle the velvet is not shed until the days begin to lengthen. Finally, on a 3-month cycle the growth phase has been displaced to a position on the curve opposite that of the natural 12-month cycle. Perhaps this has something to do with the fact that deer exposed to such accelerated light cycles are unable to grow antlers every time around.

Deer are normally in velvet for about one-third of the year. When the duration of that "year" is changed by altering the frequency of the day length curve, the absolute lengths of time in velvet or with bony antlers are profoundly affected, as can be seen by an inspection of figure 4. In artificially prolonged light cycles the velvet phase increases in absolute length but drops to only about one-quarter of the total cycle. Under abnormally short light cycles, the absolute duration in velvet progressively decreases to about half the normal length. However, since the intervals between successive rounds of antler growth diminish even more, the average relative durations in

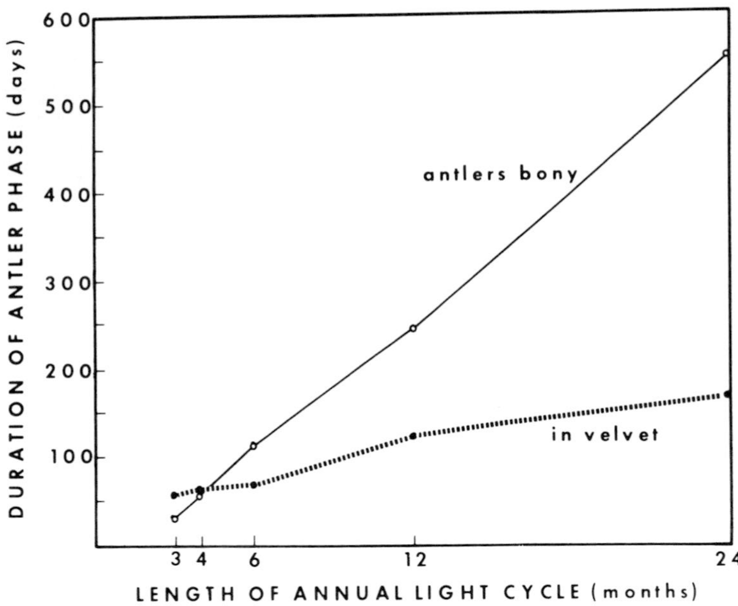

Fig. 4: Effect of varying lengths of "years" on the duration of the antler cycle. The non-growing phase, when antlers are bony, bears a linear relationship to the length of the light cycle. The time spent in velvet appears to approach a mimimum of about 2 months on abbreviated years. This sets a limit of about 3 months to the shortest possible light cycle to which a deer can react. (From Goss (1969a), by permission of The Wistar Institute Press.)

velvet increase to 38 percent of the 6-month cycles, 54 percent of the 4-month cycles and 65 percent of the 3-month cycles. It may be significant that this last figure is just the reverse of the normal situation, and the 3-month cycle marks the minimal length of time in which deer can sustain successive episodes of antler replacement.

The above experiments clearly indicate that the antler growth cycle readily locks onto reversed or accelerated light cycles within a surprisingly wide range of frequencies. But the expression of the natural rhythm when the artificial ones become too extreme (i.e., 2-month cycles) suggests the existence of an inherent mechanism that coincides with the sidereal year.

B. CONSTANT PHOTOPERIODS

To explore this possibility further, the method of choice is to eliminate all changes in day length. The most obvious way to achieve this is to hold groups of deer under simulated equatorial lighting conditions (12L/12D). Again, the results depend on the age of the deer. Yearlings introduced to these conditions prior to the winter solstice fail to replace their antlers the following spring and for an indefinite period thereafter. Adults, however, do replace their antlers in the spring even when introduced to unchanging photoperiods of 12L/12D as early as the autumnal equinox. Such animals would appear to be programmed to replace their antlers well in advance of the event. During subsequent years, however, the antler growth cycle is not expressed. Thus, in some cases these deer have retained one set of antlers for several years in a row.

It might be concluded from these findings that in the absence of changes in day length the antler growth cycle tends to be held in abeyance. Such is not the case, however. Other groups of sika deer have been held under constant photoperiods of 8L/16D, 16L/8D, and 24L/OD (fig. 5). In all three of these situations, the deer shed and regrow their antlers repeatedly. They do not do so in unison, nor do they replace their antlers at regular intervals. Nevertheless, under these unchanging artificial day lengths an irregular antler replacement cycle, with a frequency averaging about 10 months, expresses itself. Since such cycles are not exactly a year long and can occur in the apparent absence of a driving oscillation in the environment, they fit the definition of a circannual rhythm.

It is a curious thing that such a rhythm should be expressed under conditions of constant day length when the light:dark ratio is unequal to one, but not when the days and nights are of equal length. Somewhere between the equal and unequal ratios of light to dark there must be a point of transition between the suppression of a circannual rhythm and its expression. To determine where this switch takes place, a series of tests have been carried out on groups of deer exposed to lighting regimes that differed from 12L/12D by increments of 15-minute intervals. Thus, a control group was exposed to 12L/12D, while groups of experimental deer were kept at 12-1/4L/11-3/4D, 12-1/2L/11-1/2D, 12-3/4L/11-1/4D, and 13L/11D. In all cases, yearling sika deer were used rather than adults because they had previously been shown not to replace their antlers in the spring when held at 12L/12D since the preceding autumn.

Typically, sika deer grow their first

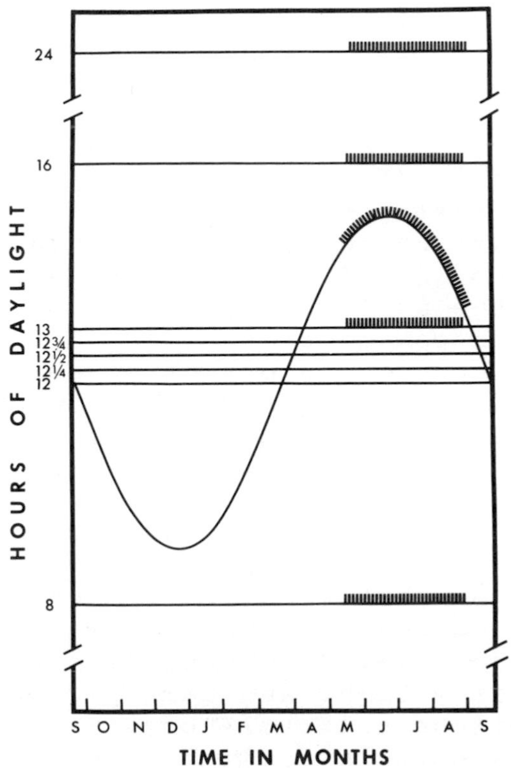

Fig. 5: Summary of experiments on the reactions of antler growth cycles to constant day lengths of different durations. The sine curve represents the natural changes in day lengths throughout the year at about 42° N. latitude. Antlers are normally in velvet from early May to early September. Yearling sika deer held under conditions of 8L/16D, 13L/11D, 16L/8D and 24L/0D since the autumnal equinox all shed and regrow antlers in the spring in synchrony with those outdoors. Animals exposed to light for 12, 12¼, 12½ and 12¾ hours per day, however, failed to replace their antlers the following year.

set of spike antlers during their second summer. The yearlings used in these experiments were all beginning their second year of life having just completed the growth of their first set of antlers. They were introduced to the experimental lighting conditions at about the autumnal equinox and were kept under observation until the following August. Temperatures were held at 20-22°C throughout the study. As expected, the 8 deer held on 12L/12D failed to replace their antlers the next spring. Similarly, none of the 9 yearlings exposed to constant day lengths up to 12-3/4L/11-1/4D replaced their antlers in the spring or summer. Of those maintained on constant lighting conditions of 13L/11D, however, 3 out of 4 yearlings shed their old antlers and grew new ones the following spring. These results, as illustrated graphically in figure 5, may be taken to indicate that when the difference between the lengths of day and night is 1-1/2 hours or less (6.25 percent of a day), the circannual rhythm of antler replacement fails to be expressed. When there is a 2-hour difference (8.33 percent of a day) between the daily light and dark periods, however, the circannual cycle cuts in. In terms of the earth's latitude, this would correspond to a zone approximately 14° to 18° from the equator, latitudes at which the differences between the summer and winter solstice are about 1-1/2 to 2 hours (fig. 6).

The foregoing results do not resolve the question of whether the circannual rhythm of antler replacement cannot be expressed at 12L/12D because the light:dark ratio equals one or because the absolute lengths of the light and dark periods are 12 hours. One way to distinguish between these

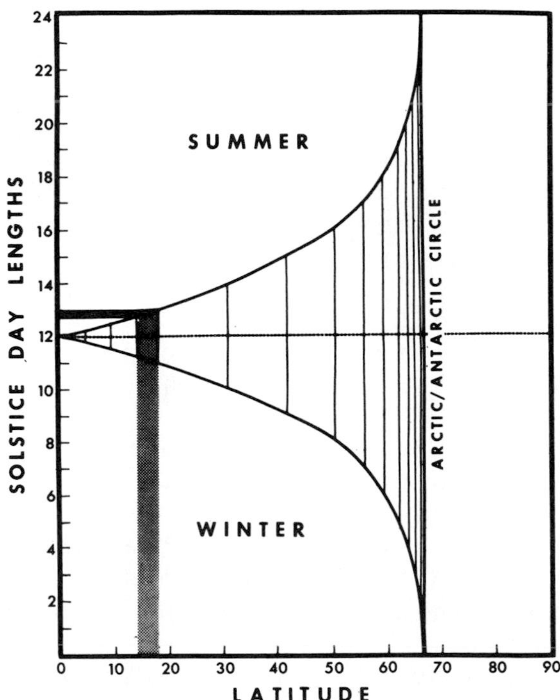

Fig. 6: Graph of the maximum and minimum lengths of natural days as a function of latitude. The first 4 vertical lines from the left represent solstice day lengths of ±15, 30, 45 and 60 minutes. Hourly intervals are marked off thereafter until, at the arctic and antarctic circles, the days are either 0 or 24 hours long at the solstice. Shaded area indicates the zone between 14° and 18° N. latitude where the length of day at the summer solstice is between 12¾ and 13 hours.

alternatives would be to hold the light at 12 hours while varying the dark period, or vice versa. Another approach would involve holding the periods of light and dark equal, but letting their lengths be something other than 12 hours. For example, if yearling deer were to be maintained on a light regime of 6L/6D/6L/6D from the autumnal equinox the ratio of light to dark would still be one, but the individual light and dark periods would no longer be 12 hours long. Hence, if the circannual rhythm of antler replacement is suppressed under conditions of 12L/12D simply because these intervals are equal then one would expect the antlers not to be replaced in the spring following a year on alternating light and dark periods of 6 hours. But if circannual suppression occurs because the intervals of illumination are 12 hours long, then antler renewal should take place in the spring on a regime of 6L/6D/6L/6D. Pending the outcome of experiments along these lines, comparable studies could be carried out at alternating light and dark periods set at the many other intervals into which our 24-hour day can be divided.

It is possible that abolition of the circannual rhythm at 12L/12D might be attributable to the fact that these intervals are of 12 hours' duration. In order to learn if deer can follow a light cycle that never reaches 12 hours in length, several sika deer have been exposed to 4-hour photoperiods alternating with ones 8 hours long at 2 month intervals. Previous experiments (Goss, 1969a, b) have shown that the antler cycle will follow similar frequencies in which photoperiods less than 12 hours (e.g., 9 or 11 hrs.) interchange with ones longer than 12 hours (e.g., 15 or 13 hrs.) every

2 months. In the present experiment, deer tended to replace their antlers in coordination with alternating photoperiods of 4L/20D and 8L/16D. These findings indicate that the antler cycle locks onto changes in day lengths *per se* even if these changes do not involve light intervals that fluctuate across the 12-hour mark.

V. LIGHT AND LATITUDE

From what has been learned, it might be predicted that if deer native to the temperate zone were transported to the equator, they would fail thereafter to replace their antlers and would remain in a state of year-round fertility. Unhappily, there are no cases on record of this having been put to the test. Therefore, one can only resort to an examination of wild populations native to the tropics (Goss, 1963).

All known species of tropical deer replace their antlers on a yearly basis. Some of them do so in synchrony with a particular season. Others, however, are capable of growing antlers at any time of year. Although each individual may grow a new set of antlers at regular 12-month intervals, they do not do so in unison. Such deer are fertile throughout the year, even when in velvet, and fawns may be produced at any time. Evidently, the schedule of antler replacement is determined by when the deer happened to have been born.

Records show that these deer are adhering to an annual, not circannual, rhythm. They replace their antlers at the same time each year but are out of phase from the others in the same population. Even when kept in zoos outside the tropics, these species are still unsynchronized in their antler growth cycles. Indeed, it is not uncommon to find tropical deer, living in temperate zone zoos, growing antlers in the middle of winter; unseasonal births among such animals can be very inconvenient. These species of tropical deer, therefore, fail to respond to the changing day lengths of temperate zones, yet their antler growth cycles are remarkably precise. If such cycles are not inherent, they must be entrained to an environmental cue which is different for each individual and independent of latitude (Goss and Albach, 1974; Goss and Rosen, 1974).

How could such a condition have evolved? It is probably safe to conclude that antlers arose in the temperate zone and that their annual replacement evolved in correlation with their breeding seasons. Now if some of these species expanded their range toward the equator their more tropical descendants would be subject to less severe environmental constraints on their breeding seasons: Hence, it is not difficult to imagine how a fairly restricted breeding season at high latitudes might expand throughout the year at lower, less seasonal, latitudes where selective pressures do not favor one time of year over another.

Should this have been the case, it might be possible to trace such a transition if there were one kind of deer whose habitat extended from the temperate zone to the equator across an uninterrupted land mass. Inasmuch as there are no deer native to tropical Africa, and since the Asiatic mainland does not cross the equator, the western hemisphere is the only place in the world where there is a continuous land bridge between the northern and southern hemisphere that is populated with deer (fig. 7). This same area is inhabited by various species

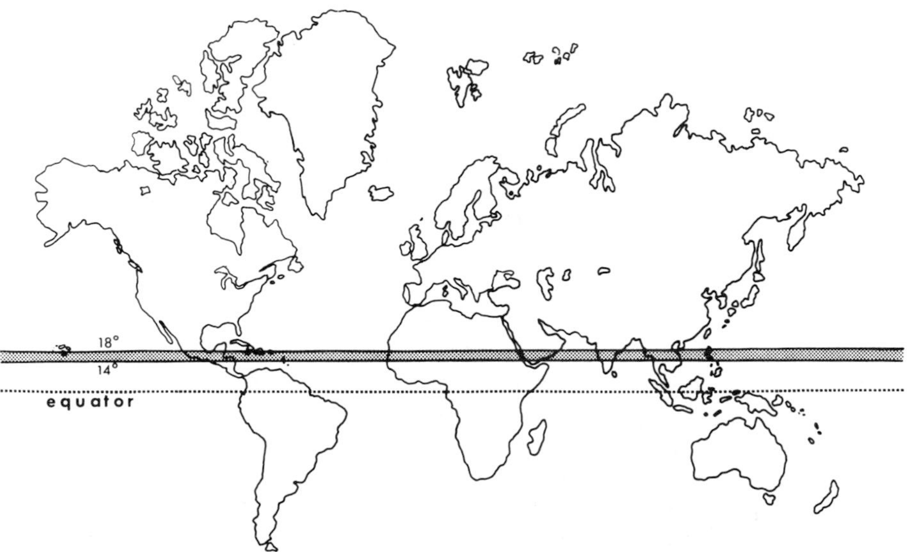

Fig. 7: Showing the relation of N. latitudes 14° and 18° to the world's land masses. Only in Central and South America is there a land bridge between the hemispheres inhabited by deer. The shaded zone is believed to approximate the transitional region between the seasonal pattern of reproduction in the temperate zone and aseasonal reproduction in the tropics.

of white-tailed deer in the genus *Odocoileus*. In South America and part of Central America, these deer adhere to the tropical mode of aseasonal, unsynchronized birth and antler growth cycles. In North America and much of Mexico, these cycles are seasonal. Predictably, there must be a zone of transition, as yet unstudied, somewhere in Central America where the phylogenetic shift from the temperate zone to the tropics may still be preserved in the reproductive and antler growth cycles of the existing population.

VI. DISCUSSION

It helps to put things in perspective by comparing what happens in one system, such as the antler growth cycle, with the reactions of other biological processes under constant conditions of light and dark. In view of the sometimes long-range effects of lighting conditions, it is important that any rhythms studied under constant photoperiod should be observed for more than one cycle since the first time around may have been programmed by conditions prevailing before the beginning of the experiment.

Thus, although Wolfson (1949) noted that the spring breeding condition was reached by juncos held on equal days and nights, subsequent cycles, including molting, were absent over a period of 20 months. Similarly, cycles of molting failed to occur for 3 years in chickens held at 12L/12D, and their egg laying was continuous in the absence of the seasonal declines that characterize normal lighting conditions (Greenwood, 1962). On the other hand, Gwinner (1967) held warblers on 12L/12D for up to 27 months and noted that their body weight changes and molting cycles were of normal frequency. Their activity rhythms, however, were less regular, but nevertheless persisted at periods averaging 9.6 months.

Mammals can also express circannual rhythms. Various studies of the ground squirrel held at constant temperature and light (12L/12D) have confirmed that these animals hibernate at intervals that are usually slightly less than a year (Pengelley and Fisher, 1957, 1963; Pengelley and Kelly, 1966; Pengelley and Asmundson, 1969, 1970; Heller and Paulson, 1970; Mrosovsky and Lang, 1971). In sheep, wool growth and shedding follow a normal annual cycle under constant lighting conditions of 12L/12D (Bennett, *et al*, 1962; Hutchinson, 1965). According to Hafez (1952), sheep living on the equator can breed all year round. Radford (1961) found this to be the case with some sheep, but reported that others were seasonal. Thwaites (1965) noted that breeding adhered to the normal cycle during the first year's exposure to 12L/12D, but that it became sporadic thereafter. In cattle living on the equator, cycles of hair growth are not evident (Bonsma, *et al*, 1943).

Judging from the above records, it is clear that some circannual rhythms can be expressed under constant conditions of equal light and dark, while others are abolished. Similar discrepancies are found if one examines the reactions of various animals to constant but unequal light/dark cycles as reviewed by Goss (1969b). In all, there are more cases on record in which the cycles persist in constant photoperiods than there are in which they disappear. This testifies to the validity of circannual rhythms but does little to explain why some of them disappear under certain conditions while others do not. In our present state of ignorance, we can only continue to gather data on this intriguing problem and hope that some day the bewildering profusion of facts may fall into place.

VII. SUMMARY

The replacement of deer antlers is triggered by the annual cycle of day length changes. Reversal of the light cycle shifts antler growth 6 months out of phase. If the day lengths are made to increase and decrease at frequencies other than once a year, the antler cycle will lock onto these experimental driving oscillations. Antlers may be replaced as often as 2, 3 or 4 times a year, or only every other year, depending on how much the "annual" light cycle has been accelerated or decelerated. If the daily photoperiods are held constant, antler replacement can still occur, but only if the ratio between the light and dark periods deviates in excess of 12.55 percent from unity (e.g., > 12-3/4L/11-1/4D). Under these conditions new antlers grow at irregular intervals averaging about 10 months. When deer are maintained under conditions of illumination that simulate the equator, however, this circannual cycle cannot be expressed and their antlers may not be replaced for at least several years in a row. It is estimated that a zone between 14° and 18° N. latitude may represent the region of transition among wild populations of deer from the aseasonal pattern of reproduction characteristic of tropical species to the seasonal rhythms of temperate zones.

ACKNOWLEDGEMENTS

Supported by NSF grant GB-20166 and NIH grants HD-00192 and GM-18805. The cooperation of the Southwick Animal Farm, Blackstone, Massachusetts, and the Roger Williams Park Zoo, Providence, Rhode Island is gratefully acknowledged, as is the invaluable assistance of Lois T. Brex and Angela M. Black.

REFERENCES

BENNETT, J. W., HUTCHINSON, J. C. D., & WODZICKA-TOMASZEWSKA, M. (1962). Annual rhythm of wool growth. Nature 194, 651-652.

BILLINGHAM, R. E., MANGOLD, R., & SILVERS, W. K. (1959). The neogenesis of skin in the antlers of deer. Ann. N. Y. Acad. Sci. 83, 491-498.

BONSMA, J. C., & PRETORIUS, A. J. (1943). Farming in South Africa 18, 101-120.

GOSS, R. J. (1963). The deciduous nature of deer antlers. In "Mechanisms of Hard Tissue Destruction" (R. Sognnaes, ed.) AAAS Publ. No. 75, Washington, D. C. Pp. 339-369.

GOSS, R. J. (1968). Inhibition of growth and shedding of antlers by sex hormones. Nature 220, 83-85.

GOSS, R. J. (1969a). Photoperiodic control of antler cycles in deer. I. Phase shift and frequency changes. J. Exp. Zool. 170, 311-324.

GOSS, R. J. (1969b). Photoperiodic control of antler cycles in deer. II. Alterations in amplitude. J. Exp. Zool. 171, 223-234.

GOSS, R. J. (1970). Problems of antlerogenesis. Clin. Orthopaed. 69, 227-238.

GOSS, R. J. & ALBACH, D. A. (1974). How antler growth cycles are influenced by day lengths and latitude. Internat. Zoo Yearbook 14. (In press).

GOSS, R. J. & ROSEN, J. K. (1974). The effect of latitude and photoperiod on the growth of antlers. J. Reprod. Fert. Suppl. No. 19. (In press).

GOULD, S. J. (1973a). The misnamed, mistreated, and misunderstood Irish Elk.

Nat. Hist. 82, 10-19.

GOULD, S. J. (1973b). Positive allometry of antlers. Nature 244, 375-376.

GREENWOOD, A. W. (1962). An experiment with a constant environment for the domestic fowl. Anim. Prod. 4, 80-90.

GWINNER, E. (1967). Circannuale Periodik der Mauser und der Zugunruhe bei einem Vogel. Naturwiss. 15/16, 447.

HAFEZ, E. S. E. (1952). Studies on the breeding season and reproduction of the ewe. J. Agric. Sci. 42, 189-265.

HELLER, H. C. & POULSON, T. L. (1970). Circannian rhythms. II. Endogenous and exogenous factors controlling reproduction and hibernation in chipmunks (*Eutamias*) and ground squirrels (*Spermophilus*). Comp. Biochem. Physiol. 33, 357-383.

HUTCHINSON, J. C. D. (1965). Photoperiodic control of the annual rhythm of wool growth. In "Biology of the Skin and Hair Growth" (A. G. Lyne and B. F. Short, eds.), Angus and Robertson, Sydney. Pp. 565-573.

JACZEWSKI, Z. (1954). The effect of changes in length of daylight on the growth of antlers in the deer (*Cervus elaphus L.*). Folia Biol. 2, 133-143.

JACZEWSKI, Z. (1973). Photoperiodic regulation of the antler cycle in deer. Przeglad Zoologiczny 17, 229-235.

JACZEWSKI, Z. & GALKA, B. (1967). Effects of administration of testosteronum propionicum on the antler cycle in red deer. Finnish Game Research 30, 303-308.

MOLELLO, J. A., EPLING, G. P. & DAVIS, R. W. (1963). Histochemistry of the deer antler. Amer. J. Vet. Res. 24, 573-579.

MROSOVSKY, N. & LANG, K. (1971). Disturbances in the annual weight and hibernation cycles of thirteen-lined ground squirrels kept in constant conditions and the effects of temperature changes. J. Interdisciplinary Cycle Res. 2, 79-90.

NOBACK, C. V. (1929). The internal structure and seasonal growth changes of deer antlers. Bull. N. Y. Zool. Soc. 32, 34-40.

PENGELLEY, E. T. & ASMUNDSON, S. M. (1969). Free-running periods of endogenous circannian rhythms in the golden-mantled ground squirrel, *Citellus lateralis*. Comp. Biochem. Physiol. 30, 177-183.

PENGELLEY, E. T. & ASMUNDSON, S. J. (1970). The effect of light on the free-running circannual rhythm of the golden-mantled ground squirrel, *Citellus lateralis*. Comp. Biochem. Physiol. 32, 155-160.

PENGELLEY, E. T. and FISHER, K. C. (1957). Onset and cessation of hibernation under constant temperature and light in the golden-mantled ground squirrel, *Citellus lateralis*. Nature 180, 1371-1372.

PENGELLEY, E. T. & FISHER, K. C. (1963). The effect of temperature and photoperiod on the yearly hibernating behavior of captive golden-mantled ground squirrels, *Citellus lateralis tescorum*. Can. J. Zool. 41, 1103-1120.

PENGELLEY, E. T. & KELLY, K. H. (1966). A "Circannian" rhythm in hibernating species of the genus *Citellus* with observations on their physiological evolution. Comp. Biochem. Physiol. 19, 603-617.

RADFORD, H. M. (1961). Photoperiodism and sexual activity in Merino ewes. I. The effect of continuous light on the development of sexual maturity. Austral. J. Agric. Res. 12, 139-146.

TACHEZY, R. (1956). Uber den Einfluss der Sexualhormones auf das Geweihwachstum der Cerviden. Saugetierkundliche Mitteilungen 4 (3), 103-112.

THWAITES, C. J. (1965). Photoperiodic control of breeding activity in the Southdown ewe with particular reference to the effects of an equatorial light regime. J. Agric. Sci. 65, 57-64.

WALDO, C. M. & WISLOCKI, G. B. (1951). Observations on the shedding of the antlers of Virginia deer (*Odocoileus virginianus borealis*). Amer. J. Anat. 88, 351-396.

WALDO, C. M., WISLOCKI, G. B. & FAWCETT, D. W. (1949). Observations on the blood supply of growing antlers. Amer. J. Anat. 84, 27-61.

WHITEHEAD, G. K. (1972). "Deer of the World." The Viking Press, New York.

WISLOCKI, G. B. (1942). Studies on the growth of deer antlers. I. On the structure and histogenesis of the antlers of the Virginia deer (*Odocoileus virgianus borealis*). Amer. J. Anat. 71, 371-416.

WISLOCKI, G. B. (1956). In "Ageing in Transient Tissues" (G. E. W. Wolstenholme and E. C. P. Millar, eds.) Ciba Foundation Colloquia on Ageing, 2, 176-187.

WISLOCKI, G. B., AUB, J. C. & WALDO, C. M. (1947a). The effects of gonadectomy and the administration of testosterone propionate on the growth of antlers in male and female deer. Endocrinology

40, 202-224.

WISLOCKI, G. B., WEATHERFORD, H. L. & SINGER, M. (1947b). Osteogenesis of antlers investigated by histological and histochemical methods. Anat. Rec. 99, 265-296.

WOLFSON, A. (1949). The relation of day length and gonadal regression to post-nuptial molt in the white-throated sparrow and other Fringillids. Anat. Rec. 105, 522.

CREDITS

Figure 4. Reproduced by permission of the Wistar Institute Press, from the Journal of Experimental Zoology, Vol. 170, pp. 311-324, 1969.

ASPECTS OF CIRCANNUAL RHYTHMS IN MAN

ALAIN REINBERG, M.D., Ph. D.

Equipe de Recherche de Chronobiologie
Humaine (C.N.R.S. No. 105)

Laboratoire de Physiologie
Fondation A. de Rothschild
29, rue Manin 75019 Paris - France

INTRODUCTION

Circannual rhythms in human beings have been reported since Hippocrates' time (circa 400 B.C.) and even before:

> There is a season for everything
> And a time for every purpose under
>   the heaven;
> A time to be born, and a time to die;
> A time to plant and a time to reap...
>
> Ecclesiastes

The existence of circannual rhythms is considered as a time-honored fact even if a large number of these rhythms still await a statistically validated detection. An impressive number of publications, with gradual increase after WW II, has been devoted, more or less specifically, to this topic. Let us quote, among many other reported circannual rhythms, changes in: heart rate, temperature, certain patterns of sleep (Kleitman *et al*, 1937; Kleitman, 1963); blood haematocrit, haemoglobin, erythrocyte sedimentation rate, plasma protein and blood chloride (Renbourn, 1947); blood pH (Petersen, 1947), circulating blood eosinophils, urinary 17-ketosteroids (Watanabe *et al*, 1956; Hamburger 1954), protein bound iodine in the serum (Du Ruisseau, 1965); potassium urinary excretion (Ghata, Reinberg, 1954), 11-oxycorticoids urinary excretion (De Pergola *et al*, 1962; Sotgiu, Lodi, 1962); blood cholesterol (Jellinek *et al*, 1936); timing of menarches (Valsik, 1965) etc. Reviews dealing with circannual rhythms in man were also published by: Hughes (1931), Kleitman (1949, 1963), Fitt (1941), Sargent (1954), Reinberg, Ghata (1964), Valsik (1963), Bunning (1963), Halberg, Engeli, Hamburger, Hillman (1965), Halberg, Reinberg (1967), Luce (1970), Klinker *et al* (1972), Reinberg (1973), Smolensky *et al* (1972), Tromp (1964, 1971) etc.

As a matter of fact recent and future progress in the study of circannual rhythms, as well as progress in chronobiology, are related to basic and methodologic considerations:

1. Physiologic functions in any living organism, including man, are not constant as a function of time; regular and predictable variations with periods of about 24 hours, about 7 days, about 30 days, about 1 year, etc. can be detected objectively; thereafter

circadian, circaseptan, circamensual, circannual biologic rhythms can be characterized and qualtified. 2. Both direct and indirect experimental arguments favor an intrinsic origin (at least in part) of such biologic rhythms. It is widely admitted, among chronobiologists, that rhythmic changes in environmental factors -- the so-called synchronizers (Halberg *et al*, 1954) or zeitgebers (Aschoff, 1954) -- do not create biologic rhythms even if they are able to influence certain of their characterizing parameters (Halberg, 1960, 1969; Aschoff, 1963; Pittendrigh, 1958; Halberg, Reinberg, 1967: Reinberg, Boissin, Assenmacher, 1970). 3. A specific methodology has to be used. To characterize any detectable bioperiodic phenomenon four parameters have to be quantified: the period, τ, the acrophase, ϕ, the amplitude, A, and the mesor, M, or the rhythm-adjusted level. Statistical analyses, the cosinor method among others (Halberg, Tong *et al*, 1967, Halberg, Engeli *et al*, 1965) allow the estimation of τ, ϕ, A and M with their respective 95% confidence interval. However, the characterization of a circannual rhythm is not restricted to the quantification of circannual τ, ϕ, A and M; the possible relationship, interference, modulation etc. between circannual, circamensual and circadian rhythms has to be taken into consideration in the study of any physiologic variable.

Therefore to begin this review paper several methodologic considerations will be summarized; then several aspects of circannual rhythms in man will be presented; they were selected of course with regard to their medical (practical) and/or their biological interest, but also with regard to the method of data gathering and to the mean of time serie analyses that were used. The aim of this review is more to give a living picture of recent and validated acquisitions in this very old branch of knowledge, than to give a tedious compilation of facts that can be sometimes questionable.

METHODOLOGIC REQUIREMENTS FOR CIRCANNUAL RHYTHM STUDIES

A. HIGH STANDARDS IN PHYSICAL MEASUREMENTS, CHEMICAL DETERMINATIONS, ETC. ARE DESIRABLE FOR EACH DATUM IN A TIME SERIES.

These experimental conditions are a necessity, but in themselves are not sufficient to validate a bioperiodic phenomenon. Even if the acquisition of each datum requires a major expenditure of time, effort and money, data collected as a function of time -- during one or several days, weeks, months or years -- cannot lead to a description of a biologic rhythm without including some additional specific requirements in the experiments and in the analyses of time series (Halberg, Reinberg, 1967; Halberg, 1969; Reinberg, 1971).

B. STATISTICAL METHODS MUST BE USED TO ANALYSE THE COLLECTED OR RECORDED TIME SERIES.

A simple and widely used procedure is to plot, as a function of time, the monthly or seasonal mean (± 1 standard error) of a given physiologic variable. From the thus obtained chronogram, one then tries to detect a statistically significant difference between annual peak and trough, i.e. with the help of Student's "t" test, or others.

More precision and objectivity can be reached with the cosinor and related methods (Halberg, Engeli *et al*, 1965; Halberg, Tong *et al*, 1967). To validate and to approximate any bioperiodic phenomenon one tries to find the best fitting cosine function (least squares method), making it possible to determine whether or not a statistically significant rhythm is detected ($p < 0.05$) for the period, τ, such as 24 hours, 7 days, 30 days, one year, etc.

For detectable bioperiodic phenomena of each given period, the cosinor method also allows the estimation of parameters such as the acrophase, ϕ, the amplitude, A, and the rhythm adjusted level or mesor, M. τ, ϕ, A and M are the parameters to be considered for rhythm characterization and objective quantification. The estimate of each parameter is given with its 95% confidence interval when a rhythm is detectable. The acrophase, ϕ, is the peak of the best fitting sine function approximating all data, taking into account a phase reference which can be a certain clock hour (i.e. midnight) for circadian rhythms, as well as a certain clock hour of a certain day for ciramensual and/or circannual rhythms as well. The bathyphase is the reverse of the acrophase; they differ from each other by 180^0; the bathyphase is the trough of the best fitting sine function used to approximate the rhythm.

The rhythm-adjusted level or mesor corresponds to the mean level when data are obtained at equal intervals for the considered period.

The results thus obtained can be either tabulated or displayed on a figure with a choice between a clockwise presentation (figure 1), that of the best fitting function (figure 2), that of the acrophases (with their respective confidence interval)

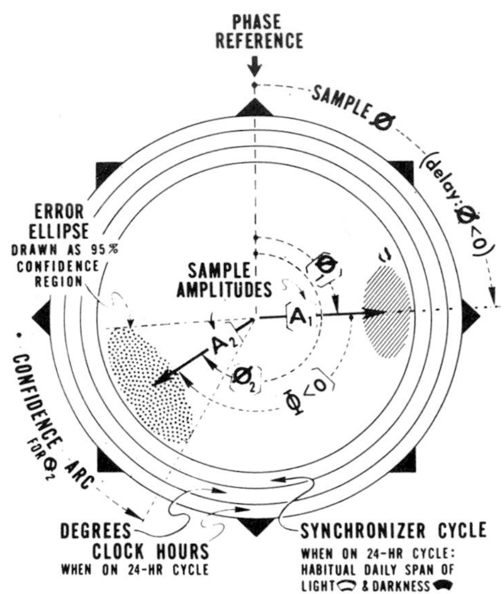

Fig. 1: Cosinor display
Results of the analyses of bioperiodic phenomena are given as polar coordinates.
The circumference represents the considered period τ.τ is equal to 360°, irregardless of the length of the period, be it 24 h, 7 days, 30 days, 1 year, etc. The synchronizer cycle can also be represented with its timing.
The error ellipse surface depends essentially upon the dispersion of phase (Ø) and amplitude (A) computed from time series on the one hand, and from the statistical p value.
When the error ellipse overlap the pole it means that no rhythm is detected with a) the considered period, τ, (or frequency $1/\tau$); b) with similar phases and amplitudes. (The number of samples may be too small; the dispersion of Øs and As may be too wide or the biologic period of the rhythm does not correspond to that which was taken for the analysis.
On the contrary a statistically significant rhythm is detectable when the error ellipse does not overlap the pole with 95% as security limits.
Then, the rhythm amplitude (A) is given by the length of the vector. The error ellipse gives the amplitude confidence interval where it intersects the vector and the prolongation of the vector.
The acrophase (Ø) is given by the location of the prolonged vector on the circumference. It can be then expressed as a phase angle with respect to a phase with respect to a phase reference, or expressed in units of time as a fraction of the considered period τ. The confidence arc of the acrophase is given by the tangents of the error ellipse drawn from the pole. (By courtesy of F. Halberg *et al:* 1967).

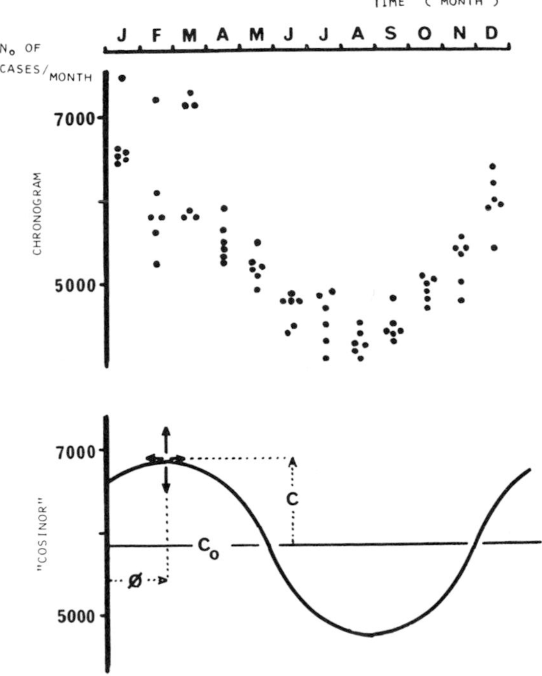

Fig. 2: Circannual rhythm of deaths in France from cerebral vascular lesions is taken here as illustrative example for the cosinor method of time series analysis.
Chronogram (Top). For macroscopic inspection, monthly data are plotted as a function of time for each of the considered years. Circannual variation is apparent, yet the task remains to objectively describe the rhythm and to quantify its characteristics. "Cosinor" (Bottom). The rhythm is approximated by the least squares method with the 365.25-day cosine function best fitting all data. A statistically significant circannual rhythm ($p < 0.005$) is detected; it can be characterized by the point estimation of several of its parameters with their respective 95% confidence limits. "Acrophase". Ø (peak of the best fitting cosine function. Ø = Feb. 20 (Feb. 9 to March 2).
"Rhythm – adjusted level". C_0 (or "Mesor". M) (here monthly mean of death) = 5837 ± 133.
"Amplitude". C(or A) = 1057 (685 to 1430). In other words, at the time of the circannual acrophase the monthly number of deaths from cerebral vascular lesions in France averages 6894 ± 373, whereas six months earlier or later, the corresponding value is 4780 ± 373.
(A. Reinberg, P. Gervais *et al:* 1973).

for a given period, etc.

There is a relationship between the duration of a sampling span, T, the number of subjects, N, the sampling interval between consecutive samples, Δt, and the period of rhythm, τ. Δt of course, should be smaller than τ. If for any reason T is fixed, it is possible to manipulate N or vice versa. The longitudinal profile (or sample) consists of repeated observations or measurements made on one and the same individual over a span, T, much longer than the period, τ, of a given physiologic rhythm under study. For example, T = 16 years in Hamburger's study on a circannual rhythm in 17-ketosteroid urinary excretion (Hamburger, 1954; Halberg, Engeli *et al*, 1965) the transverse profile (or sample) consist of "short" time series, each covering a span that is not much longer than τ, of a physiologic rhythm under study. A transverse analysis, which corresponds to the most common sampling procedure of circannual rhythm studies, requires the use of a relatively large number of subjects.

It has been demonstrated, for circadian rhythms, that with comparable subjects and experimental conditions, longitudinal and transverse studies lead to similar results (Halberg, 1965; Halberg, Reinberg, 1967; Halberg, Tong *et al*, 1967; Reinberg, 1971). This statement can be extended to circannual rhythms. The circannual rhythm chronogram of the 24-hour urinary excretion of 17-ketosteroids in an adult man shows a peak in December-January (and a trough in June-July) in the transverse study of Watanabe (1964) in Niigata, Japan, a peak in January (and a trough in June) in the transverse study of Ahuja and Sharma (1971) in New Delhi, India, and a peak in November (and a trough in May) in the longitudinal study of Hamburger (1954) in Copenhagen, Denmark. Obviously to validate such comparisons one has to take into consideration other potentially influencing factors such subjects' sex (Reinberg, Smolensky, 1973), age (Descovitch, Montalbetti *et al*, 1974), ethnic origin, (Simpson, Bohlen, 1973), geographical location (latitude) even for subjects living in the same hemisphere (Ghata, Reinberg, 1954; Smolensky, Halberg, Sargent, 1972; Batchelet *et al*, 1973; Simpson, Bohlen, 1973). As a matter of fact when referring to geographical location we are dealing with the circannual subjects' synchronization rather than a possible but small climatic influence (Smolensky, Halberg, Sargent, 1972; Reinberg, 1972). This point will be discussed later. However, it leads us to consider an other methodologic aspect of prime importance, namely the specific methodology.

C. SPECIFIC METHODOLOGY IN DATA ACQUISITION

This important aspect has been discussed by Reinberg and Smolensky (1973) with reference to circamensual rhythms in healthy women. Basically the same methodological considerations have to be kept in mind in the study of any low-frequency rhythm (or infradian rhythm) i.e. circannual, circamensual as well as circaseptan rhythms. The interpretation of findings finally depends upon the experimental design used to gather data. In the past, most investigators, in researching circannual rhythms neglect the possible influence (or interference) of other periodicities, especially circadian and circamensual rhythms, which may be characterized by larger amplitude than the circannual rhythm. Thus, by sampling once daily at different clock hours -- or for that matter, even at the same time of day in subjects adhering to non-standardized societal routine -- collected data may reflect, to a greater degree, circadian (or circamensual) rather than circannual rhythmicity.

Moreover, let us underline that not only the circadian mesor, M, (time serie mean level) but also the circadian acrophase, ϕ, and the circadian amplitude, A, of a physiologic variable can be the subject of a circannual rhythm. This possibility, suspected by Ghata and Reinberg (1954) in experiments on healthy mature men, was demonstrated experimentally by Haus and Halberg (1970) in mice.

Circannual variations in serum corticosterone levels in Balb/C mice were demonstrated in pooled data of profiles involving at least 6 time points of sampling per 24-hour span. In Minnesota, high values are found during the winter months and low values in late spring and summer (figure 3). Cosinor analysis of the circadian rhythm of serum corticosterone during different times of the year shows a change in the circadian acrophase from about 43° in February to 95° in May. (24 h = 360°; 1 h = 15°; acrophase reference, mid-light span). This change is evident after a seven-day standardization span at relatively constant temperature and a regimen of LD = 12:12 with light from 0600 to 1800 (figure 3).

Evidence for circamensual changes in circadian A, M, ϕ (and/or possibly τ) values of healthy adult women, with spontaneous menstruation, come from recently completed studies (Smolensky, Reinberg *et al*, 1973); on rhythms in cutaneous reactivity to histamine (Reinberg, Smolensky, Ghata,

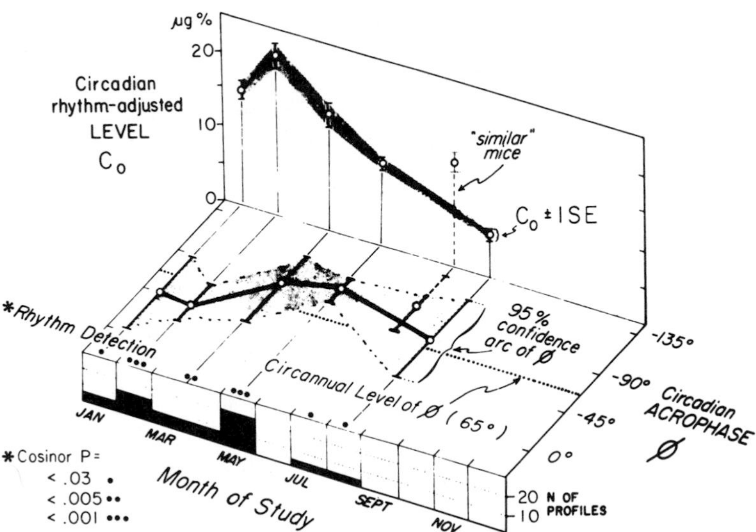

Fig. 3: Serum corticosterone circannual rhythm in mice.

The vertical plane of the graph represents the circannual change in the circadian rhythm adjusted level of the variable.

The horizontal plane of the graph shows the circannual change in the circadian acrophase of plasma corticosterone.

95% confidence limits as well as rhythm detection obtained from cosinor analysis are also represented.

Mice were standardized, for at least 7 days prior to killing, in artificial light from 0600 to 1800. They were kept previously under largely natural lighting conditions.

(By courtesy of E. Haus and F. Halberg: 1970).

Gervais, 1973); on rhythms in urine 17-hydroxycorticosteroids excretion, Proccaci *et al*, (1972) and on rhythms in pain threshold to radiant heat.

Evidence for circannual changes in circadian A, M, ϕ (and/or τ) values of healthy subjects will be presented here. For example a statistically significant circannual change in circadian ϕ was demonstrated in the urinary excreted epitestosterone glucuronide of healthy young men by Lagoguey, Dray, Chauffournier and Reinberg (1973).

Therefore, it is necessary to consider the possibility of circannual modulation of either one or all of the circadian (or other) rhythms characterizing parameters besides the mesor (time series mean level). When circannual modulation of the circadian ϕ (and/or τ) occurs, sampling once daily may not only fail to reveal such events, but may lead to erroneous findings (Smolensky, Reinberg *et al*, 1973). In such cases, sampling once daily -- eventually at the same clock hour all the year long -- may reveal a biphasic rhythm; while a more frequent circadian sampling would reveal a monophasic wave form with a change in circadian ϕ during the circannual rhythm.

Thus, in order to more thoroughly study circannual and other low frequency rhythms as well, it is strongly suggested, when appropriate, that further works utilize a rhythmometric approach including the use of profiles of 24-hour duration, scheduled at least at weekly intervals throughout the menstrual cycle and at monthly (or every other month) intervals throughout the year.

Obviously in research on circadian modulation of circannual rhythms, clock-hours of lights-on and of lights-off (sleep-wakefulness) have to be recorded as precisely as possible since the prominent (circadian) synchronizer in man is of socio-ecological origin (Apfelbaum, Reinberg *et al*, 1969; Ghata *et al*, 1969; Halberg *et al*, 1959; Reinberg, 1971).

The persistance of a biologic rhythm, when the studied organism is isolated from the influence of known synchronizers, can be taken as an excellent argument in favor of the intrinsic origin of the rhythm. This method has been and is still widely used for circadian rhythm studies in plants and animals including man. An impressive number of circadian rhythms -- and eventually circamensual rhythms (Halberg, Reinberg, 1967; Reinberg, 1967) -- have been demonstrated to persist in many experiments with men (or women) during several weeks or months of isolation in laboratory apartments (Aschoff, 1969), or underground in natural caves (Kleitman, 1949, 1963; Mills, 1966, 1973; Ghata *et al*, 1969; Apfelbaum, Reinberg *et al*, 1969; Halberg, Reinberg, 1967; Halberg, Reinberg, Haus *et al*, 1970). The method of isolation, in almost constant and controlled environmental conditions, has been used to demonstrate the persistance of circannual rhythms in birds, (Benoit *et al*, 1956; Assenmacher *et al*, 1970) as well as in small mammals (Pengelley, 1971); for obvious reasons human experimental isolation for a span of time as long as one year has not been performed as yet. Thus the evidence of an intrinsic component of circannual rhythms in man can be given only by indirect arguments.

CIRCANNUAL ENDOCRINE AND RELATED RHYTHMS

The available information deals mainly with bioperiodic activities of the adrenal, the testis and the ovary measured more or less directly by indexes such as urinary excretion; 17-ketosteroids, 17-hydroxycorticosteroids, potassium, catecholamines, testosterone glucuronides, etc.; serum testosterone; timing of menarches, menstrual cycle duration, etc.

From an endocrinological point of view these indexes are not necessarily the best that can be selected for their specificity, their precision, etc. However one must understand that the authors of such long term studies had to compromise between what was scientifically desirable and what was realistically possible.

URINARY 17-KETOSTEROID EXCRETION

A longitudinal study with data covering 16 years (with breaks) of 24 hour samples, with determination of urine volume and 17-ketosteroid excretion, was performed by Hamburger on himself and analysed with the help of cosinor and related methods by Halberg, Engeli, Hamburger and Hillman (1965). On the chronogram (figure 4) a circannual variation with a trough in May and a peak in November can be demonstrated for 17-ketosteroid urinary excretion.

Then the entire series of data on 17-KS and urine volume, as well as sections of these series, each covering several years, were analyzed by a special computer program (least squares method); certain spectral windows restricted to relatively narrow regions with trial periods chosen linearly in period were then computed for the data as a whole, and, thereafter for separate sections.

"The window on the right of figure 5 indeed visualizes the 24-hour-synchronized circadian component of 17-KS excretion. The reader should note that the amplitude scale of this window differs from that used in the other two windows of the same figure, in order to accomodate the relatively large amplitude of the circadian rhythm next to the much smaller amplitudes of about - weekly or about - yearly rhythms..."

"An amplitude crest of urinary 17-KS in the spectral region around one year can be seen in the window on the left of figure 5. The location, in this window, of a precise one-year period is indicated by a vertical line and by an arrow in heavy print at the pertinent point of the abscissa. The amplitude crest of urine volume in the about -yearly region lies right on the line, above the arrow. The amplitude crest of 17-KS is found to the left of the line corresponding to a trial period of precisely 365 days, the crest lying at the trial period of 378 days. The average "circannual" period seems to be somewhat longer than a year, but this circannual amplitude crest of 17-KS is relatively broad..."

"One can suggest merely that the deviation from 365 days of the circannual component in 17-KS excretion is probably not a methodologic artifact; a precise annual (average) period can be seen indeed in the spectral windows of 17-KS. The different results obtained in spectra computed for the two physiologic functions in the circannual domain represent, probably, differences between the behaviour of rhythms in the two time series..."

The window in the middle of figure 5 reveals the about-weekly component in the

Daily (07-07) 17-Ketosteroid Excretion (KS) as a Function of Month of Year [A Healthy Mature Man]

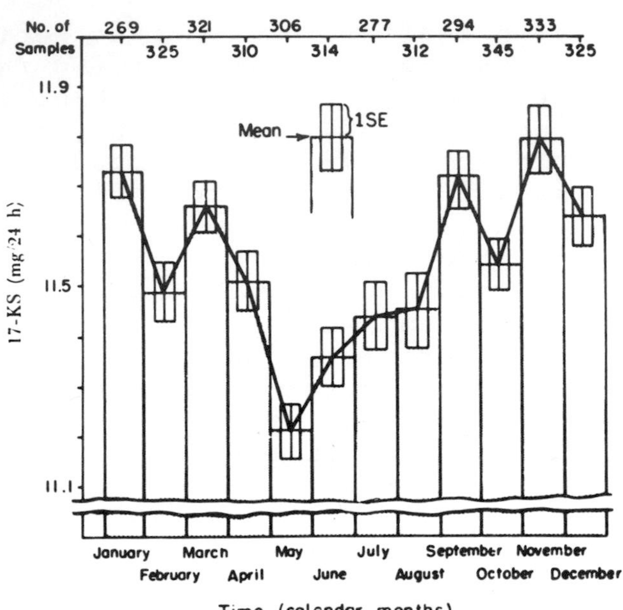

Fig. 4: Circannual rhythm in the urinary 17-ketosteroid excretion (KS). Chronogram of a longitu dinal study covering more than 15 calendar years. Trough in May and peak in November. (By courtesy of F. Halberg, M. Engeli, C. Hamburger and D. Hillman: 1965).

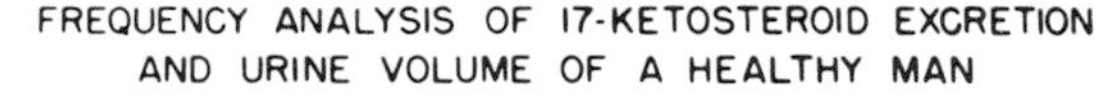

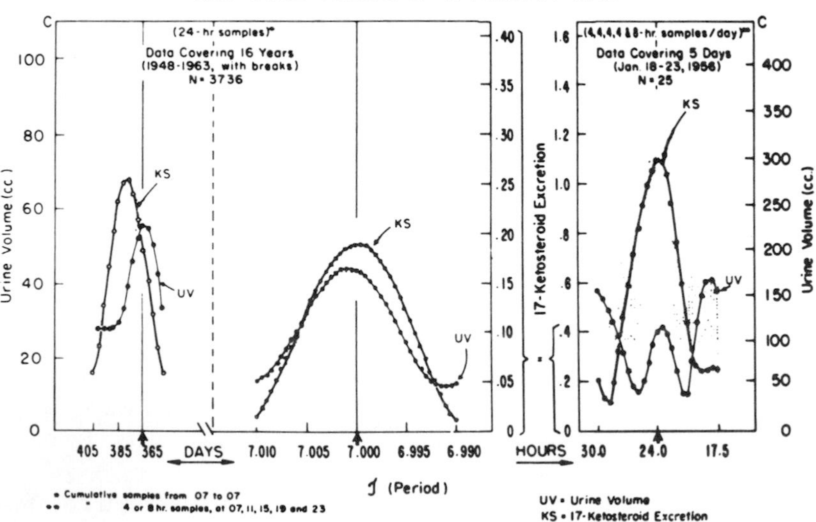

Fig. 5: From left to right circannual (about 1 year), circaseptan (about 7 days) and circadian (al 24 h) rhythm detection is shown. (see text).
(By courtesy of F. Halberg, M. Engeli, C. Hamburger and D. Hillman: 1965).

spectrum of 17-KS excretion. On the basis of the time series as a whole, this rhythm appears to have been a precise 7-day period; however, a slight yet statistically significant deviation from precisely 7 days can be detected in one of the 4 serial sections restricted to a 4 year span of time. This rhythm can be then desynchronized from a societal 7-day routine. The desynchronization taking place at a time during which the subject was given a testosterone treatment. It is suggested by the authors that: "the long-continued frequent administration of androgen had eventually desynchronized [without damping of the approximate 7-day rhythm amplitude] one or both of the gland pairs contributing to the urinary 17-KS-adrenal and/or testes - from their superimposed pituitary and hypothalamic controls."

These results suggest also that infradian (or low frequency) rhythms in man are able to free-run after appropriate manipulation of the organism.

The integrated 24-hour urine sampling, which obscured the circadian rhythms, about once a month during a year in the transverse studies of both Watanabe (1964) and Ahuja and Sharma (1971) also show a circannual variation. A good agreement between the results of these three groups working independently is not surprising in spite of the facts that (a) the day of the week and/or the month was not taken into consideration either by Watanabe or by Ahuja and (b) geographical location, and climate were changing. The amplitude of both the 7 day and 30 day rhythms of 17-KS urinary excretion is smaller than that of circadian and circannual rhythms; therefore the risk of a circannual rhythm oblitaration by a methodologic defect in data gathering was minimized. Climatic changes (i.e. that of ambient temperature) even if they are well correlated with 17-KS excretion do not "explain" the circannual rhythm of this variable since annual change of climate in Niigata, Japan; New Delhi, India; and Copenhagen, Denmark are less than comparable for the given year of the transverse study as well as the 16 years of the longitudinal study in Copenhagen. A better understanding of such circannual changes in a physiologic variable is given by the existence of an intrinsic component than by an extrinsicly (and exclusively) induced rhythm.

PLASMA CORTISOL

The plasma concentration of 11-hydroxycorticosteroids was documented by Mills and Waterhouse (1973) in a longitudinal study of one year with sampling twice a day (blood collected on rising and just before retiring) at 12 different days corresponding roughly to different months. The highest concentration took place in the morning except for two days, May 26th and October 29th, when the concentration was high at bedtime. These findings are coherent with a circannual phase shift in the circadian acrophase of plasma corticosterone demonstrated by Haus and Halberg (1970) in mice. It corresponds also to a phase shift in the circadian acrophase of urinary potassium as shown in the results published by Mills and Waterhouse (1973).

Since circadian rhythms in plasma cortisol, circulating blood cells etc. were not documented round the year by other authors, the results they obtained are questionable from a methodologic point of view.

URINARY 17-HYDROXYCORTICOSTEROID EXCRETION

Circannual rhythms dealing with the 24-hour 17-OHCS excretion can be considered for discussion, in spite of the fact that circamensual changes, demonstrated by Reinberg, Smolensky *et al* (1974) in healthy women, were ignored when the following works were published.

A circannual variation of mean 24-hour urinary 17-OHCS was reported by Watanabe (1964) with a peak in February, by Ahuja *et al* (1971) with a peak in January and a trough in July, while Okamoto *et al* (1964) reported a statistically significant difference between a high mean in January and a low mean in September.

Thus a circannual rhythm in the circadian mean level of 17-OHCS excretion can be theorized with a winter peak and a summer trough in the northern hemisphere.

URINARY POTASSIUM EXCRETION

Let us remember (see review paper by Reinberg, 1971) that biological rhythms of potassium metabolism can be demonstrated in man as well as in other animals. These rhythmical and predictable variations have been studied extensively (a) in healthy human subjects under various experimental situations and/or conditions, e.g. new-borns raised in self demand, adults following their usual daily routine of activity and rest, or desynchronized by environmental manipulations such as underground isolation without time cue, and in intercontinental flight across several time zones (transmeridian flight), and (b) in patients suffering from various diseases such as adrenal insufficiency, Addison's disease, adrenal hyperactivity, Cushing's syndrome, etc.

From experimental facts it appears that potassium excretory bioperiodicity is a well documented example of an intrinsic rhythm in man. Several of its parameters can be changed by the manipulation of synchronizers, nevertheless synchronizers do not create the rhythm.

Several attempts have been made to find a specific relationship between the potassium rhythm and a physiologic function (Conroy and Mills, 1970). Obviously the adrenal glands, the kidneys and other organs or tissues with rhythmic activity play a role in inducing the K^+ excretory rhythm; but no one of these organs or tissues alone can be considered as the clock or as the inducer of this rhythm. The potassium excretory rhythm is likely to be the result of several component cyclic variations involving rhythmical activities in the adrenals, kidneys, nervous system, etc. Even if the adrenal seems to be a very important factor in sodium and potassium excretory rhythms, other factors must not be neglected. In addition, rhythms of adrenal secretions are themselves associated with adrenal activity rhythms such as circadian variations in mitoses, metabolic processes, response to ACTH stimulation, etc. which can be considered also as component rhythms.

In spite of these qualifications, potassium excretory rhythms can be summarized in a paragraph dealing with metabolic reflections of corticoadrenal activities. From chronogram examination of data collected in 1952-1954, on urinary volume and potassium excretion, Ghata and Reinberg (1954) reported that a circannual rhythm seemed to occur. These time series were reanalyzed for a well documented discussion about circannual rhythms in mice (Haus and Halberg, 1970). Circannual rhythms may be expressed, among other ways, as changes in level and/or in parameters (ϕ or A) of each of the rhythms in the several frequency domains. Ghata and Reinberg's data suggest the occurence of a circannual shift in circadian acrophase of urinary volume and K^+ excretion by human subjects. A circadian acrophase was computed by a least squares fit of a 24 hr cosine curve to data on 6-8 subjects, each sampled at 4 hr intervals over a 24 hr span during either September-October in Paris (1952, 1954) and Copenhagen (1953), or during March-April (1954) in Paris. In order to focus upon possible circannual changes in circadian acrophase, the site where the profiles were obtained can be ignored. By ignoring further the circular distribution (in view of the location away from 360° of the acrophase and their proximity one to another), an F-test was then used to compare the 3 series obtained in September-October with that obtained in March-April. A statistically significant difference (below the 5% level) was thus detected. The circadian acrophase (with their standard errors) expressed as delay from local 0000 were located at $-189^\circ \pm 8^\circ$ for urine volume and at $-184^\circ \pm 6^\circ$ (12 hr 16 min.) for urinary potassium excretion during September-October. The corresponding values for March-April were located at -261° and at -233° (15 hr 32 min.) respectively.

There was no statistically significant seasonal change in the 24 hr mean value of excreted potassium, a result which is in good agreement with that of Mills and Waterhouse (1973), Ahuja *et al* (1971), Simpson and Bohlen (1973) among others.

A circannual shift in the circadian

acrophase of the potassium excretory rhythm was also reported by Mills and Waterhouse (1973) and by Simpson and Bohlen (1973) and Bohlen (1971).

The studies performed by Simpson and Bohlen are interesting to consider since they deal not only with circannual and circadian changes of several variables (temperature, urinary excretion for water potassium and sodium) but also with the latitude at which each of these bioperiodic phenomena was documented.

Summer (June-July) circadian rhythm parameters were obtained from Hugh Simpson and his wife who made eight urine collections daily for one week while on holiday in St. Lucia (Lat. 14° N) in 1968 and a further series in Devon Island (Lat 76° N) in 1969 for three weeks. Comparison can thus be made between data from two latitudes (near the Equator and in the High Arctic), derived from the same two subjects at the same time of year and nearly the same longitude and only one year apart. The sleep/awake routine was reasonably similar in both cases, the midsleep time in St. Lucia being 0240 whereas in Devon Island it was 0300. Diet was not controlled however. Similar daily potassium excretion rates in the two locales were recorded.

Regarding the circadian rhythm in potassium excretion, a large difference in amplitude (both absolute and relative) is found between the equatorial and polar study amounting to a 50% damping in the latter; the acrophases for the two subjects, on average, occur 3.1 hr later from midsleep in the Arctic.

An acrophase shift of about 3 hr in the same direction was found by Ghata and Reinberg (1954) in the potassium circadian rhythm between similar groups of subjects (7-8) living in Paris (lat 49° N) and in Copenhagen (Lat 56° N) at the same time of the year (September-October).

According to Simpson and Bohlen (1973) in low equatorial latitude, and due to the regularity of a mean LD = 12:12 photoperiod, one might expect a relatively high-amplitude circadian rhythm phased in a precise relation to it or to the habitual routine which will normally be based on it.

In the middle latitudes, the smaller 24 hr amplitude of light intensity might result in a later phase in the circadian system if brightness level in the morning were important; however social factors (including the effect of putting on the electric light when the alarm clock rings) might counteract this.

In high latitudes, particularly near the poles where there is no 24 hr periodicity in light, one might expect increasing lateness of phase as the photo-synchronizer decreased in amplitude, and the circadian system tended to resume a circadian period longer than 24 hr which is characteristic of cave isolation studies.

Moreover the change of daily photoperiod length with season could well result in effects on the circadian system. Dawn (Simpson and Bohlen, 1973) as well as sunrise (Mills and Waterhouse, 1973) might have a more effective synchronizing effect at some phases of the circadian system than at others. If these factors are important one would expect seasonal variations in phase and amplitude of circadian rhythms, such as potassium excreted in urine and others.

Bohlen made a study of eight adult Eskimos (5M + 3F) in Wainwright, Alaska (Lat. 71°N). Urinary excretion rates for potassium and sodium and oral temperature among other variables were documented in order to evaluate the relative prominence of any rhythmic circadian and/or circannual components. The subjects were studied intensively (sampling interval $\Delta t = 2$ hr/by "day") for about one month in each of the four seasons i.e. the solstices and equinoxes because of the difficulty of continuous sampling through out the year.

Cosinor analysis for potassium excretion (2237 urine samples collected from 8 Eskimos) show first that a circadian rhythm was demonstrable (fig. 6) at all four seasons with phase and frequency synchronized between individuals. They are similar to findings made by J. Bohlen, H. Simpson and their Caucasian colleagues on themselves when sojourning in the High Arctic. Second, cosinor analysis shows the potassium acrophase for the Wainwright Eskimo group to be as follows: winter 11.8 hr, spring 10.1 hr, summer 12.0 hr, and autumn 10.1 hr, after midsleep.

Cosinor analysis for oral temperature also reveals (1) statistically significant circadian rhythm at all seasons and (2) a circannual change in the circadian acrophase (fig. 7).

URINARY CATECHOLAMINE (EPINEPHRINE AND NOREPINEPHRINE) EXCRETION

Circannual rhythms in the 24 hr urinary mean level of epinephrine (E)and norepinephrine (NE) excretion have been reported by Hale *et al* (1968) in presumably healthy young males with a peak for NE in summer; while Johanson *et al* (1969), in a large group of men and women, reported an increased excretion of E in winter and no change in

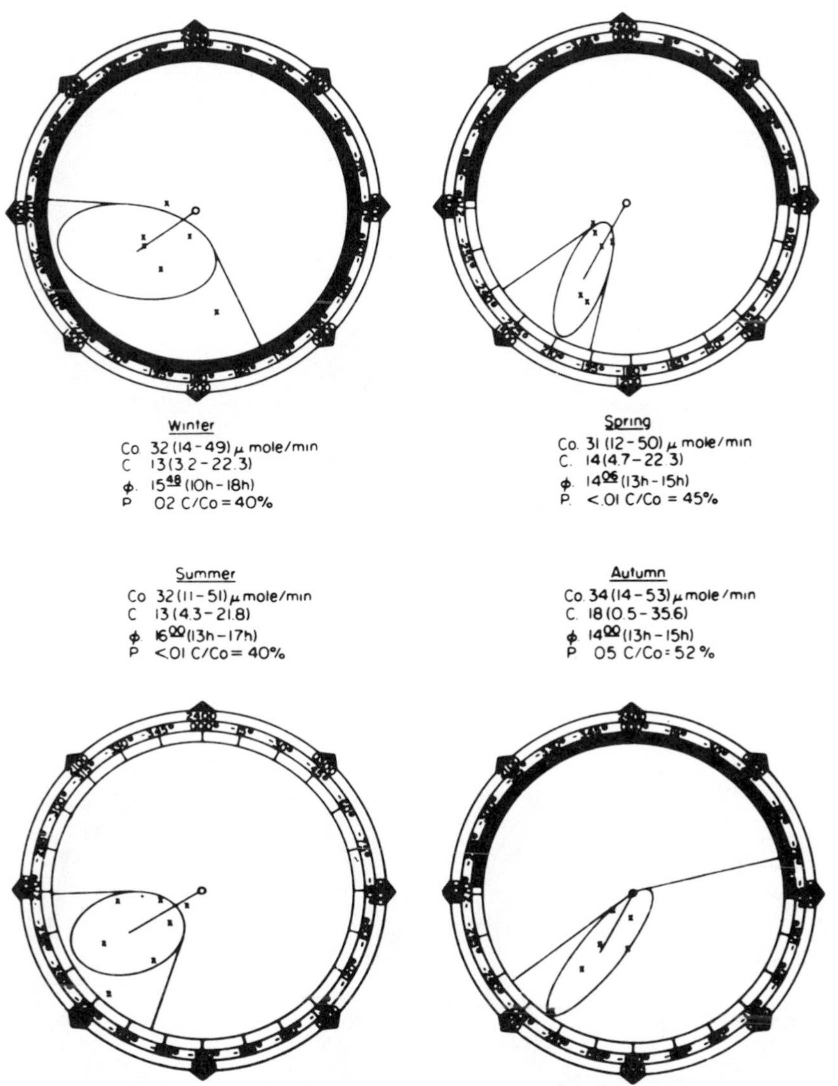

Fig. 6.
Cosinor summaries of urinary potassium circadian rhythm in Wainwright Eskimos. Crosses represent individual subjects. Ellipses and parentheses indicate 95% confidence limits for the group and P the probability that there is no phase and frequency synchronized rhythm for the group. Ø is acrophase in clock hours; to use midsleep as reference substract 4 hr. Black portion of inner circle indicates hours of darkness.
The subjects (8 adult Eskimos : 5 M + 3 F) were studied intensively for one month at the four seasons i. e. the solstices and equinoxes.
(By courtesy of H. Simpson and J. Bohlen: 1973).

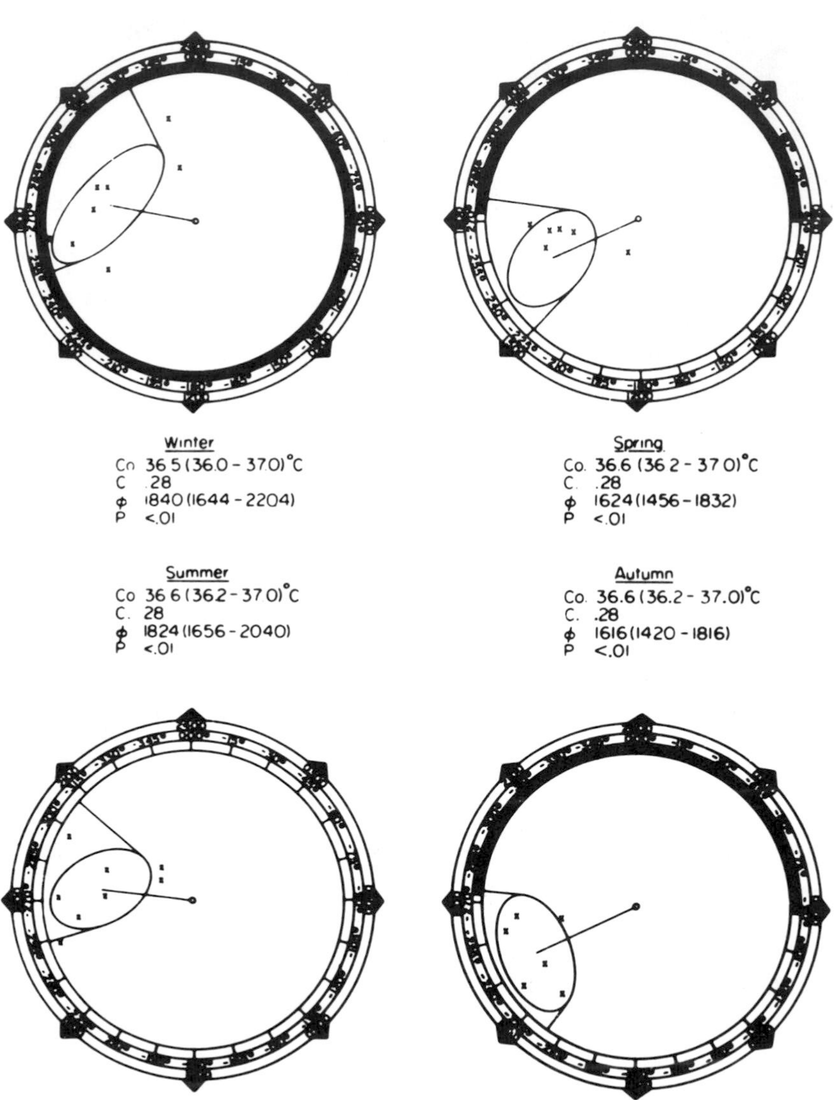

Fig. 7:
Cosinor summaries of oral temperature rhythm in Wainwright Eskimos. Conventions as in fig. 6.
(By courtesy of H. Simpson and J. Bohlen: 1973).

NE excretion.

Since the chronobiologic methodology of both of these studies is questionable the results are difficult to comment on and to compare.

Descovich, Montalbetti *et al* (1974) were able to demonstrate objectively a circannual rhythm for both urinary E and NE, in analyzing serially independent sampling from presumably healthy human subjects of both sexes. The aim of the study was to quantify circadian and circannual rhythms and to document the effect of aging.

The 106 subjects were synchronized for 4 days before and during the 3 days of sampling (Δt = 6 hr), with light on at 0600 and light off at 2200.

A statistically significant circadian rhythm was detected in both E and NE with an acrophase located about 12 hr after midsleep. (This result is in good agreement with that of comparable studies: Reinberg, Ghata *et al*, 1970; and others). Moreover, Descovich, Montalbetti *et al* (1974) also demonstrated a circadian rhythm in NER (NE/Ne+E ratio) with an acrophase coinciding with midsleep. NER reflects perhaps, *inter alia*, the relative conversion of NE into E.

Circadian rhythms in E and NE in the age group 20-30 years, and in the age group of 66-99 years have a similar timing but their mesors and amplitudes differ, decreasing with age. The amplitude of NER rhythms also decrease with age ($p < 0.001$) while the NER mesor increases with age ($p < 0.001$).

The circamensual changes in urinary catecholamines of spontaneously menstruating women were not taken into account for the circannual change. However a statistically validated circannual rhythm was detected for the three considered variables with the following acrophase timing: May-June for both E and NE and October for NER (table 1).

PLASMA AND URINARY TESTOSTERONE IN MATURE AND HEALTHY YOUNG HUMAN MALES

Both circadian and circannual rhythms of urinary testosterone glucuronide (TG) and epitestosterone glucuronide (ETG) have been documented by Lagoguey, Dray, Chauffournier and Reinberg (1972, 1973).

Fourteen apparently healthy adult men (19 to 32 years of age at the time of the test) were volunteers for this transverse study. Each one of the subjects was synchronized for at least one week before the test by a nocturnal rest from midnight to 0600, and a diurnal activity from 0600 to midnight. Meals (self selected diet without restriction) were taken at fixed hours: 0700, 1300 and 2000.

Each subject made a urine collection, every 6 hr (0600, 1200, 1800, 2400) for 36 hr, starting at 0600 on day 1. Urinary volume was measured immediately, and after the addition of mercurothiolic acid each sample was preserved at $-25^{o}C$. Determinations were made serially. The study began in May 1969 and ended in November 1971.

TG and ETG determinations in urines were made according to a double isotope derivative assay (Dray, 1970) using as internal standard 1,2-H^3 radio-testosterone and 1,2-H^3 radio-epitestosterone.

A statistically significant (cosinor method) circadian rhythm ($p < 0.05$) is detectable in urinary excretion of water, TG and ETG. In relation to the subjects' synchronization the acrophase is 1339 (from 0921 to 1756) for water, 0917 (from 0400 to 1433) for TG and 1043 (from 0546 to 1539)

Table 1

Circannual variation in urinary epinephrine (E), norepinephrine (NE) and NE/ (NE + E) ratio (NER) in 106 presumably healthy adults.

Variable	P	Mesor ng/min.	1SE	Amplitude ng/min.	1SE	Acrophase* (.95 Conf. int.)
E	0.03	4.6	0.3	1.4	0.44	-168 (-134,-202)
NE	0.01	16.7	0.4	3.9	0.47	-179 (-163,-195)
NER	0.58	8.0	0.8	1.3	1.22	- 308

* One year = 360 ; one month = 30 ; acrophase reference: December 22 at 0000.

Data from: Descovich, Montalbetti *et al* (1974).

Table 2

Statistically significant shift of circadian acrophase in epitestosterone glucuronide* urinary excretion between April-July (7 subjects) and November-January (7 subjects)

	Mid-April Mid-July ± 1 SE	End of November Early January ± 1 SE
24 hr urinary excretion		
Water (ml)	1212 ± 211	1243 ± 154
Testosterone glucuronide (ug)	76.4 ± 13	94.3 ± 10
Epitestosterone glucuronide (ug)	118.3 ± 24	136.8 ± 24
Circadian acrophase ϕ (in degrees) ($1\ hr = 15^o$; $24\ hr = 360^o$)		
Water	-197 ± 24	-162 ± 26
Testosterone glucuronide	-138 ± 20	-110 ± 15
Epitestosterone glucuronide*	-174 ± 14	-87 ± 23

* In ETG excretion a statistically significant difference, of about 5 hours, is detected between the circadian acrophase of 7 subjects studied in April-July ($\phi = -174^o \pm 14^o = 11^{36} \pm 56$ mi) when compared with the circadian acrophase of 7 subjects studied in November-July ($\phi = -87^o \pm 23^o = 05^{48} \pm 92$ mi). No statistically significant difference is detected between the groups, for the 24 hr urinary excretion of water, TG and ETG nor for circadian ϕ of water and TG.

for the ETG excretory rhythms. The amplitude represents about 25 per cent of the mean level for the TG as well as the ETG circadian rhythms, while the amplitude of diuresis is about 18 per cent of the mean level of water excretion.

The study took place from mid-April to mid-July for 7 subjects and from mid-November to the beginning of January for the other 7 subjects. Between these two groups a statistically significant shift in phase of about 5 hr is detectable in the circadian acrophase for ETG urinary excretion. In April-July, $\phi = -174^{\circ} \pm 14^{\circ}$ (1136 ± 56 min.); in November-January, $\phi = -87^{\circ} \pm 23^{\circ}$ (0548 ± 92 min.). There is no statistically significant change in the circadian acrophase for water and TG between the two groups. A circannual change in total urinary excretion was not detected for water and ETF (table 2).

Most of the TG excreted appears to arise from plasma testosterone (Southren, Tochimoto *et al*, 1965; Dray, 1970). Therefore the circadian rhythms for urinary TG probably reflect the circadian rhythm of plasma testosterone demonstrated earlier (Dray, 1970; Dray, Reinberg and Sebaoun, 1965; Southren *et al*, 1965).

The origin of plasma and urine epitestosterone is still unknown. The occurence of a circannual rhythm in the ETG circadian acrophase favors an adrenocortical origin (at least partially) of this metabolite. In fact, circannual shifts in phase of circadian acrophase have been described so far only for physiologic variables directly or indirectly influenced by adrenocortical activity: e.g. circannual rhythms in plasma corticosterone of mice (Haus and Halberg, 1970) and circannual rhythms of urinary potassium excretion in men (Ghata and Reinberg, 1954).

In both of these experiments as well as in results here reported, the circadian acrophase occurs earlier during the winter months than it does during summer months.

In spite of the fact that no statistically significant difference occurs between urinary TG in late spring-early summer (76.4 ± 13 ug/24 hr) and in winter (94.3 ± 10), a rise in the latter was suspected to exist. (Both small size of samples and group difference might have contributed in obliterating a seasonal change). To check this possibility and thus quantify a possible annual change in gonadal activity of the human male, Lagoguey and Reinberg (1974 - as yet unpublished data) are exploring both circadian and circannual rhythms in plasma free testosterone (radio-immunologic assay). Every other month, during the course a year, 5 mature healthy young males volunteered to document circadian changes (T = 28-32 hr; Δt = 4 hr) in a set of physiologic variables including plasma testosterone. A statistically significant circadian rhythm is detectable, with acrophase and amplitude values in good agreement with previous findings (Dray *et al*, 1965; Southren *et al*, 1965).

A circannual rhythm is detectable in the mesor circadian rhythm of plasma testosterone. The 24 hr rhythm adjusted level (M ± 1 SE) shows a peak in November for subjects living in Paris (M = 0.938 ± .034 ug of plasma testosterone/100 ml) and a trough in May (M = 0.672 ± .033 ug/100ml) the difference being statistically significant.

Therefore, the existence of a circannual change in testicular activity of the human mature male, with a peak in late autumn-early winter and a trough in late spring, in the northern hemisphere, emerges from a coherent body of knowledge, i.e. urinary 17-ketosteroids and may be TG excretion, and the mean circadian level of plasma testosterone.

Circannual changes in plasma testosterone concentration of male reindeer and caribou have been reported by Whitehead and McEwan (1973). The plasma testosterone levels of reindeer increased from 1 ng/ml in August to 30-60 ng/ml in mid-September. From both a comparative endocrinological and a methodologic point of view, the circannual amplitude of plasma testosterone in these cervid species is presumably larger than the circadian one, especially when the concentration has declined to barely detectable levels (late October).

In the human male, both circadian and circannual rhythms of plasma testosterone seems to have the same amplitude.

Let us add that from indirect arguments, which will be presented in the next paragraph, it is also possible to presume that the pituitary-ovarian system in the mature human female has a circannual rhythmicity with a peak in late autumn-early winter and a trough 6 months earlier or later, in the northern hemisphere.

THE CIRCANNUAL CYCLE OF MENARCHES AND THE POSSIBLE INFLUENCE OF LIGHT ON THE HUMAN FEMALE MENSTRUAL CYCLE

According to several authors a statistically significant seasonal change is detectable for the beginning of menstrual function in young healthy girls.

Engle and Shelesnyak (1934) studied both menarches and further menstruations of 250 healthy girls recorded in the Hebrew

Orphan Asylum of New York City. The menarche occured in winter (December to February) in 29.6% of the group and in summer (June to August) in only 18%, the difference being statistically significant. Moreover, the occurence of temporary amenorrheas (cycles over 57 days in length) were studied. The amenorrheas occuring in July were twice as numerous as those initiated in any other two months of the year.

Breipohl (1938) reported that the annual peak in menarches of 600 young healthy German girls was timed in December-January. The author suggested a relation between this circannual change and that of the mean duration of the day (from sunrise to sunset) in the course of the year.

Grimm (1952) recorded the menarches of 2685 young healthy German girls. He found a circannual peak in December and January in Halle, in January in Berlin but also a secondary peak in June, in Berlin; from this latter finding Grimm considered that Breipohl's interpretation was questionable.

The results of a statistical analysis of menarcheal seasonal changes dealing with 1560 Cuban girls (white, 651; black, 417; mestiza, 492), was published by Pospisilova-Zuzakowa *et al* (1965). The annual peak occured in December for the mestiza and in January for both white and black girls.

Valsik (1965) reported, in a well documented review paper, his own findings since 1934, showing a statistically significant winter peak in menarches of young girls living in Praha, Brno, Bratislava, etc. and in mountain villages located more than 600 meters above sea level. In the villages situated in plains, as well as in Brno and Bratislava, a secondary peak occurs in summer. The author discusses the possible influence of such factors as altitude, nutrition, urbanisation, socio-economic conditions, etc. but surprisingly he did not consider the possible effect of seasonal and geographical changes in light.

Before discussing this important point, additional information related to lighting and the human menstrual cycle has to be given.

Zacharias and Wurtman (1964) have attempted to investigate the effect of light on the human ovary by comparing the age of menarche in girls with and without visual function and light perception. Four groups of girls were studied: 181 prematurely born girls who were blind by retrolental fibrosis (among them 101 girls had light perception remaining, 80 had none); 99 prematurely born girls with normal vision; 54 girls born at full term, who were blind at birth or became blind; 336 girls born at full term with normal vision. It was found that, among prematurely born girls, blindness is associated with an earlier onset of menarche than that observed in girls with normal vision. This effect is increased when light perception as well as vision is absent.

The effect of altitude studied by Valsik *et al* (1963) can be summarized as follow: the menarches of 1866 girls living in the Northwestern Slovakian mountains were recorded. The higher the girls lived (from 400 to 650 m above sea level) the later occured the mean age of menarche. Moreover, in the mountain villages the percentage of menstruating young girls decreases with altitude. These differences are statistically significant. Let us remember that the intensity of the light as well as a change in spectral distribution of solar radiation (amount of short wavelength radiation) is relatively higher in the mountains than at sea level (Gates, 1966; Koller, 1964).

In addition to these descriptive statistical studies on human population a physiologic approach can now be given regarding other indirect evidence of lighting on the human menstrual cycle. Circamensual as well as circadian rhythms in self-recorded rectal temperature (among other physiologic variables) of a young healthy woman (26 years of age) were analyzed before, after and during about 3 months isolation underground in a natural cave (Reinberg, Halberg *et al*, 1966; Halberg and Reinberg, 1967; Reinberg, 1970). Ambient temperature was almost constant (6^{0} C ± 2). During her 88 day isolation, she was without time cue or other known synchronizer; the intensity of artificial light was very small (not exceeding 50 lux at 1 meter from source). Light was available on a self-selected schedule. Least squares spectral analysis of rectal temperature series revealed that several spectral components persist during isolation. The period of the circadian rhythm lengthens during isolation from 24 to 24.5 hr, whereas a circamensual spectral component of about 29 days shortens during isolation to 25.9 days on the average; menstrual cycle length undergoes changes similar to those of the circamensual component of body temperature.

The influence of light on the human menstrual cycle and its circannual variations, can be viewed from two different but complementary points of view:

(1) the facts reported above yield indirect tentative evidence suggesting that the environmental lighting regimen (among other environmental factors) which is known to affect gonadal function in small mammals and

birds, also influences the ovaries of women, although one cannot differentiate between a possible stimulating effect of darkness and/or an inhibition of ovarian activity by light.

(2) Bunning (1936) in plant physiology as well as Hamner (1964) in bird physiology consider the photoperiodic response-rhythms. "We may ... proceed with the assumption that the periodic timing mechanism of the house finch utilizes a circadian rhythm as a biological clock to temporally regulate the breeding season" says Hamner. Changes in the effects of light as a function of its timing in the 24 hr scale is also considered by Mills (1973), by Simpson and Bohlen (1973), and by Reinberg (1970) in human chronobiology.

Therefore, both circannual changes in light quality and intensity (Koller, 1964; Gates, 1966) and circannual changes in the daily photofraction, and its timing, possibly play the role of an annual synchronizer in men, among other cyclic environmental factors.

CIRCADIAN AND CIRCANNUAL RHYTHMS IN HUMAN NATALITY

Two main sources of up to date information have been used for this aspect: a review paper from Smolensky, Halberg and Sargent (1972) and a paper from Batschelet, Hillman, Smolensky and Halberg (1973).

An unequal distribution of spontaneous birth along the 24-hour time scale has been recognized since Buek (1829). The cosinor analyses of birth data by Smolensky *et al* (1972) represents an extension of an earlier investigation (i.e. Kaiser and Halberg, 1962).

The hourly incidence of 207,000 spontaneous labors, 2,082,453 natural, 30,493 induced births and 12,081 still births were studied. A statistically significant circadian rhythm was detected in each of these variables.

The circadian acrophase for labor (spontaneous initiation of painful contractions and/or rupture of fetal membranes) occurs at 0100 (from midnight to 0148, with the 95% confidence interval). For natural human birth a circadian acrophase was determined at 0416 (from 0324 to 0508).

The incidence of induced births (deliveries assisted by chemical, mechanical or operative means) and of still births (pregnancies terminating in a non-viable neonate following an average gestation) is non-ran dom. The circadian acrophase of induced births occurs at 1336 (from 1232 to 1624). This phase angle difference from the natural birth circadian acrophase presumably reflects the preferred diurnal working schedules of physicians. The still birth circadian acrophase is timed at 1708 (from 1044 to 2120). This ϕ value differs from that of natural live births by almost 180^{0} (reflecting those still births following long, difficult labors as well as those resulting from the side effects of labor inducement) which implies an increased neonatal survival for those deliveries close to the expected acrophase of birth (Kaiser and Halberg, 1962).

Quetelet (1825) was one of the earliest investigators to report circannual rhythms in human birth. The cosinor summary published by Smolensky, Halberg and Sargent (1972) reveals a rhythm with an overall sample circannual acrophase at -50^{0} or about 50 days (February 10 in the northern hemisphere from the chosen phase reference: midnight of the average longest night of the year, December 22 in the northern hemisphere and June 22 in the southern hemisphere). However, samples of human birth data from countries with widely differing geographic locations have a circannual acrophase that varies considerably from this ϕ of -50^{0}.

When ϕs are replotted simply as a function of absolute geographic latitude, as in figure 8, the timing of the acrophases for human natality appears to vary in such a pattern that on the average the greater the latitude north (or south) of the equator, the later the ϕs occur in the year. The winter solstice is used as ϕ reference (ϕ = 0^{0}).

Data gathered by Gauquelin (1968) concerning simian birth at different latitudes were reanalized by Batschelet *et al* (1973) and incorpored in figure 8.

However, in human data from countries as diverse as Australia, Cuba, India, South Africa and the U.S.A. there is an unexplained "leveling off" of the relation between birth acrophases and latitude.

The variation in acrophase with latitude suggests an environmental influence upon human fertility patterns. Birds, sheep, ferrets and other forms of life in their natural surroundings may be synchronized in terms of circannual breeding rhythms by the predictable changes in photofraction (Amoroso and Matthews, 1955; Farner, 1959; Hafez, 1952, 1959; Halberg, 1970; Hamner, 1964).

"The possibility that today man may be influenced (even if only indirectly and partially) in a similar manner is intriguing, since we regard ourselves largely liberated from our natural milieu through the creation

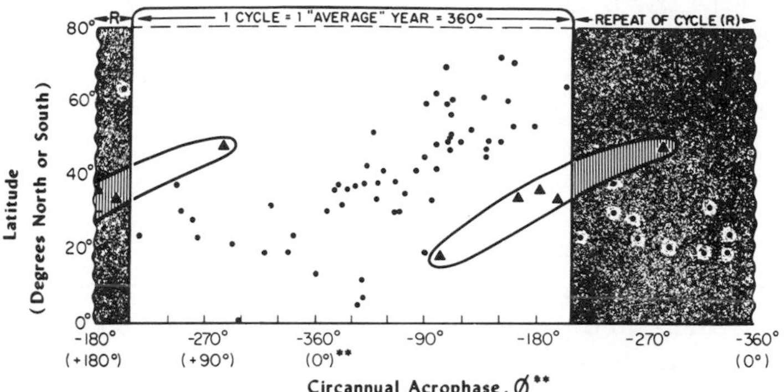

Fig. 8:
Development of a coefficient for computing correlations between an angular and a linear variate allows demonstration of a statistically significant correlation between the timing of a circannual rhythm in human and simian natality and geographic latitude. The greater the latitude north or south from the equator, the later from winter solstice the circannual Ø. This impression is gained by plotting according to geographic latitude circannual acrosphases computed for time series summarizing the monthly incidence of natural simian (triangle) and human (dot) births.
(By courtesy of E. Batschelet, D. Hillman, M. Smolensky and F. Halberg: 1973).

Table 3

Circannual rhythm in human birth rate

mple nd and size	No. of time series	Amplitude, A^1 (0·95 confidence	Acrophase, ϕ^2 interval)
ing‡ birth control devices* (20 x 10^6)	9	3·9(0·2 to 7·6)	-120°(-45° to -198°)
t using‡ birth control devices** (39 x 10^6)	20	5·8(1·9 to 9·6)	-16°(-320° to -57°)
ited States** (12 x 10^6)	48	4·8(4·4 to 5·1)	-218°(-201° to -232°)
l † ε** (11 x 10^7)	58	3·2(0·5 to 5·8)	-50°(-4° to -94°)

Amplitude expressed relative to M(M = 100 per cent).
ϕ reference = 00^{00} of the average longest night of the year, December 22, in the northern hemisphere and June 22 in the southern hemisphere (1° = 1·01 d).
48 time series for U.S. births represented by A and ϕ values from a cosinor summary.
Rhythm description** = $0·001<P<0·01$, * = $0·01<P<0·05$.
Undocumented presumption made for majority of a given population.

of comfortable artificial environments."

"Against this background, the degree to which a relationship holds between the circannual birth ϕ and latitude is most interesting. One may ask, of course, whether latitude may represent societal and thus only indirectly geophysical influences. This relationship, if proved, may reflect with possibly yet greater likelihood, differences between the more urban and industrial populations of the middle and higher latitudes as compared to the more agrarian and gaming ones of the equatorial regions. Moreover, even at similar latitudes, the circannual acrophase of birth may differ drastically because of varying customs." (Batschelet *et al*, 1973).

Table 3 (Batschelet *et al*, 1973) summarizes differences in the timing of human birth rhythms for various subsets of all samples. In populations where a majority are believed not to practice birth control, an average acrophase of -16° is estimated; but for countries in which contraceptives are freely available and their use encouraged, the circannual ϕ occurs at -120°. For the United States the timing of ϕ is much later, at -248°.

A circannual rhythm in birth for regions inhabited mainly by populations relying primarily on natural methods of family planning must not be interpreted as expressing an endogenous human fertility rhythm. On the contrary, such a periodicity, and in particular its timing, apparently reflects a societal influence. For example, any widespread practice of sexual abstinence and delay of marriage during spans involving certain kinds of religious observance may influence if not synchronize human circannual birth rhythms.

In populations with sexual abstinence during Lent the number of marriages increases after Easter, thus influencing births about nine months later. Assuming an average human gestation of 266 days from fertilization, the calendar date of a high number of conceptions can be computed from the overall circannual birth ϕ if this ϕ in populations presumed not to use birth control devices is reasonably well approximated by the fitted sinusoid: as noted earlier, a ϕ of -16° from the day with the longest night is found for data from such populations. As also noted, the day with the longest night is used as ϕ reference. This choice of ϕ reference constitutes an obscuring factor, since religious schedules, by contrast to geophysical ones, do not show a hemispherical disparity. Nonetheless, the confounding effect is seemingly overridden; thus, a ϕ found at -16° (7 January) for births (in relation to December 22) in the northern hemisphere corresponds to a high number of conceptions occuring about 15 April. This date coincides with the weeks immediately following Easter, the occurence of which varies from one year to another between late March and April.

In addition to these remarks from Batschelet *et al* (1973) let us remember that the circannual peak of human gonadal activity is found in November (northern hemisphere) and thereafter, the circannual peak of natality could be expected nine months later if these phenomena are correlated; this does not seem to be the case in humans.

CIRCADIAN, CIRCASEPTAN AND CIRCANNUAL RHYTHMS IN SPONTANEOUS NUTRIENT AND CALORIC INTAKE OF 4 ± 1.5 YEAR OLD HEALTHY CHILDREN

A circannual rhythm in the spontaneous intake of food was reported by Campbell in the albino rat (1945) and by Haberey, Dantlo and Kayser (1966-1967) in *Eliomys quercinus*, a hibernating mammal. The hexose-monophosphate-shunt (HMS) in glucose consumption exhibits a circannual rhythm in 3 different species of hibernating mammals; the HMS is less important in December and March than in October and June. This circannual change in the pathway of glucose metabolism is associated with a circannual change in the spontaneous food intake with an increase in carbohydrate consumption in the autumn (and a decrease in lipid intake) while the protein intake remains unchanged throughout the year.

A circannual rhythm in feeding behavior of 6 human children (initial age: 6 to 10 months) was also reported by Sargent (1954). Each of these infants was maintained completely on a self-selected diet for an observational span of time covering 36 to 57 months -- longitudinal study. This "seasonal periodicity" in humans is only one example among many others gathered by Sargent from a large body of previously isolated facts. Such circannual changes are discussed by Sargent with reference to Edmund Hughes' hypothesis of "vestigial physiology" (1931). According to Hughes and to Sargent the seasonal variations in physiological functions and metabolic diseases of human beings are attributable to "a vestigial mechanism the function of which is analogous to that of hibernation in lower animals."

A transverse study was performed by Debry, Bleyer and Reinberg (1973-1974). Food intake (proteins, lipids, carbohydrates and calories) was measured individually in 128 children. Four groups of 34 to 37 children were formed consecutively, one for each season of the year (a core sub-group of 8 children was maintained from group to group). These groups can be considered homogeneous regarding the sex (about 60% of boys); the age histogram, with 4 ± 1.5 years of age, as a mean; body weight and height; personal status (orphan, forsaken child, broken home, etc.).

The children lived constantly in two separate institutions at Nancy; both were following rigoursly similar rules all year long regarding social synchronization (light -on at 0700, light-out at 1830), timing of meals (the fixed clock hours being 0800, 1100, 1400 and 1800), timing and type of activities, the origin and nature of food availability to the children.

Each child was asked and encouraged to be on a spontaneous selected diet. At (fixed) meal time he (she) self-selected the nature, the quantity and the quality of food he (she) desired; no order in the sequence of ingestion was required. Thereafter, during a seven day span of time, a careful determination of the proteins, lipids, carbohydrates and calories was done for each meal actually eaten.

Time series thus obtained were analyzed according to the cosinor statistical method.

No statistically significant differences between boys and girls were found in biological rhythms documented in this study. This result is not surprising in itself; however it allows us to computerize each group as a whole.

Stastically significant circadian rhythms are detectable in each of the 4 studied variables for each day of the week and for each of the 4 seasonal groups (figure 9).

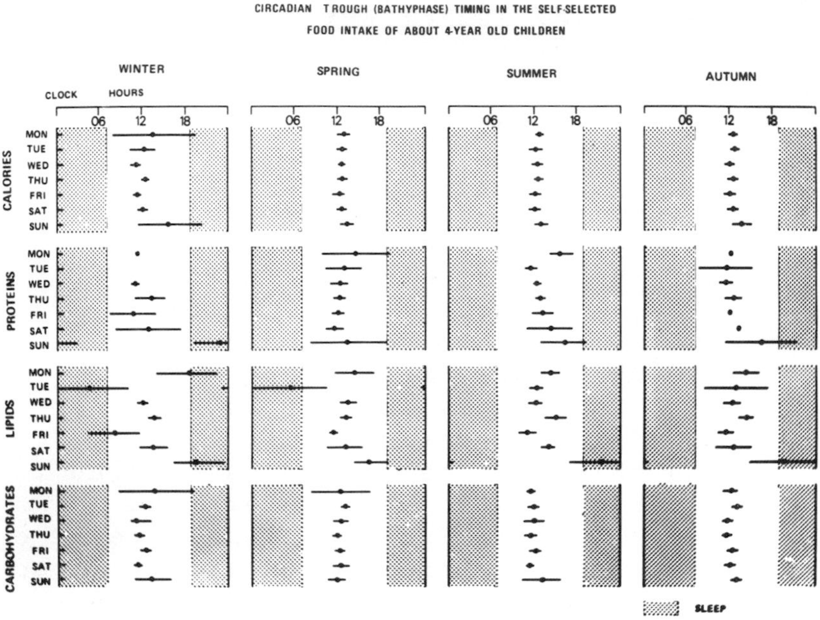

CIRCADIAN TROUGH (BATHYPHASE) TIMING IN THE SELF-SELECTED FOOD INTAKE OF ABOUT 4-YEAR OLD CHILDREN

Fig. 9: Circadian trough (bathyphase) timing in the self-selected food intake of about 4-year old children.

Food availability at fixed clock hours only : (meal timing 0800, 1100, 1400 and 1800). Subjects' social synchronization light-on at 0700, light-off at 1830.

When a statistically significant circadian rhythm is detected ($p < 0.05$) the corresponding bathyphase is given with its 95% confidence limits.

The bathyphase occurs around noon in almost all the studied circumstances. This means that, as a rule, larger self-selected food intake takes place at 0800 (breakfast) and at 1800 (supper). However the bathyphase of both lipid and protein spontaneous intake was clustered around 1800 on Sunday only.

(G. Debry, R. Bleyer and A. Reinberg: 1974).

The bathyphase (trough) occurs around noon-midday in almost all the studied circumstances. In other words larger meals took place usually at 0800 (breakfast) and at 1800 (supper) for these children on self-selected diets.

The bathyphase of both lipid and protein spontaneous intake was clustered around 1800 on Sunday only. For the other days of the week the circadian trough in lipid and protein intake occured around 1200.

There are no statistically significant changes in both amplitude and mean level of circadian rhythm studied here.

Cosinor analyses of individual seven-day time series (of each variable of each season) lead to the detection of statistically significant circaseptan rhythms for glucides spontaneous intake. Circaseptan rhythms of other variables are detectable in the summer and to a lesser extent in autumn and winter (figure 10).

Acrophases of studied rhythms usually occur on Saturday, Sunday or Monday but almost never on Wednesday (figure 10). However the timing of the acrophase in any one season, is able to shift from variable to variable; i.e. in summer, circaseptan ϕ for spontaneous protein intakes occurs on Monday (1800) (from Monday (0700) to Tuesday (1900) with 95% confidence intervals). Circaseptan ϕ for lipid spontaneous intake occurs on Tuesday (1800) (from Monday (0800) to Thursday (0200)), while circaseptan ϕ for protein and calorie spontaneous intake occur respectively on Sunday (0900) (from Saturday (0700) to Sunday (midnight)) and on Sunday (2000) (from Saturday (2200) to Monday (1600)).

Statistically significant circaseptan changes in amplitude were not detected.

Changes in the weekly mean adjusted

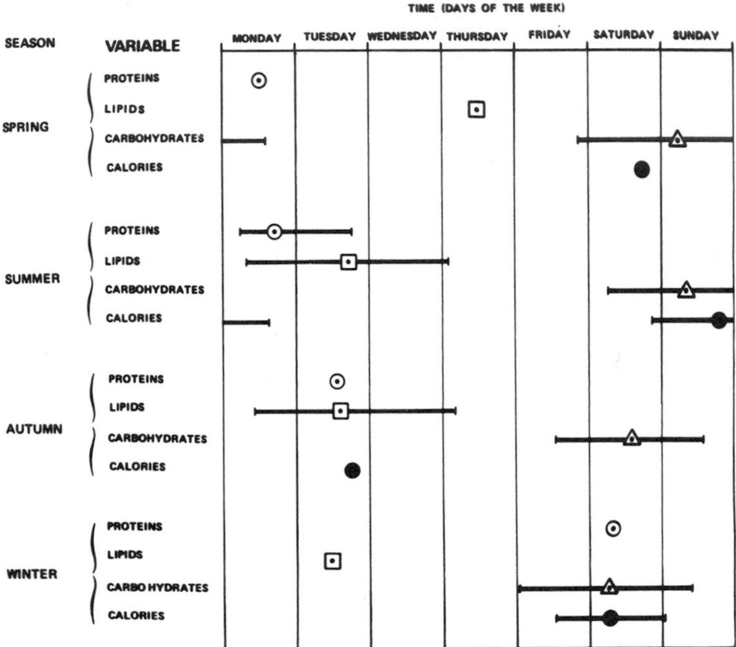

Fig. 10: About-7-day peak (circaseptan acrophase) timing in the self-selected food intake of about 4-year old children.
No change from day to day in subjects' social synchronization, type of activity (same choice at each meal, same meal timing at fixed clock hours each day, etc).
When a statistically significant circaseptan rhythm is detected ($p < 0.05$) the corresponding acrophase is given with its 95% confidence limits.
Circaseptan acrophase usually occur on Saturday, Sunday or Monday and almost never on Wednesday; however the timing of the acrophase is able to vary, in any one season, from variable to variable.
(G. Debry, R. Bleyer and A. Reinberg: 1974).

levels obtained by this method are as follows: (1) demonstrate circannual variations in spontaneous intake of lipids, carbohydrates and calories (protein changes are not statistically significant) see figure 11; (2) show that the peak of lipid spontaneous intake occurs in spring, while both peaks of carbohydrate and calorie spontaneous intake occur in summer; (3) show also that protein, lipid and carbohydrate spontaneous intake remain "equilibrated" for their respective proportion (roughly the proportion of protein, lipid and carbohydrate daily intake -- expressed in calories -- is respectively 1: 2: 4).

Another conventional statistical method has been used. The overall daily mean for each nutrient (and calories) in each season leads to the same result: a circannual change in the spontaneous intake of lipids, carbohydrates and calories with a statistically significant trough in winter and no circannual change in spontaneous protein intake; (differences between annual peak and trough were t-tested by the student method); t value is equal to 3.1 for carbohydrates, 2.39 for calories, 2.21 for lipids and 1.98 (not significant) for proteins, respectively.

It is also possible to check the results thus obtained from a transverse sampling, by those obtained from a longitudinal sampling since a core-sub group of 8 children was studied from season to season. For the daily calorie intake, i.e. the following mean figures (± 1 SE) were found: 1268 (± 52) in winter, 1331 (± 65) in spring, 1409 (± 30) in summer and 1395 (± 44) in autumn.

The existence of circadian rhythm in spontaneous food intake of 4 year old children is coherent with the existence of such

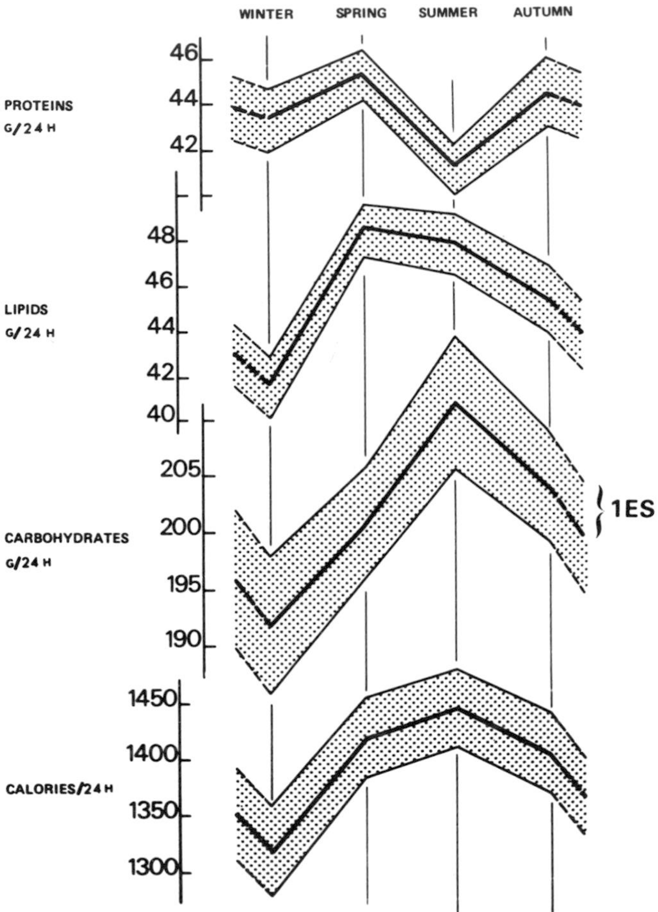

Fig. 11: Circannual changes in the 7-day mean level of the spontaneous nutrient and caloric intake of groups of about 4-year old children, during a self-selected diet.
Statistically significant differences are detected for lipid, carbohydrate and caloric intake between winter (trough) and summer (peak). (The peak might be located in spring time for lipids). The seasonal change in protein intake is not statistically significant.
(G. Debry, R. Bleyer and A. Reinberg: 1974).

bioperiodic phenomenon in rodents (Le Magnen, 1966, 1971; Collier *et al*, 1972, 1973; Haberey *et al*, 1966, 1967) as well as in the human new-born (Halberg, 1969). In healthy animals, when food is available *ad libitum*, food ingestion takes place during the active span of the 24 hours. These circadian changes in hormonaly active secretions (Reinberg, Apfelbaum and Assan, 1974; Haus and Halberg, 1966; Apfelbaum, Reinberg *et al*, 1972), and circadian changes in the response of certain CNS structures as shown by Margules *et al* (1972). These authors have demonstrated a circadian rhythm in the response of hypothalamic centres to l-noradrenalin in respect to the temporal feeding pattern of rats.

To come back to circannual rhythms, a striking similarity occurs between our results and those of Haberey *et al* (1966, 1967) on *Eliomys quercinus* showing circannual rhythms in spontaneous food intake with an increase in carbohydrate consumption in autumn while the protein intake remained unchanged during the course of the year.

Circannual rhythms in spontaneous food intake of infants were also reported by Sargent (1954) with (a) weight gain late in the summer and early in the autumn; (b) a tendancy to consume more fat in spring-summer rather than in autumn-winter; (c) the consumption of carbohydrates tends also to be maximal in spring-summer; (d) however considerable individual variability is evident.

An autumnal accentuation in the weight curve is characteristic in the case of groups of newborn infants, growing children and tubercular adults, according to Fitt (1941). These autumnal peaks are associated with a trough in weight gain -- and even a tendancy for weight loss -- in the spring (Fitt, 1941).

In human adults circannual rhythms in food intake were also reported by Sargent (1954). Soldiers in training camps, urban families and mill workers consume significantly more calories in the fall-winter than in the spring-summer; this is also the case in young and adult rats (Campbell, 1945).

Thus circannual changes in feeding behavior might be different in growing children and in mature men with regard to peak and trough timing for carbohydrate, fat and calorie intake.

The problems of intrinsic components of these rhythms in mammals (including man) has to be discussed now.

For each biological rhythm one has to consider on the one hand, an intrinsic component genetically inherited and, on the other hand, an extrinsic component, synchronizer among others, which influence but do not create the rhythms.

Moreover, according to Collier *et al* (1972, 1973) "... research on feeding has been dominated by a search for those physiological factors (level of circulating metabolites or hormones, size of reserves, gastrointestinal load, etc.) which covary with the nutritional state of the animal and which are associated with the initiation and termination of feeding ..." The view derived from free feeding experiments and the types of feeding patterns that are encountered in the field suggests that the momentary state is irrelevant and the critical class of variables is that relating to the total food economy of the animal. Thus the appropriate unit of analysis is the feeding cycle rather than the ingestive responses, *per se*. Variables which alter the metabolic state of the animal such as growth, pregnancy and environmental temperature are probably reflected in the modification of feeding patterns in terms of long-term regulation of energy balance.

With this background two remarks have to be made on results presented in this paper. (1) A spectrum of rhythms, with different periods, can be demonstrated in the feeding behavior of 4 year old children; (2) within certain limits both acrophase (or bathyphase) and mean level of the studied rhythms are able to change from variable to variable (carbohydrate, lipid, protein, calorie intake) in each spectral domain (about 24 hours, about 7 days and about 1 year). These facts can be considered as indirect evidence of an endogenous component of the studied rhythms.

In his attempt to understand seasonal rhythms in man Sargent (1954) developed the theory of vestigial physiology first proposed by Hughes (1931). According to this theory, regular seasonal variation in man should be qualitatively similar to the changes observed in hibernating mammals. The implicit mechanism is metabolic in nature, and seasonal variations would be governed by the nutritional demands and expectations of the tissues with annual changes in daily photoperiod serving as the appropriate informative stimulus. According to Sargent the annual rhythm is biphasic; the anaphase (approximately April to October) corresponds to the original period of activity and preparation for inactivity, and the cataphase (approximately October to April) to the reversed situation. During the anaphase the tissues and organs "expect" to use carbohydrate as the chief source of fuel and to store fat against the expected period of inactivity. During the cataphase (autumn-winter) this stored fat should be metabolized either by direct utilization or by gluconeogenesis.

For several reasons "vestigial physiology" seems now to have been an unfortunate terminology. (1) "Vestigial" has no objective meaning. From a neo-darwinian point of view of evolution it is preferable to avoid statements concerning advantage, usefulness, etc....i.e. a genetic character can be both an advantage for the species and a disadvantage for individuals (Dobzhansky and Boesinger, 1968). Moreover we are reluctant to reopen the at least partially academic discussion leading to the decision as to whether or not human infants are genetically related to hibernating mammals. (2) Vestigial physiology deals only with circannual rhythms and not with rhythms of other periods (or frequency).

In spite of these qualifications Sargent's hypothesis concerning the alternations in lipid and carbohydrate metabolisms in the course of the year is a very stimulating one. On the basis of one circadian period, rhythmical changes in carbohydrate and other metabolisms as well as in energy balance has been demonstrated: persistance of circadian rhythms, including that of glycogen deposition in the liver of fasting mice (Haus and Halberg, 1966); persistance of circadian rhythms in oxygen consumption and in respiratory quotient (among others) of healthy women during a severe caloric restriction (Apfelbaum, Reinberg *et al*, 1972; Reinberg, 1973). These facts lead to the conclusion that the organism is able to favor glycolysis during a certain part of the 24 hours and to favor gluconeogenesis during another part of the day. There are also circadian changes in glucose tolerance and metabolism (Bowen *et al*, 1967; Jarrett *et al*, 1970; Lestradet *et al*, 1970; Sensi *et al*, 1970).

According to Sargent's terminology it would be possible to consider for human circadian changes of metabolism an anaphase taking place roughly in the morning and a cataphase taking place during the night for subjects with a routine of diurnal activity and nocturnal rest. By the same token hormonal circadian rhythms such as that of the human growth hormone, insulin, glucagon, cortisol and catecholamine must be taken into consideration since (1) they play a role in metabolic regulation and (2) they persist during starvation or during a low calorie diet. (Chossat, 1843; Haus and Halberg, 1966; Reinberg, Apfelbaum *et al*, 1974). A few experiments also suggest the existence of metabolic circannual rhythms in addition to those reported or quoted by Sargent. According to Hommel *et al*(1971), the effect of a fixed dose of insulin shows a seasonal variation in glucose uptake and glycogen content of eviscerated nephrectomized rabbits, as well as in isolated perfused hearts of rats. The observed effect is at its peak in December and at its trough in July, the difference being statistically significant. Annual changes in diabetes have also been reported with a decreased need for insulin therapy from late spring to early fall (Chrometzka, 1940) or with a summer minimum of clinical manifestations (Gamble *et al*, 1969).

This discussion bringing together a coherent body of facts and concepts leads to the conclusion that circadian, circaseptan and circannual rhythms in spontaneous nutrient and calorie intake, demonstrated in 4 year old children, are at least partially, inherited. However most of the work remains to be done since only indirect evidence is in favor of this conclusion.

CIRCANNUAL RHYTHMS IN HUMAN MORTALITY AND MORBIDITY -- (CIRCANNUAL CHRONOSUSCEPTIBILITY)

It has long been considered that cyclical changes, i.e. seasonal changes, in morbidity and mortality are related mainly, if not exclusively, to cyclical changes in environmental factors. Whereby one postulates that the organism's susceptibility (or its resistance) to potentially noxious agents is constant throughout the day, the year, etc. An increasing number of facts and experimental evidence leads chronobiologists to reject this hypothesis from a biologic, physiologic and pharmacologic as well as a toxicologic point of view.

Thus when attempting to resolve rhythms in morbidity and mortality one has to consider, on the one hand, changes in environmental factors (climatic, meteorological *inter alia* and, on the other hand circadian, circannual and other cyclic changes in the organism's susceptibility (or resistance) which are inherent in the organism itself. This susceptibility can be called chronosusceptibility.

Biologic responses to various agents, (including chemical substances such as drugs or poisons, physical agents such as heat, cold, noise and x-rays) are neither constant as a function of time nor are they subject solely to random variations. A rapidly growing number of circadian rhythms (Halberg, 1960, 1969; Halberg *et al*, 1955; Reinberg, 1967; Reinberg and Halberg, 1971), circamensual rhythms (Smolensky, Reinberg *et al*, 1973), and circannual rhythms (Hommel *et al*,

1971; Altounyan, 1969) of susceptibility have been reported for various life forms including human beings.

These experimental facts lead to a new concept, i.e. the organisms' *tempus minoris resistantiae* (time of least resistance) or chronosusceptibility. With reference to circadian rhythms in susceptibility, processes of changing responsiveness are now investigated as problems of circadian chronotoxicology and circadian chronopharmacology.

As a matter of fact, regular and predictable circadian changes in biologic susceptibility can now be viewed as rather common phenomena, even in low frequency (i. e. circannual) rhythms.

The occurence of circadian rhythms in susceptibility to several potentially harmful physical agents has also been reported; mortality of *Drosophila* from x-rays (Rensing, 1969), mortality of mice from x-rays, assessed as LD 50 (Haus and Halberg, 1970) or by other endpoints (Nelson, 1966: Pizzarello *et al*, 1964), mortality of rats and of CBA mice from gamma irradiation (Grigoryev *et al*, 1969) or body weight loss from partial body x-rays (Garcia-Sainz, Halberg and Moore, 1968).

Circadian and other chronopharmacologic effects can be demonstrated at all levels of organization and for an already wide variety of chemical agents and animal species (Aron, Roos and Asch, 1967; Cardoso and Carter, 1969; Scheving and Vedral, 1966; Scheving, Vedral and Pauly, 1968; Stroebel, 1969), including man (Reinberg, Sidi and Ghata, 1965; Reinberg and Sidi, 1966; Reinberg, Zagula-Mally *et al*, 1967, 1969; Reindl *et al*, 1969; Rutenfranz and Singer, 1967).

Changes in environmental factors and in particular the manipulation of a known synchronizer can influence circadian and other susceptibility rhythms. Suppression of known synchronizers - or of their transducers (maintenance in continuous darkness (DD) or in continuous light (LL), or bilateral optic enucleation in experimental animals) - does not obliterate several circadian susceptibility rhythms. They persist with eventual changes in the parameters τ, ϕ and A (Halberg, 1960; Scheving and Vedral, 1968). A phase shift of synchronizer is followed by a phase shift of susceptibility rhythms with a delay of varying length, depending upon phase systems, species, agents and other conditions of study (Halberg, 1967; Halberg and Stephens, 1959; Reinberg and Halberg, 1971).

Let us remember that the administration of certain drugs can influence some physiologic circadian and other rhythms, as do certain manipulations of the synchronizer. (Reindl *et al*, 1969; Reinberg *et al*, 1971).

Death records in a Parisian hospital -- Hopital Fernand Widal -- (1957-1967) and in France as a whole (1962-1967) have been analyzed in order to detect and to characterize circannual, circadian and other rhythms of risk in various diseases (Reinberg, Gervais *et al*, 1973). Results thus obtained were compared to those of Smolensky, Halberg and Sargent (1972) since a similar chronobiologic methodology was used.

Circannual acrophases in the rhythm of deaths recorded at the Hopital Fernand Widal are in good agreement with those of deaths recorded in France (fig. 2 and 12) as well as with circannual acrophases computed from time series recorded in Minnesota and other parts of the United States (Smolensky, Halberg and Sargent, 1972), see fig. 13.

Circannual chronograms (fig. 14) of

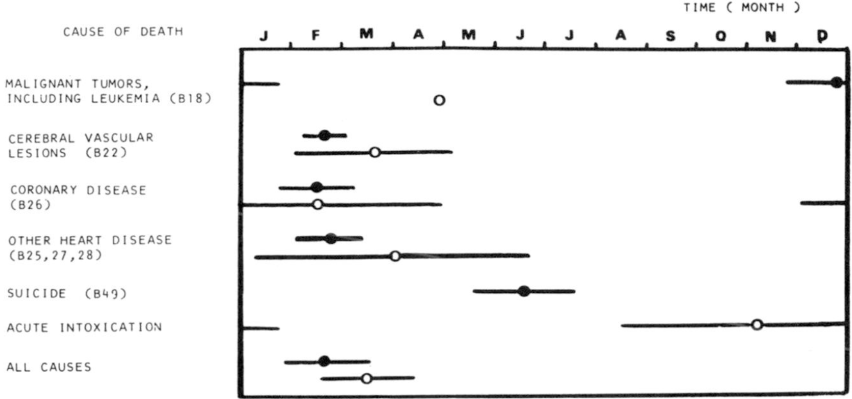

Fig. 12
Circannual acrophase chart. ∅ is given with its 95% confidence interval.
(A. Reinberg, P. Gervais *et al:* 1973).

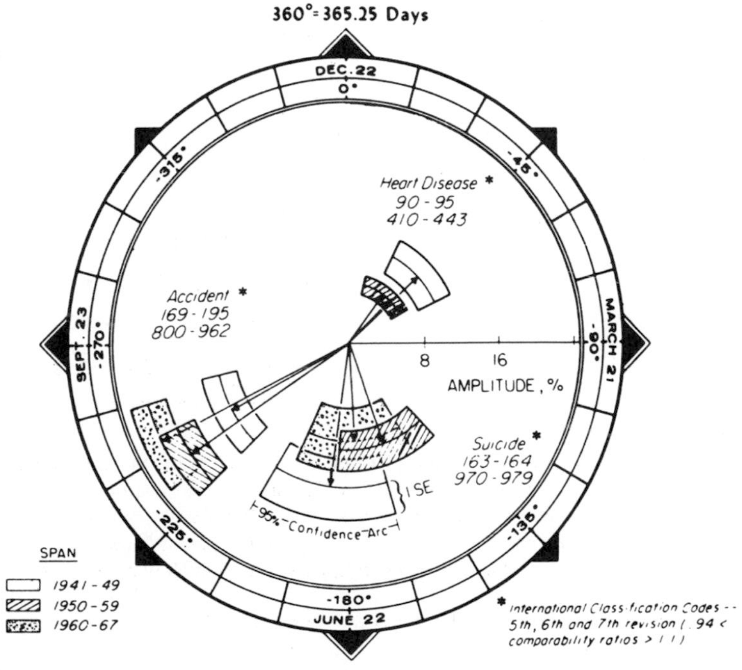

Fig. 13:
Circannual rhythm of mortality in Minnesota during 1941-1967. Note reproducibility of findings in consecutive decades for death from heart disease, suicide and accident.
(By courtesy of F. Halberg: 1973).

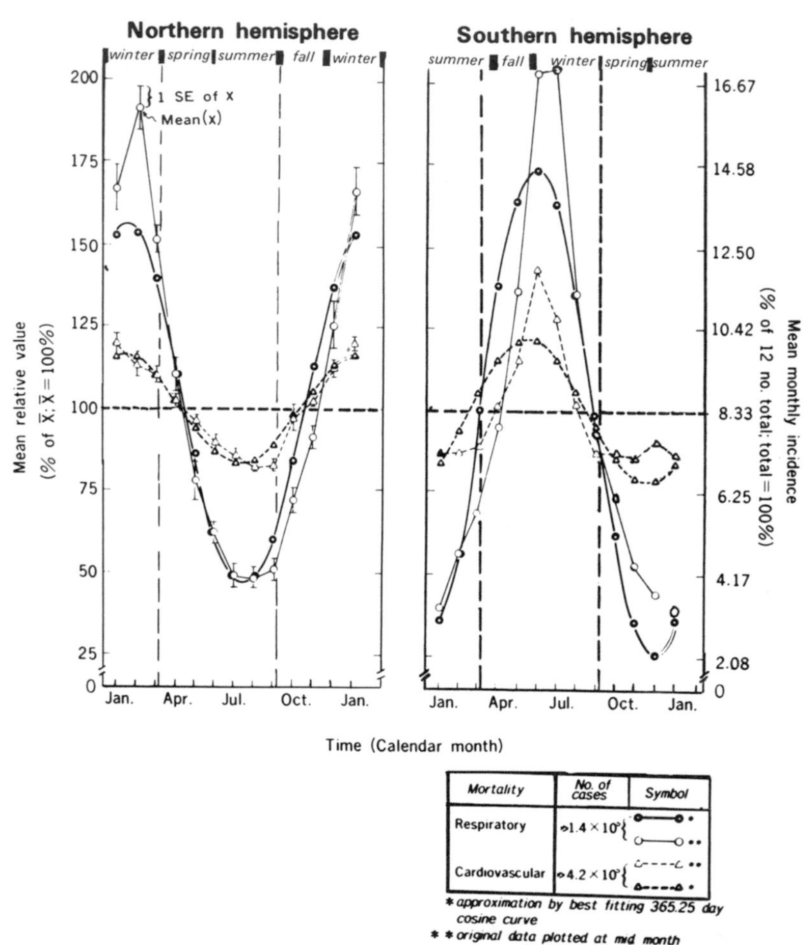

Fig. 14:

Circannual chronogram of human respiratory and cardiovascular deaths. Chronogram of human mortality rhythms reveal highest incidences during winter irrespective of hemisphere. Least squares approximation of the monthly means by a 365.25-day cosine curve accurately reflects waveforms for the northern hemisphere. Slight discrepancies between original data and approximating curve for the southern hemisphere due to small number of samples available.

(By courtesy of M. Smolensky, F. Halberg and F. Sargent: 1972).

human respiratory and cardiovascular deaths (Smolensky, Halberg and Sargent, 1972) reveal highest incidences during winter irrespective of hemisphere.

Results concerning circadian rhythms of deaths are restricted to cosinor analyses recorded at Hopital F.W. (fig. 15). Let us point out that the acrophase in the rhythm of deaths from infectious diseases occurs at 0533 (0114 to 0952), which corresponds roughly to the end of the sleep span (in darkness) for adult men; in mice the acrophase of the circadian rhythm in susceptibility to bacterial toxin also occurs toward the end of the sleep span (in light) (Halberg and Stephens, 1968).

Ultradian rhythms (0.5 hr $< \tau <$ 20 hr) have been detected in the record of death from heart disease. Halberg reanalyzing time series published by Bock and Kreuzenbeck (1966) was able to detect a disease, while no circadian rhythm was detectable in the same data.

From Hopital F.W. records we can conclude that there is: (1) a statistically significant circadian rhythm for deaths from coronary disease (B 26-576 cases); (2) no circadian rhythm detectable for deaths from other heart diseases (B25, 27, 28 - 810 cases) and (3) a statistically significant ultradian rhythm for $\tau = 6$ hr in mortality record for all heart diseases (B25, 26, 27, 28 - 1386 cases) with $\phi = 250^{0}$ (-320 to -99); ($\tau = 6$ hr $= 360^{0}$), and A = 10 (.2 to 19.4).

We may now consider four arguments in favor of an intrinsic component in rhythms of death.

The first argument comes from the experiments in chronotoxicology.

The second one is the similarity of the circannual rhythm characteristics

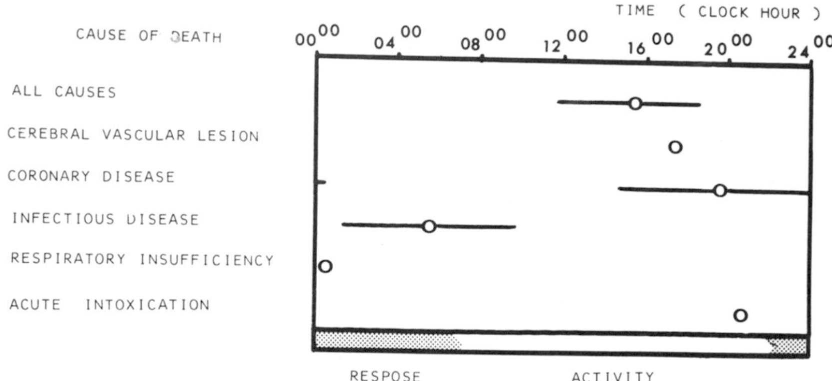

Fig. 15:
Circadian acrophase, Ø, in mortality rhythms for several causes in a Parisian hospital. Ø is given with its 95% confidence interval. It is postulated that subjects' rhythms were synchronized, with diurnal activity and nocturnal rest.
(A. Reinberg, P. Gervais *et al:* 1973).

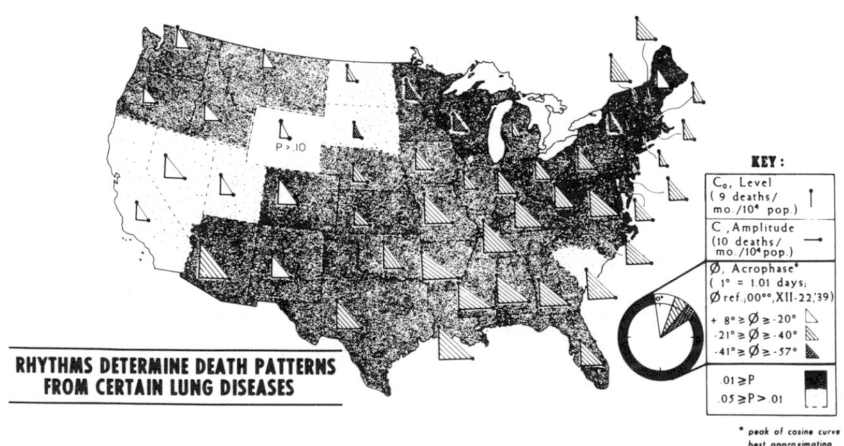

Fig. 16:
Month of changing risk for death from pneumonia and influenza. Summarized by the fit of a 365.25 day cosine curve to 1940. Human mortality by state in USA.
Circannual rhythms in death from pneumonia and influenza in 48 states of the USA have been analyzed. A statistically significant circannual rhythm is detectable in 47 states. The acrophase stands between the end of December and the end of February for all the states, the acrophase occurs earlier (December) mainly for western states, a fact which could be related to climatic factors.
(By courtesy of M. Smolensky, F. Halberg and F. Sargent (II): 1972).

observed when comparing "short" time series (T = 3 to 5 years) with each other and to the total span of time considered (T = 10 years and more). In particular the circannual rhythms of death from cardiovascular origin do not show statistically significant differences in their ϕ, A and M from short time series to others (T = 3 to 5 years) in Halberg's analyses in Minnesota (fig. 13)or in ours in France.

The third argument stems from the evidence of an ultradian (τ = 6 hr) rhythm in human mortality from certain diseases. It is easy to consider the possible existence of an environmental factor, acting as a synchronizer or as an inducer, with a 24 hr or a yearly periodic change. Such an hypothesis is difficult to sustain when dealing with ultradian rhythms which seem to depend upon the organism itself rather than an environmental factor.

The fourth argument is obtained when comparing quantified circannual rhythms in death, for a specified disease, in data gathered from different states and countries, for the same years, with important differences in weather and climate from one geographical location to another.

Circannual rhythms in death from pneumonia and influenza in 48 states of the USA have been analyzed microscopically by Halberg (fig. 16). A statistically significant circannual rhythm is detectable in 47 states. The acrophase stands between the end of December and the end of February for all the states; the acrophase occurs earlier (December) mainly for western states, a fact which could be related to climatic factors.

Nevertheless, we have to point out that the circannual peak occurs at almost the same time for populations enjoying a mild climate all year long (Florida, Louisiana, Texas) and for populations exposed to hard winters and hot summers (northern states). If the risk of mortality for certain lung diseases is higher between late December and late February - in the northern hemisphere, it is not necessarily because cold and stormy weather occurs at this time but, rather because the human organism is then more susceptible to this type of infection than at any other time.

Comparison of the acrophases of circannual rhythms of deaths for various diseases in Minnesota and in France (mild climate as compared to that of Minnesota) leads to the same conclusion. Since the intrinsic component in periodic changes of human susceptibility have been largely ignored or considered of no consequence we felt it worth while to present this set of arguments.

If, for instance, in the northern hemisphere mortality for certain diseases reaches its peak in January and February, this fact need not necessarily and exclusively be related to perceptive poetical remarks aimed literally and literarily to nurture, such as "winter months are killers of poor people" (J. Richepin); the intrinsic susceptibility (or resistance) of the human organism also varies rhythmically with a period of about 1 year, as well as with a period of about 24 hours, among others.

We also believe that a set of intriguing circannual rhythms in human morbidity will reach better understanding with the help of a chronobiologic approach; e.g. circannual changes in bronchial response to standard aerosol histamine challenge (Altounyan, 1969); circannual variation of congenital dislocation of the hip with peak incidence in winter months (Cohen, 1971);

circannual variation in the birth of children with aneuploid chromosome abnormalities (Nielsen *et al*, 1973); etc.

Aims for future investigations should be a better quantification of valid data and an estimation of the respective importance of intrinsic factors (susceptibility rhythms, *tempus minoris resistantiae*) and of extrinsic factors (changes in climate, meteorological conditions, potentially noxious environmental agents, etc.).

SUMMARY

The existence of circannual rhythms in man is considered as a time-honored fact since at least Hippocrates' time (about 400 B.C.). A large number of these bioperiodic phenomena still await a validated objective detection from both a methodologic and a statistical point of view.

As a matter of fact recent and future progress in the study of circannual rhythms, as well as progress in chronobiology, are related to basic and methodologic considerations:

1. Physiologic functions in any living organism, including man, are not constant as a function of time: regular and predictable variations with period, τ, of about 24 hours (circadian), about 7 days (circaseptan) about 30 days (circamensual), about 1 year (circannual), etc. can be detected objectively.

2. Each detectable rhythm can then be characterized by the estimation of such parameters as: the acrophase ϕ (peak time of the best fitting mathematical function i.e. cosine used to approximate all the time series' data); amplitude, A, and mesor, M, (rhythm-adjusted mean level). The least

squares method and special computer programs (F. Halberg's cosinor) are useful for rhythm detection as well as for the estimation of τ, ϕ, A and M with their respective 95% confidence interval.

3. Circannual rhythm studies require a specific methodology. (a) High standards in physical measurements, chemical determinations, etc. are desirable for each datum in a time series; however, this experimental necessity is not in itself sufficient to validate a bioperiodic phenomenon. (b) Statistical methods must be used to analyse the collected or recorded time series. (c) The interpretation of findings finally depends upon the experimental design used to gather data. In researching circannual rhythms one must not neglect the possible influence (or interference) of other periodicities especially circadian and circamensual rhythms. By sampling only once daily, at different or even at the same clock hour, or sampling in subjects adhering to non-standardized societal routine, may reflect, to a greater degree, circadian (or circamensual) rather than circannual rhythmicity. (d) Obviously other potentially influencing factors such as the subjects' sex, age, ethnic origin, geographical location (latitude) etc. have to be taken into consideration. (e) For understandable reasons human experimental isolation for a span of time as long as one year have not been performed as yet; thus the evidence for an intrinsic component of circannual rhythms in man are deducted only from indirect but consistent arguments.

The presented results were selected in consideration with the desiderata summarized above. They deal with (1) endocrine and related rhythms; 17-ketosteroids, 17-hydroxycorticosteroids and potassium urinary excretion, plasma cortisol, catecholamine (epinephrine, norepinephrine) urinary excretion; plasma and urinary testosterone in mature young males, timing of menarches and menstrual cycle length in mature young females, etc. (2) feeding behavior (mainly in 4 ± 1.5 year old children). (3) chronosusceptibility; circannual rhythms in human mortality and morbidity.

The intrinsic origin of these rhythms, as well as several hypothesis such as vestigial physiology (E. Hughes) long-term versus short-term regulation, climatic influences, annual changes in daily light intensity and timing as possible synchronizer, etc. are discussed.

ACKNOWLEDGEMENTS

I wish to express my deep gratitude to Drs. Franz Halberg, Michael Smolensky, Hugh Simpson, Joe Bohlen, Norberto Montalbetti and Erhard Haus for their pertinent advice; to Dr. Charles Abulker and Mr. Jean Dupont for their help in computer analyses; to Miss Jocelyne Clench for help in the preparation of this manuscript, and to Miss Annonciade Nicolai for her technical assistance.

REFERENCES

AHUJA, M. M. S., SHARMA, N. J. (1971). Adrenocortical function in relation to seasonal variations in normal Indians. Indian J. Med. Res. 59, 1893-1905.

ALTOUNYAN, R. E. C. (1969). Changes in histamine and atropine responsiveness as a guide to diagnosis and evaluation of therapy in obstructive airways disease. In : Disodium cromoglycate in allergic airways disease. Editors: J. Pepys & A. W. Frankland. Butterworth. London, pp. 47-53.

AMOROSO, E. C., MATTHEWS, L. H. (1955). The effect of external stimuli on the breeding cycle of birds and mammals. British Medical Bulletin, 11, 87-92.

APFELBAUM, M., REINBERG, A., NILLUS, P., HALBERG, F. (1969). Rythmes circadiens de l'alternance veille-sommeil pendant l'isolement souterrain de sept jeunes femmes. Presse Medicale, 77, 879-882.

APFELBAUM, M., REINBERG, A., ASSAN, R., LACATIS, D. (1972). Hormonal and metabolic circadian rhythms before and during a low protein diet. Israel J. Medical Sciences, 8, (6), 867-873.

ARON, CL., ROOS, J. and ASCH, G. (1967). Donnees nouvelles sur le role joue par la ponte provoquee dans les phenomenes de reproduction chez la ratte. Ann. Endocrinol. (Paris), 28 : 19-30.

ASCHOFF, J. (1954). Zeitgeber der tierischen Tagesperiodik. Naturwissenschaften, 41, 49-56.

ASCHOFF, J. (1963). Comparative physiology; diurnal rhythms. Am. Rev. Physiol. 25, 581-600.

ASCHOFF, J. (1969). Desynchronization and resynchronization of human circadian rhythms. Aerospace Med. 40, 844-849.
ASSENMACHER, I., BOISSIN, J. (1970). Rythmes circannuels et circadiens du fonctionnement cortico-surrenalien et thyroidien en relation avec le reflexe photo-sexuel. Neuroendocrinologie Colloque du C.N.R.S., N° 927. C.N.R.S. Publ. Paris, pp. 405-423.
BATSCHELET, E., HILLMAN, D., SMOLENSKY, M., HALBERG, F. (1973). Angular - linear correlation coefficient for rhythmometry and circannually changing human birth rates at different geographic latitudes. Internat. J. Chronobiol. 1, 183-202.
BENOIT, J., ASSENMACHER, I., BRARD, E. (1956) Apparition et maintien de cycles sexuels non saisonniers chez le Canard domestique place pendant plus de trois ans a l'obscurite totale. J. Physiol. Paris, 48, 388-391.
BOCK, K. D., KREUZENBECK, W. (1966). In : Antihypertensive Therapy. Principles and Practice. F. Gross (ed.) Springer Verlag, Berlin, 224-237.
BOHLEN, J. G. (1971). Circumpolar chronobiology. Ph.D. Thesis. University of Wisconsin.
BOISSIN, J., ASSENMACHER, I. (1970). Circadian rhythms in adrenal cortical activity in the quail. J. Interdiscip. Cycle Res. 1, 251-265.
BOWEN, A. J., REEVES, R. L. (1967). Diurnal variation in glucose tolerance. Archive Internal Medicine, 119, 261-264.
BREIPOHL, W. (1938). Uber die Beziehungen der juvenilen Ovarialfunktion zu klimatischen. Arch. Gynak, 166, 202-204.
BUEK, H. W. (1829). Nachrichte von dem

Gesundheitszustand der Stadt Hamburg; Gersons und Julius Magazin der Auslandischen Litterature der Gesammten Heilkunde (Anahng Hamburg) pp. 347-354.

BUNNING, E. (1936). Die endogene Tagesrhythmik als Grundlage der photoperiodism Reaktion. Ber. Deutsch. Bot. Ges. 54, 590-607.

BUNNING, E. (1963). Die Physiologische Uhr. Springer-Verlag Berlin, 153 pp. (2nd ed.)

CAMPBELL, H. L. (1945). Seasonal changes in food consumption and rate of growth of the albino rat. Amer. J. Physiol. 143, 428-433.

CARDOSO, S. S., CARTER, J. R. (1969). Circadian mitotic rhythms as a guide for the administration of antimetabolites. Proc. Soc. Exp. Biol. (N.Y.), 131, 1403-1406.

CHOSSAT, CH. (1843). Recherches experimentales sur l'inanition. Memoire Academie Royale des Sciences de l'Institut de France, 8, 438-640.

CHROMETZKA, F. (1940). Sommer-Interrhyrhmus in menschlichen Stoffwechsel. Der Sommer-Winterrhythmud des diabetes Stoffwechsels. Klin. Wchnschr. 19, 972-976.

COHEN, PH. (1971). Seasonal variations of congenital dislocation of the hip. J. Interdiscipl. Cycle Res. 2 (4), 417-425.

COLLIER, G., HIRSCH, E., HAMLIN, P. H. (1972). The ecological determinants of reinforcement in the rat. Physiol. and Behav. 9, 705-726.

COLLIER, G., HIRSCH, E., KANAREK, R., MARWINE, A. (1973). Environmental determinants of patterns of feeding. Proceedings of the Int. Soc. Chronobiol.

Meeting. Hannover, Int. J. Chronobiology (in press).

CONROY, R. T. W. L., MILLS, J. N. (1970). Human circadian rhythms. J. & A. Churchill, London, pp. 236.

DEBRY, G., BLEYER, R., REINBERG, A. (1973). Circadian, circaseptan and circannual rhythms in spontaneous nutrient and calorie intake of 4 ± 1.5 year old healthy children. Proceeding of the Int. Soc. Chronobiology Meeting. Hannover. Int. J. Chronobiology (in press).

DEBRY, G., BLEYER, R., REINBERG, A. (1974). Circadian, circannual and other rhythms in spontaneous nutrient and calorie intake of 4 ± 1.5 year old healthy children. (in press).

DESCOVICH, G. C., MONTALBETTI, N., KUHL, J. F. W., RIMONDO, S., HALBERG, F. (1974). Age and catecholamine rhythms. (in press).

DOBZHANSKY, TH., BOESINGER, E. (1968). Essais sur l'evolution. Masson & Cie, Paris, pp. 183.

DRAY, F. (1970). Contribution a l'analyse des steroides et a l'etude metabolique de la testosterone et de l'epitestosterone, These de Doctorat d'Etat es Sciences, N° 4197, Paris.

DRAY, F., REINBERG, A. and SEBAOUN, J. (1965) Rythme biologique de la testosterone libre du plasma chez l'homme adulte sain : existence d'une variation circadienne, Comptes-rendus Academie des Sciences, Paris, 231, 573-576.

DU RUISSEAU, J. P. (1965). Seasonal variation of P.B.I. in healthy Montrealers. J. Clin. Endocr. Metab. 25, 1513-1515.

ENGLE, E. T. and SHELESNYAK, M. G. (1934). First menstruation and subsequent menstrual cycles of pubertal girls. Human Biology, 6, 431-453.

FARNER, D. A. (1959). Photoperiodic control of annual gonadal cycles in birds. In : Photoperiodism and Related Phenomena in Plants and Animals. American Association for the Advancement of Science. Publication, N° 55 Washington, pp. 717-750.

FITT, A. B. (1941). Seasonal influence on growth, function and inheritance. New Zealand Council for Educational Research Series, N° 17. Wellington. Whitcombe & Tombs, Ltd, 182 pp.

GAMBLE, D. R., TAYLOR, K. W. (1969). Seasonal incidence of diabetes mellitus. Brit. Med. J. III, 631-633.

GATES, D. M. (1966). Spectral distribution of solar radiation at the earth's surface. Science, 151, (N° 3710), 523-529.

GARCIA-SAINZ, M., HALBERG, F. and MOORE, V. (1968). Rev. Mex. Radiol. 22: 131-146.

GAUQUELIN, M. F. (1968). Le cycle annuel de reproduction du Macaque, Macaca irus. Bulletin Biol. France, 102, 261-270.

GHATA, J., HALBERG, F., REINBERG, A., SIFFRE, M. (1969). Rythmes circadiens desynchronises (17-OHCS, temperature rectale, veille-sommeil) chez deux sujets adultes sains. Ann. Endocrinol. (Paris), 30, 245-260.

GHATA, J., REINBERG, A. (1954). Variations nycthemerales, saisonniere et geographique de l'elimination urinaire du potassium et de l'eau chez l'homme adulte sain. Comptes-Rendus Ac. Sci. Paris, 239, 1680-1682.

GRIGORYEV, Y. G., DARENSKYA, N. G., DRUZHININ, Y. P., KUSNETSOVA, S. S. and SERAYA, V. M. (1969). Diurnal rhythms and ionizing radiation effects. Abstracts : Cospar XII Plenary Meeting, Prague, 163-164.

GRIMM, H. (1952). Uber jahreszeitliche Schwankungen im Eintritt der Menarche. Zentralblt. f. Gynakologie, 74, 1577-1581.

HABEREY, P., DANTLO, CH., KAYSER, CH. (1966). Methode d'exploration du comportement alimentaire d'un hibernant le Lerot *Eliomys guercinus*. C.R. Soc. de Biol. 160, 655-659.

HABEREY, P., DANTLO, CH., KAYSER, CH. (1967). Evolution saisonniere des voies metaboliques du glucose et du choix alimentaire spontane chez un hibernant, le Lerot (*Eliomys quercinus*). Arch. Sci. Physiologiques, 21, 59-66.

HAFEZ, E. S. E. (1952). Studies on the breeding season and reproduction of the ewe. I. The breeding season in different environments: Journal of Agricultural Science, 42, 189-199.

HAFEZ, E. S. E. (1959). Studies on the breeding season and reproduction of the ewe. II. The breeding season in one locality. Journal of Agricultural Science, 42, 199-265.

HALBERG, F. (1960). Temporal coordination of physiologic function. In : Cold Spring Harbor Symp. Quant. Biol. New York. Long Island Biol. Ass. N.Y. 25, 289-310.

HALBERG, F. (1964). Physiological rhythms in physiological problems in space travel. Hardy J.D. Ed. Springfield. Ill. Charles C. Thomas, 298-322.

HALBERG, F. (1965). Some aspects of biological data analysis; longitudinal and transverse profile of rhythms. In : Circadian Clocks. J. Aschoff Ed. North

-Holland. Publ. Co. Amsterdam, 13-22.

HALBERG, F. (1967). Ritmos y corteza supra-renal. IV. Simposio panamericano de farmacologia y terapeutice. Mexico 1967. Excerpta Medica Int. Cong. Ser. 185, pp. 7-39.

HALBERG, F. (1969). Chronobiology, Ann. Rev. Physiol. 31, 675-725.

HALBERG, F. (1970). Body temperature, circadian rhythms and the eye. In : La photoregulation de la reproductionchez les Oiseaux et les Mammiferes. J. Benoit et I. Assenmacher. eds. Centre National de la Recherche Scientifique, N° 172, 497-528.

HALBERG, F. (1973). Laboratory techniques and rhythmometry in biologic aspects of circadian rhythms. J.N. Mills, Editor. Plenum Press. London and New York, pp. 1-26.

HALBERG, F., ENGELI, M., HAMBURGER, C., HILLMAN, D. (1965). Spectral resolution of low-frequency, small-amplitude rhythms in excreted ketosteroid : probable androgen-induced circaseptan desynchronization. Acta Endocrinol. Supplement, 103, 54 pp.

HALBERG, F., HALBERG, E., BARNUM, C. P., BITTNER, J. J. (1959). Physiologic 24-hour periodicity in human beings and mice, the lighting regimen and daily routine. In photoperiodism and related phenomena in plants and animals. AAAS publ. Washington, D.C., 55, 963-978.

HALBERG, F., HALBERG, E. and MONTALBETTI, N. (1969). Premesse e sviluppi della cronofarmacologia. Quad. Med. Quant. Sperimentazione Clin. Controllata, 7: 5-34.

HALBERG, F., REINBERG, A. (1967). Rythmes

circadiens et rythmes de basses frequences en physiologie humaine. J. Physiol. Paris, 59, 117-200.

HALBERG, F., REINBERG, A., HAUS, E., GHATA, J., SIFFRE, M. (1970). Human biological rhythms during and after several months of isolation underground in natural caves. Bull. Nat. Speleological Soc. 32 (4) : 89-115.

HALBERG, F. and STEPHENS, A. N. (1959). Susceptibility to ouabain and physiologic circadian periodicity. Proc. Minn. Acad. Sci. 27 : 139-143.

HALBERG, F., SPINK, W. W., ALBRECHT, P. G., GULLY, R. (1955). Resistance of mice to brucella somatic antigen, 24-hr periodicity and the adrenals. J. Clin. Endocr. Metab. 15, 887.

HALBERG, F., TONG, Y. L., JOHNSON, E. A. (1967). Circadian system phase : an aspect of temporal morphology; procedures and illustrative examples. In the Cellular Aspects of Biorhythms. H. Von Mayersbach, Ed. Springer, Berlin-Heidelberg, New York, pp. 20-48.

HALBERG, F., VISSCHER, M. G., BITTNER, J. J. (1954). Relation of visual factors to eosinophil rhythm in mice. Amer. J. Physiol. 179, 229-235.

HALE, H. B., ELLIS, J. P., WILLIAMS, E. W. (1968). Urinary creatinine-based ratios in relation to season. Aerospace Med. 39, 1048-1051.

HAMBURGER, C. (1954). Six years' daily 17-ketosteroid determination on one subject; seasonal variations and independence of volume and urine. Acta Endocr. 17, 116-127.

HAMNER, W. E. (1964). Avian photoperiodic response-rhythms : evidence and inference. In : Circadian Clocks. Proceedings of the Feldafing Summer School, Sept. 1964. J. Aschoff. Ed. Amsterdam, North-Holland, Cy, p. 379-384.

HAUS, E., HALBERG, F. (1966). Persisting circadian rhythms in hepatic glycogen of mice during inanition and dehydration. Experientia, 22, 113-115.

HAUS, E., HALBERG, F. (1970). Circannual rhythm in level and timing os serum corticosterone in standardized inbred mature C-mice. Environmental Research, 3, 81-106.

HOMMEL, H., FISHER, U. (1971). Jahresperiodik der Insulinwirkung am eviszerierten Kaninchen und am isolierten perfundierten Rattenherzen. Diabetologia, 7, 6-9.

HOWE, P. E., BERRYMAN, G. H. (1945). Average food consumption in the training camps of the United States Army (1941-1943). Amer. J. Physiol. 144, 588-594.

HUGHES, E. (1931). Seasonal variation in man. H. K. Lewis and Co. Ltd, London, 126 pp.

JARRETT, R. J., KEEN (1970). Further observations on the diurnal variation in oral glucose tolerance. British Medical Journal, 4, 334-337.

JELLINEK, E. M., LOONEY, J. M. (1936). Studies in seasonal variation of physiologic functions. I. The seasonal variation of blood cholesterol. Biometric Bull. 1, 83-95.

JOHANSSON, G., FRANKENHAEUSER, M., LAMBERT, W. W. (1969). Note on seasonal variations in catecholamine output. Perceptual and Motor Skills, 28, 677.

KAISER, H., HALBERG, F. (1962). Circadian aspect of birth. Ann. N.Y. Acad. Sci. 98, 1056.

KLEITMAN, N. (1949). Biological rhythms and

cycles. Physiol. Rev. 29, 1-30.
KLEITMAN, N. (1963). Sleep and wakefulness. Chicago Univ. Press. pp. 552 (2d Ed.)
KLEITMAN, N., MULLIN, F. J., COOPERMAN, N. R., TITELBAUM, S. (1937). Sleep characteristics. Chicago Univ. Press. pp. 86.
KOLLER, L. R. (1964). Ultraviolet radiation. J. Wiley & Sons, Inc., New York.
LAGOGUEY, M., DRAY, F., CHAUFFOURNIER, J. M., REINBERG, A. (1972). Etude des rythmes circadiens et circannuels des glucuronides de testosterone et d'epitestosterone chez l'homme adulte sain. Comptes-Rendus Ac. Sci. Paris, 274, 3435-3437.
LAGOGUEY, M., DRAY, F., CHAUFFOURNIER, J. M., REINBERG, A. (1973). Circadian and circannual rhythms of urine testosterone and epitestosterone glucuronide in healthy adult men. Internat. J. Chronobiology, 1, 91-93.
LAGOGUEY, M., REINBERG, A. (1974). unpublished data.
LE MAGNEN, J., TALLON, S. (1966). La periodicite spontanee de la prise d'aliment "*ad libitum*" du rat blanc. J. Physiol. Paris, 58, 323-349.
LE MAGNEN, J. (1971). Advances in studies of the physiological control and regulation of food intake. In Progress in Physiological Psychology. E. Stellar & J. M. Spragues Ed. New York, Academic Press, 4, 204-261.
LESTRADET, H., LABRAM, C., ALCALAY, D. (1970) Diabete du soir : espoir ! ou l'importance de l'horaire dans l'interpretation d'une courbe d'hyperglycemie provoquee. Presse Medicale, 78, 1481-1482.
LUCE, G. (1970). Biological rhythms in

psychiatry and medicine. Public Health Service Publication, N° 2088, U. S. Government Printing Office. Washington, D.C. pp. 183.

MARGULES, D. L., LEWIS, M. J., DRAGOVITCH, J. A., MARGULES, A. (1972). Hypothalamic norepinephrine : circadian rhythms and the control of feeding behavior. Science, 178, 640-643.

MILLS, J. N. (1966). Human circadian rhythms. Physiol. Rev. 46, 128-171.

MILLS, J. N. (1973). Transmission processes between clock and manifestations. In : "Biological aspects of circadian rhythms". J.N. Mills Editor. Plenum Press. London. New York, pp. 27-84.

MILLS, J. N., WATERHOUSE (1973). Circadian rhythms over the course of a year in a man living alone. Internat. J. Chronobiology 1, 73-79.

NELSON, R. F. (1966). Variation of radiosensitivity of mice with time of day. Acta Radiol. (Stockholm), 4 : 91.

NIELSEN, J., BRUUN PETERSEN, G., THERKELSEN, A. J. (1973). Seasonal variation in the birth of children with aneuploid chromosome abnormalities. Humangenetik, 19, 67-74.

OKAMOTO, M., KOHZUMA, K., HORIUCHI, Y. (1964). Seasonal variations of cortisol metabolism in normal man. J. Clin. Endocr. Metab. 24, 470-471.

PENGELLEY, E. T., ASMUNDSON, S. J. (1971). Annual biological clocks. Scientific American. 224 (4), 72-79.

PETERSEN, W. F. (1947). Man, weather and sun. Charles C. Thomas, Springfield, Ill.

PITTENDRIGH, C. S. (1960). Circadian rhythms and the circadian of living systems in : Cold Spring Harbor. Symp. Quant.

Biol. Long Island Biol. Assoc. N.Y. 25, 159-182.

PIZZARELLO, D. J., ISAAK, D., CHUA, K. E. and RHYNE, A. L. (1964). Circadian rhythmicity in the intensity of two strains of mice to whole body radiation. Science, 145, 286-291.

POSPISLOVA-ZUZAKOVA, V., STUKOVSKY, R. and VALSIK, J. A. (1965). Die Menarche bei Weissen, Negrinnen und Mulattinnen von Habana (Cuba). Zschr. arztl. Forbild. 59 Jg, H.9, p. 508-515.

PROCACCI, P., BUZZELLI, G., PASSERI, I., SASSI, R., VOLGELIN, M. R., ZOPPI, M. (1972). Studies on the cutaneous pricking pain threshold in man. Circadian and circatrigintan changes. Res. Clin. Stud. Headache, 3, 260.

QUETELET, L. A. J. (1826). Memoire sur les lois des naissances et de la mortalite a Bruxelles. Nouveaux Memoires de l'Academie Royale des Sciences et Belles Lettres de Bruxelles, 3, 495-512.

REINBERG, A. (1967). The hours of changing responsiveness or susceptibility. Persp. Biol. Med. 11 : 111-128.

REINBERG, A. (1971). Biological rhythms of potassium metabolism. Proc. of the 8th Colloquium of the International Potash Institute Uppsala, Sweden, June 1971. Published by the Int. Potash Institute. Berne, Switzerland. pp. 160-180.

REINBERG, A. (1971). Methodologic considerations for human chronobiology. J. Interdisciplinary Cycle Res. 2, 1-15.

REINBERG, A. (1971). La chronobiologie. La Recherche, 2, 242-250.

REINBERG, A. (1972). Biological rhythms and human biometerology, with special reference to mortality rhythms and chronotoxicology. 6th Internat. Biometeorological Congress. Noordwijk, Netherlands. Internat. J. Biometeorology. Supp. to vol. 16, 97-112.

REINBERG, A. (1973). Biological rhythms and energy balance in man. In "Energy balance in man" (Paris. Sept. 1971). M. Apfelbaum, Ed. Masson, Paris, pp. 285-295.

REINBERG, A., APFELBAUM, M., ASSAN, R. (1974). Chronophysiologic effects of restricted diet (220 cal/24 hr as casein) in young healthy but obese women. Intern. J. Chronobiology (in press).

REINBERG, A., GERVAIS, P., HALBERG, F., GAULTIER, M., ROYNETTE, N., ABULKER, C. and DUPONT J. (1973). Rythmes circadiens et cercannuels de mortalite des adultes dans un hopital parisien et en France. Nouvelle Presse Medicale 2 : 289-294.

REINBERG, A., and GHATA J. (1964). Les rythmes biologiques. Presses Universitaires de France (Que Sais-Je?), 1 vol. Paris, 1964 (seconde edition). Biological Rhythms. Walker, New York, 138 pp.

REINBERG, A., HALBERG, F. (1971). Circadian chronopharmacology. Annual Review of Pharmacology, 11, 455-492.

REINBERG, A., HALBERG, F.,GHATA J. et SIFFRE, M. (1966). Spectre thermique (rythmes de la temperature rectale) d'une femme adulte saine avant, pendant et apres son isolement souterrain de trois mois. Comptes-Rendus Acad. Sc. 262, 782-785.

REINBERG, A., SIDI, E., and GHATA J. (1965). Circadian reactivity rhythms of human skin to histamine or allergen and the adrenal cycle. J. Allergy, 36, 273-283.

REINBERG, A., SIDI, E. (1966). Circadian changes in the inhibitory effects of an antihistaminic drug in man. J. Invest. Dermat., 46 : 415-419.

REINBERG, A., SMOLENSKY, M. H. (1973). Circatrigintan secondary rhythms related to hormonal changes in the menstrual cycle : general considerations. In : Biorhythms and human reproduction. Wiley. Interscience, New York (in press).

REINBERG, A., SMOLENSKY, M., GHATA, J., GERVAIS, P. (1974). Chronobiologic approach in the study of menstrual changes of healthy women with spontaneous menstruation. Internat. J. Chronobiol. (in press).

REINBERG, A., ZAGULLA-MALLY, Z., GHATA, J. and HALBERG, F. (1967). Circadian rhythm in duration of salicylate excretion referred to phase of excretory rhythms and routine. Proc. Soc. Exp. Biol. (N.Y.), 124, 826-832.

REINBERG, A., ZAGULLA-MALLY, Z., GHATA, J. and HALBERG, F. (1969). Circadian reactivity rhythm of human skin to house dust, penicillin and histamine. J. Allergy, 44, 292-306.

REINDL, K., FALLIERS, C., HALBERG, F., CHAI, H., HILLMAN, D. and NELSON, W. (1969). Circadian acrophases in peak expiratory flow rate and urinary electrolyte excretion of asthmatic children: phase shifting of rhythms by prednisone given in different circadian system phases. Rass. Neurol. Veg. 23 : 5-26.

RENBOURN, E. T. (1947-). Variation diurnal and over longer periods of time in blood haemoglobin, haematocrit, plasma protein, erythrocyte sedimentation rate and blood chloride. J. Hyg. London, 45, 455-467.

RENSING, L. (1969). Ein circadianer Rhythmus der Empfindlichkeit gegen Rontgenstrahlen bei Drosophila. Z. Vergl. Physiologie, 62 : 214-220.

RUTENFRANZ, J. and SINGER, R. (1967). Untersuchungen zur Frage einer Abhangigkeit der Alkoholwirkung von der Tageszeit. Int. Z. Angew. Physiol. 24, 1-17.

SARGENT (II), F. (1954). Season and the metabolism of fat and carbohydrate : a study of vestigial physiology. Meteorological Monographs. 2 (3), 68-80.

SCHEVING, L. E. and VEDRAL, D. (1966). Circadian variation in susceptibility of the rat to several different pharmacological agents. Anat. Rec., 154 : 41.

SCHEVING, L. E. and VEDRAL, D. F. (1968). Daily circadian rhythm in rats to D-Amphetamine Sulphate : effect of blinding and continuous illumination on the rhythm. Nature (Lond.), 219 : 621-622.

SCHEVING, L. E., VEDRAL, D. F. and PAULY, J. E. (1968). A circadian susceptibility rhythm in rats to pentobarbital sodium. Anat. Rec. 160 : 741-750.

SENSI, S., CAPANI, F., CARADONNA, P., POLICICCHO, D., CAROTENUTO, M. (1970). Diurnal variation of insulin response to glycemic stimulus. Biochimica e Biologia Sperimentale, 9 (3), 153-156.

SIMPSON, H., BOHLEN, J. (1973). Latitude and the human circadian system in : Biological aspects of circadian rhythms J. N. Mills, Editor. Plenum Press. London. New York, pp. 85-120.

SMOLENSKY, M., HALBERG, F., SARGENT (II), F. (1972). Chronobiology of the life sequence. In : Advance in Climatic

Physiology. S. Ito, K. Ogata, H. Yoshimura Ed. Igaku Shoin, Ltd, Tokyo, pp. 281-318.

SMOLENSKY, M. H., REINBERG, A., LEE, R. E., McGOVERN, J. P. (1973). Secondary rhythms related to hormonal changes in the menstrual cycle : special reference to allergology. In Biorhythms and human reproduction. Wiley. Interscience. New York (in press).

SOUTHREN, A. L., TOCHIMOTO, S., CARMODY, N., and ISURUGI, K. (1965). Plasma production rates of testosterone in normal adult men and women and in patient with the syndrome of feminizing testes. Journal of Clinical Endocrinology and Metabolism, 25, 1441-1447.

STROEBEL, C. F. (1969). Biologic correlates of disturbed behavior in the rhesus monkey. In Circadian Rhythms in Non Human Primates. F.H. Rohles (Ed.) Karger, Basel, 91-105.

TROMP, S. W. (1963). Medical Biometerology. Elsevier Publ. Comp. Amsterdam, 919 pp.

TROMP, S. W. (1964). Weather, climate and man. In : Adaptation to the Environment. D.B. Dill (ed). Handbook of physiology, feet. 4. Williams and Wilkins, Baltimore.

TROMP, S. W. (1971). Monthly and yearly fluctuations in mortality from arterosclerosic heart diseases and cerebral apoplexy in the Netherlands. J. Interdisc. Cycle Res. 2 : 315.

VALSIK, J. A. (1965). The seasonal rhythm of menarche : a review. Hum. Biol., 37 : 75-90.

VALSIK, J. A., STUKOVSKY, R., BERNATOVA, L. (1963). Quelques facteurs geographiques et sociaux ayant une influence sur l'age de la puberte. Biotypologie.

24, 109-123.

WATANABE, G.I. (1964). Seasonal variation of adrenal cortex activity. Arch. Environmental Health. 9, 192-200.

WATANABE, G.I., AOKI, S., NAGAI, T. (1956). Climatic effect on circulating eosinophil count. J. Appl. Physiol. 9, 461.

WATANABE, G.I., YOSHIDA, S. (1956). Climatic effect on output of netral 17-ketosteroids. J. Appl. Physiol. 9, 456.

WHITEHEAD, P.E., McEWAN, E.H. (1973). Seasonal variation in the plasma testosterone concentration of reindeer and caribou. Canad. J. Zool. 51, 651-658.

ZACHARIAS, L. and WURTMAN, R.J. (1964). Blindness: its relation to age of menarche. Science, 144 (N° 3622), 1154-1155.

CREDITS

Figure 1. Reproduced by permission from Journal de Physiologie, 59: 117-200, 1967. Masson, Paris, France.

Figures 2,12 and 15. Reproduced by permission from Nouvelle Presse Medicale, 2: 289-294, 1973. Masson, Paris, France.

Figure 3. Reproduced by permission from Environmental Research 3: 81-106, 1970. Academic Press and Dr. Erhard Haus.

Figures 4 and 5. Reproduced by permission from Acta Endocrinologica, Supplement 103: 50, 1965. Periodica, Copenhagen, Denmark and Dr. Franz Halberg.

Figures 6 and 7. Reproduced by permission from Biological Aspects of Circadian Rhythms, Editor J.N. Mills, 1974. Plenum Press, London, England.

Figure 8. Reproduced by permission from International Journal of Chronobiology 1: 183-202, 1973. John Wiley and Sons Ltd., London, England.

Figures 13,14 and 16. Reproduced by permission from Advances in Climatic Physiology, Editors S. Ito, K. Ogata and H. Yoshimura, 1972. Igaku Shoin Ltd., Tokyo, Japan and Dr. Michael Smolensky.

CIRCANNUAL RHYTHMS IN CIRCADIAN PERSPECTIVE

MICHAEL MENAKER

Professor of Zoology

Patterson Laboratories 30

The University of Texas at Austin

Austin, Texas 78712

As is indicated by the name chosen to describe the phenomenon, the study of circannual rhythmicity rests very near its conceptual base on an analogy with circadian rhythms. Sometimes this analogy is explicitly drawn; more often it is implied, but it is almost always present in the thinking of those who study endogenous annual biological rhythms and its influence on the design and interpretation of their experiments is obvious. Perhaps the most useful contribution that I can make to these proceedings is to examine critically that analogy and to ask: has it been fruitful? and will it continue to be? I will argue that the answer to the first question is clearly yes, whereas the answer to the second is, with some qualifications, no.

Let us examine the analogy in some detail. Circadian rhythms are innate (i.e. inherited) and self-sustained (their continued motion does not require a periodic input from the environment). They can be entrained (synchronized) by environmental cycles of light, temperature and occasional other factors and their formal behavior in response to such entraining signals bear remarkable resemblances to the behavior of physical self-sustained non-linear oscillators. Finally, their period lengths are remarkably well buffered from environmental influences, especially from the effects which one would expect changes in environmental temperature to exert on the rate of any biochemically based process. Although each of these properties, and certainly all of them together, may have been most thoroughly studied in circadian systems, there is every indication that they occur in other kinds of biological oscillators, especially in the much more rapid oscillations of nerve cells and networks and in certain biochemical oscillations that can be produced *in vitro*. What then are the unique features of the circadian system which makes it attractive as an analogue of circannual rhythmicity?

It seems to me that there are two such features, a minor and a major one. The minor one involves simply the unexpectedly long time constants of the circadian oscillation. If, as seems almost certain, most living cells contain (or are) circadian oscillators, then in many cases the periods of the rhythms which they generate represent a significant fraction of their lifetimes. This fact serves as a dramatic reminder of the extent to which biological organization is history dependent - not only in the obvious evolutionary sense but also in the lives of individual cells and organisms. The cell at noon is not the cell it was at 6:00 A.M. nor yet the cell that it will become at midnight. If the time constants involved in circadian oscillations are surprisingly long how much more so are those in circannual rhythms. Here a single cycle commonly occupies a third or more of the life of complex and relatively long-lived organisms. Indeed perhaps the major difficulty in the study of circannual rhythms is a consequence of the ratio of the period length of a single circannual cycle to the length of the productive life of the biologist! Interest in both circadian and circannual oscillations has to a large degree been fueled by the challenge of attempting to understand the internal mechanism by which, in living things normally characterized by rapidity of turnover and response, a regular series of events is caused to recur once a day or once a year. Beyond this, however, the fundamental message is the same. Life is a dynamic process with respect to time every bit as much as it is with respect to space. An individual animal or plant is not the same, in many ways which we understand and in many more which we do not, in May as it is in December.

The major feature of circadian oscillations which all but forces upon us the analogy between them and circannual rhythms has to do with their adaptive significance.

They are used by organisms to "measure" time - not simply biological time but astronomical (environmental) time. They are, in fact, clocks - internal clocks constructed to be sure of "flesh and blood" but used by organisms, in several extensively documented situations and presumably in many others, in exactly the way (at least in principle) as we use our kitchen clocks or wristwatches. It was the perception of this fact in the 1950's primarily by Pittendrigh, which led to the reawakening of interest in what are now called circadian rhythms, the major features of which had been known for more than 200 years. To my mind at least, this remains the most interesting fact about circadian rhythms. As we learn more about the temporal organization of living things we will certainly discover other oscillations that are inherited, self-sustained, entrainable and temperature compensated, but most will not serve to relate internal biological events to astronomical time. It is precisely because circannual rhythms seem to fall into this very limited class of biological oscillations that they are of particular interest and that the analogy between them and circadian rhythms seems so natural and attractive.

Our analogy can now be restated more precisely in terms of the question: Do some organisms have internal oscillators which operate as biological calendars, as most organisms have internal oscillators which operate as biological clocks? The fact that the answer to this question is yes is, on strictly *a priori* grounds, most surprising. First because of the long time constants already discussed but second and more importantly because the general line of argument which supports the assumption of adaptive significance for biological clocks will not support a similar assumption concerning biological calendars. Reduced to its bare bones, this argument is that it is useful for organisms to be able to initiate events at phase points (times) in the real day which are not marked by obvious environmental reference points. If it is, for whatever reason, of adaptive significance to do something in the middle of the night an internal clock becomes useful as a means of recognizing that particular point in time. In its most general form this line of argument does not apply to the potential adaptive significance of circannual rhythms simply because there is, over most of the earth's surface, a clear and accessible environmental reference point for every phase of the astronomical year - the length of the day or photoperiod. Further, it is evident from a now very large literature that many organisms are able to measure daylengths (often with the aid of their circadian clocks) with great enough accuracy so that if they were to employ this method to determine time of year they would never be in error by more than a week or two. While it is true that there is potential ambiguity in such a method because the same daylength occurs twice each year, from our armchairs it is easy to imagine a solution to this difficulty as well. If organisms can measure daylength accurately surely they can tell whether the photoperiod is lengthening or growing shorter. (However, I know of no case in the literature concerned with photoperiodic time measurement in which it has been well documented that an organism responds to progressive changes in daylength as distinct from responding to fixed photoperiods of particular durations. Considering the wide variety of organisms which have been studied and the fact that photoperiodic time measurement has evolved independently many times, I find it remarkable that no organism appears to have "learned" how to make this discrimination which would so usefully distinguish spring from fall and more generally provide access to a complete exogenous calendar.) There is clearly something wrong with this chain of *a priori* reasoning because it is very difficult to imagine that there is no adaptive significance to a phenomenon which seems as improbable as endogenous circannual rhythmicity. The argument, however, does serve to emphasize both the importance and the difficulties of accounting for the evolution of this particular biological oscillation and of assessing critically its putative role as an endogenous calendar.

Even if their use as calendars has not yet been fully documented, circannual rhythms are remarkably similar, in many of their general properties, to circadian oscillations and it is safe to say that much of the research which has led to the delineation of the similarities has been stimulated by the widely accepted analogy between them. Analogies, however, are fruitful only so long as they suggest interesting and answerable questions. While some important questions suggested by this particular analogy have not yet been resolved - chiefly the identity of the environmental agents responsible for synchronizing circannual rhythms with the annual cycle - I would argue that most, if not all, have been asked. I believe that, if the analogy is pushed much farther, answers to the resulting questions will require unreasonable amounts of time to obtain and may well turn out to be biologically meaningless.

If it is overdrawn, the analogy between circadian and circannual rhythms unproductively biases the kinds of questions which people studying the latter ask. Take as an example the recent discussion in the circannual literature of the question of whether circannual rhythms are "really" rhythms or are "merely" a series of sequental steps (Mrosovsky, 1970; Pengelley and Asmundson, 1972). This discussion arises rather naturally out of the very large amount of physiological information which is available about events within a single circannual cycle (much more, at least potentially, than we have about the circadian system) but its emphasis is counterproductive. Such considerations have been largely ignored by students of circadian rhythms because of lack of information. This has led to the general (incorrect) impression that the circadian mechanism is some sort of rather mysterious whole thing, which does not consist of discrete identifiable functions and further to the implicit, logically peculiar conclusion that if an apparent rhythm yields to explanation as a series of sequential steps then it is not really a rhythm. Of course both circannual and circadian rhythms must consist of sequences of interdependent steps: the productive question in both fields (but perhaps more easily studied in circannual rhythms) is: at what level of organization do the crucial steps occur and can they be identified and their interactions analysed.

The kind of detailed, formal description and modeling which has proven so successful in explaining (on one level) the entrainment of circadian oscillations by environmental cycles and in uncovering other interesting properties of the circadian system, is obviously impractical to apply to the analysis of circannual rhythms. The limits imposed by the number of cycles obtainable in an experimenter's lifetime appear to me, at least, to be insurmountable. Analogy with circadian rhythms is of further use only in underlining the futility of an experimental concern with such variables as the period of a circannual rhythm in steady state [to achieve steady state may require 70-100 cycles of the circadian oscillation in *Passer domesticus* (Eskin, 1971)] or transient motion of the underlying oscillation. Formal model building in the study of circadian rhythms has depended almost exclusively on kinds of data which simply will not be available for circannual rhythms. Furthermore the history of the field of circadian rhythms should give pause to those who would emulate the formal approach. It has proven extremely difficult to extract from the system by formal means, general features which are not themselves of adaptive significance and thus object to convergent selection. It cannot be overemphasized that we do not yet know whether the very great formal similarities among the properties of circadian rhythms in diverse organisms are due to common physiological mechanisms or to convergence imposed by rigorous functional limitations. After 20 years of vigorous work by relatively large numbers of people on a system from which data are available at a rate which is at least 365 times that at which they are available for circannual rhythms, the physiological light is only now beginning to appear at the end of the formal circadian tunnel.

It appears most unlikely that the

physiological mechanisms by means of which organisms measure annual time will be found to bear more than the most general resemblance to those by means of which circadian time is kept. It therefore seems expedient for students of circannual rhythms, at least temporarily, to place the analogy between the two phenomena on the shelf (with a nod of thanks for past services rendered) and to proceed with the physiological analysis of the "series of sequential steps" of which circannual rhythms are surely composed. In that endeavor the extreme length of circannual cycles, which makes formal analysis so difficult, is something of an advantage as it allows ample time for experimental perturbations to be applied with a single cycle and for the effects of such perturbations to be measured.

In the context of such a physiological analysis, one can ask to what, if any, extent does the state of the circadian system affect circannual rhythmicity? Note that this question is completely different in kind from the questions, suggested by the analogy between the two phenomena, which have been considered above. It asks not whether and in what ways the two phenomena are similar but rather, regardless of their degree of similarity, does one play a role in the physiology of the other? In its simplest, though far from its only, form it can be restated as: do organisms generate circannual rhythms by counting and summing circadian cycles? Although the answer to this question in its simple form may well be no, my personal intuition suggests that experimental work guided by the question in its more general form will be fruitful. Such work, however, to be successful must be carried out in an unusually favorable experimental system and I would like to conclude this essay by extolling the virtues of one such system which appears to me to be very attractive, in an attempt to encourage students of circannual rhythms to work with it.

Ideally one would like, in studying the potential influence of circadian on circannual rhythms, to work with an organism for which there is extensive background data on both rhythms. Of course the more general physiological information that is available the better but, more important, the data which are available should indicate at least a potential richness of interaction among the known rhythmic parameters. In addition it would be of great advantage to be able to work with a laboratory animal which had nonetheless not lost, through artificial selection as has the white rat, much of the mechanism by which its physiology in the wild state is related to environmental cycles.

Although it does not have a documented circannual cycle, the golden hamster comes so near to being ideal for studies of this kind in other respects that the effort required to document circannual rhythmicity, which is almost certainly present in this animal, would appear to be well worthwhile. Circadian rhythms in hamsters are beautifully precise and have been more thoroughly studied than those of any other vertebrate (Burchard, 1958; Pittendrigh, 1960, 1974; DeCoursey, 1964; Aschoff, Figala and Poppel, 1973) (in fact perhaps than those of any other organism with the exception of *Drosophila pseudoobscura*). Hamsters have annual cycles of both hibernation and reproductive condition (Mogler, 1958; Vendrely *et al.*, 1971; Reiter, 1972b) about which there is a good deal of physiological information.

Either one or both of these annual cycles might, in fact, be circannual. The annual reproductive rhythm is under photoperiodic control (Gaston and Menaker, 1967), but not all of its components require direct driving by day length [i.e. there is a refractory period to the inhibitory effects of short days and a "spontaneous" recrudescence of the gonads of animals held for long time on short photoperiods (Reiter, 1972a)]. Photoperiodic time measurement is accomplished by the circadian system, indicating a relationship between a circadian and a potentially circannual rhythm (Elliott, Stetson and Menaker, 1972). The difficulty which some workers have experienced in inducing hamsters to hibernate in the laboratory could probably be surmounted, as it may well be due in large part to the fact that the most commonly used laboratory light cycles stimulate steroid hormone production which is known to inhibit hibernation in other rodents. Finally Alleva *et al.* (1971), in their studies of the control of estrous in hamsters, have produced evidence which suggests that estrous may be timed by the counting and summing of circadian cycles. Alleva's results are, to my knowledge, unique in strongly suggesting the participation of the circadian oscillation in the timing of another endogenous rhythm of longer period, and illustrate one kind of interaction for which one might look in the relationship of circadian to circannual rhythms.

ACKNOWLEDGEMENTS

It is a pleasure to thank Carl Johnson for his able assistance in reviewing the circannual literature and for valuable discussions. I should also like to thank NICHD for their support of some of the research referred to (HD03803 and HD07727) and for a Research Career Development Award (5-K04-HD09327-05) held by the author.

REFERENCES

ALLEVA, J.J., WALESKI, M.V. & ALLEVA, F.R. (1971). A biological clock controlling the estrous cycle of the hamster. Endocrinol. 88, 1368-1379.

ASCHOFF, J., FIGALA, J. & POPPEL, E. (1973). Circadian rhythms of locomotor activity in the golden hamster (*Mesocricetus auratus* measured with two different techniques. J. Comp. Physiol. Psychol. 85, 20-28.

BURCHARD, J.E. (1958). Resetting a biological clock. Ph.D. Dissertation, Princeton University.

DeCOURSEY, P.J. (1964). Function of a light response rhythm in hamsters. J. Cell Comp. Physiol. 63, 189-196.

ELLIOTT, J.A., STETSON, M.H. & MENAKER, M. (1972). Regulation of testis function in golden hamsters: A circadian clock measures photoperiodic time. Science 171, 1169-1171.

GASTON, S. & MENAKER, M. (1967). Photoperiodic control of hamster testis. Science 158, 925-928.

MOGLER, R. K-H. (1958). Das endokrine System des syrischen Goldhamsters unter Berücksichtigung des naturlichen Winterschlafs. Z. Morph. Oekol. Tiere 47, 267-308.

MROSOVSKY, N. (1970). Mechanism of hibernation cycles in ground squirrels: circannian rhythm or sequence of stages. Pennsylvania Academy of Science 44, 172-175.

PENGELLEY, E.T. & ASMUNDSON, S.J. (1972). An analysis of the mechanisms by which mammalian hibernators synchronize their behavioral physiology with the environment. In Hibernation and

Hypothermia, Perspectives and Challenges. Elsevier/Excerpta Medica/ North Holland Pub. Co., Amsterdam, pp. 637-661.

PITTENDRIGH, C.S. (1960). Circadian rhythms and the circadian organization of living systems. Cold Spring Harbor Symp. Quant. Biol. 25, 159-182.

PITTENDRIGH, C.S. (1974). Circadian oscillations in cells and the circadian organization of multicellular systems. In The Neurosciences, Third Study Program, F.O. Schmitt and F.S. Worden, eds. in chief, MIT Press, Cambridge, pp. 437-458.

REITER, R.J. (1972a). Evidence for refractoriness of the pituitary-gonadal axis to the pineal gland in golden hamsters and its possible implications in annual reproductive rhythms. Anat. Rec. 173, 365-372.

REITER, R.J. (1972b). Pineal control of seasonal reproductive rhythm in male golden hamsters exposed to natural daylight and temperature. Endocrinology 92, 423-430.

VENDRELY, E., GUERILLOT, C., BASSEVILLE, C. & DaLAGE, C. (1971). Poids testiculaire et spermatogenèse du Hamster doré au cours du cycle saisonnier. C.R. Soc. Biol. 165, 1562-1565.

Subject Index

A 4
B 5
C 6
D 7
E 8
F 9
G 0
H 1
I 2
J 3